Chapters in the Evolution of

Chromatography

Chapters in the Evolution of

Chromatography

Leslie S. Ettre

Retired Adjunct Professor
Yale University, USA

edited by

John V. Hinshaw

Serveron Corporation, USA

ICP

Imperial College Press

Published by

Imperial College Press
57 Shelton Street
Covent Garden
London WC2H 9HE

Distributed by

World Scientific Publishing Co. Pte. Ltd.
5 Toh Tuck Link, Singapore 596224
USA office: 27 Warren Street, Suite 401-402, Hackensack, NJ 07601
UK office: 57 Shelton Street, Covent Garden, London WC2H 9HE

British Library Cataloguing-in-Publication Data
A catalogue record for this book is available from the British Library.

CHAPTERS IN THE EVOLUTION OF CHROMATOGRAPHY

ISBN-13 978-1-86094-943-2
ISBN-10 1-86094-943-6

Typeset by Stallion Press
Email: enquiries@stallionpress.com

Printed in Singapore by Mainland Press Pte Ltd

Contents

Preface

The Present is the future of the Past and the past of the Future

Motto used at Technion, the Israel
Institute of Technology, in Haifa

Originally developed for the separation and characterization of naturally occurring plant pigments, in the 100 years since its invention the application of chromatography gradually extended into almost all fields of science: studies of complex natural substances and biochemical processes, investigations of the nature of petroleum and its derivates, control of chemical syntheses, and determination of trace impurities polluting the soil and air, as well as the drinking and surface water of our planet. From a technique used 100 years ago by a lonely Russian botanist, chromatography eventually became the most widely used laboratory technique.

For my generation the rise of chromatography, its broadening out into various fields, and the development of its variants, was part of our life: we had been actively involved in it. We personally knew the principal players and regularly met them at the frequently held international symposia. There we learned about each others' work and could not wait the closing of the meeting when we rushed home to try out the others' newest results, adapting them to improve our own work. We also learned about the mistakes made by others, just as they learned about our errors. We were part of the evolution of chromatography: it was an exciting time.

Today's chromatographers represent a new, young generation who did not participate in the evolution of the various branches of the technique: the events which have been observed firsthand by my generation represent the distant past to our young colleagues. Therefore, it is important for them to become familiar with the origin of the achievements they utilize in their daily work. Let us not forget the old saying: he who does not know history will repeat past mistakes!

This book represents an attempt to explore the evolution of chromatography, examine the background of the key developments — placing them in the proper historical context — and investigate the life and work of the pioneers. I hope that our readers will not only enjoy the fascinating stories how these milestones of chromatography were developed, but also learn a lesson from them, that they can utilize in their everyday's work.

May 10, 2007
Leslie S. Ettre

Acknowledgments

The chapters of this book are mostly based on historical articles published in the past 20 years in four journals: *LCGC Magazine* (as part of the Milestones in Chromatography series), *Chromatographia, Analytical Chemistry*, and the *Journal of Chromatography*. However, practically all the original articles have been extensively rewritten and some additional materials added. In each chapter reference is given to the original article or articles on which it is based. The cooperation of the editors and publishers of these journals permitting the use of parts and figures is much appreciated.

Naturally, many persons helped me in collecting the information used when writing the articles; detailed listings have been given in the original publications and it would be impossible to repeat these here. In a few cases special help was obtained from certain individuals and institutions, and these are indicated in the footnotes to the chapters' titles. I would also like to acknowledge the help of two scientists, now deceased, who were participants in the evolution of chromatography, and to whom I could always turn for advise with respect to little known events and background material; they were Professor Richard L. M. Synge (1914–1994), the co-recipient of the 1952 Chemistry Nobel Prize, and Professor Ivo Hais (1918–1996) of Prague, Czech Republic. Special gratitude is due to Ms Debra Kaufman, head librarian of Perkin-Elmer Instruments (originally in Norwalk, and since 2001 in Shelton, CT) without whose help the gathering of all the referenced publications would have been impossible, and to Mr Bernard Dudek (Wethersfield, CT) who was of great help in the preparation of some figures for publication.

Some of the original articles on which the individual chapters are based have been written with the help of co-authors who provided much needed information on the subject. Their names and affiliations are listed below.

Chapters 2 and 7	Mr Heinz H. Bussemas Gemeinschaftspraxis für Laboratoriumsmedizin Dortmund, Germany
Chapters 4 and 5	Prof. Dr Karl I. Sakodynskii† M. S. Tswett Association of Chromatographers Moscow, Russian Federation
Chapter 8	Priv.Doz. Dr Veronika Meyer Swiss Federal Laboratories for Materials Testing St Gallen, Switzerland
Chapter 9	Prof.em. Dr Robert L. Wixom Department of Chemistry, School of Medicine University of Missouri Columbia, Missouri, USA
Chapters 10 and 23	Dr Peter J. T. Morris Science Museum London, United Kingdom
Chapter 16	Prof. Dr Huba Kalász Department of Pharmacology, Semmelweis University Budapest, Hungary
Chapter 18	Prof.em. Dr Charles W. Gehrke Department of Biochemistry University of Missouri Columbia, Missouri, USA

†deceased.

Leslie S. Ettre

Introduction: One Hundred Years of Chromatography

Chromatography is 100 years old. M. S. Tswett, its inventor, started its development during the early years of the 20th century and presented the first report on his early investigations in March 1903, at a local meeting in Warsaw, to an audience of 41 including his colleagues and students of the university. At that time the technique was in a somewhat embryonic state and Tswett was not yet sure about the final methodology. It took him three more years to finalize and describe it for international audience in the famous twin papers of 1906.

At the beginning Tswett's method was ridiculed as an oddity and he was considered a parvenu who tries to intrude in a field where he does not belong. A remark by Leon Marchlewski, then an internationally respected Polish scientist from Cracow, immediately after Tswett's fundamental publications is typical: he warned that Tswett should not believe that "a simple filtration experiment" (this is how Marchlewski characterized chromatography) would be enough for him to "swing himself to the height of a reformer of chlorophyll chemistry".[1] Even the otherwise polite Richard Willstätter, professor at the University of Munich and the highest authority in chlorophyll research, considered chromatography an "odd way" to carry out pigment research.[2] It is thus not surprising that, in the first 25 years after Tswett's publications, the use of chromatography was tried only in about half a dozen laboratories. Even as late as in 1929, there were scientists who still expressed their negative opinion about the importance of Tswett's invention. It is suffice to cite here F. M. Schertz, an American agricultural chemist,[3] according to whom

> it is evident that Tswett was never at any time dealing with pure pigments, for not once were the substances crystallized.

The big breakthrough finally came during 1930–1931 in Heidelberg, in the laboratory of Richard Kuhn, followed almost immediately by Paul Karrer, in Zurich, László Zechmeister, in Pécs, Hungary, and then many others. It is worthwhile to quote here Karrer's statement during 1939, just a decade after Schertz' dictum[4]:

> It would be a mistake to believe that a preparation purified by crystallization should be purer than one obtained from chromatographic analysis. In all recent investigations chromatographic purification widely surpassed that of crystallization.

The meteoric rise of the application of chromatography in the 1930s can be best illustrated by comparing the number of publications in the decade following Tswett's invention with the decade following Kuhn's activities. Between 1906 — Tswett's twin papers — and 1914 — the outbreak of the First World War — we can find a total of only nine publications (in addition to those of Tswett) in the international literature describing applications of chromatography. At the same time, the bibliography section in the second edition of Zechmeister's chromatography book published in 1938 lists a total of 550 publications for the period of 1930–1938.[5]

One of the reasons for the delay in the acceptance of chromatography was that Tswett's method represented a radical change in the existing philosophy of how complex natural substances are investigated. Instead of obtaining a single compound in crystal form, he separated all the individual compounds from the matrix and from one another. In other words, instead of isolating one single compound and discarding the rest, chromatography provided all (or at least most) compounds present in pure form, and permitted to do it by using only a small amount of the starting complex mixture. This change in the philosophy needed to appreciate the superiority of chromatography was best characterized in 1937 by G. M. Schwab, professor at the University of Munich. According to him[6] only after biochemistry, pressed by new problems, demanded methods for the reliable separation of small quantities of similar substances, could chromatography celebrate a rapid and brilliant resurrection.

These citations, the last three within a period of less than a decade, are given here to illustrate Tswett's struggle for acceptance and then the sudden change in the attitude of the international scientific community.

In descriptions chromatography is usually considered a technique. However, it is more than a simple technique: it is an important part of science encompassing chemistry, physical chemistry, chemical engineering, biochemistry, and is cutting through different fields. When introduced 100 years ago, it represented a new paradigm, and provided the theory and practice of interactions between two different phases. Also, while we primarily consider it a laboratory method, the amounts handled by chromatography cover many orders of magnitude. It is true that most of the samples analyzed are very small in the domain of microchemistry — let us not forget that it was gas chromatography which initiated the development of microsyringes, with capacities of less than one microliter (10^{-3} g) — and today, we routinely determine amounts in the nanogram (10^{-9} g) to picogram (10^{-12} g), and even to the femtogram (10^{-15} g) level; at the same time, however, we know of real industrial plants constructed in the former Soviet Union in the 1970s, using gas chromatography columns of 15–200 cm diameter, for the production of 200–1200 metric tons/year of pure compounds,[7] and in February 2007 the 19th International Symposium on Preparative (Liquid) Chromatography was held in Baltimore, Maryland, discussing mainly applications for biochemical and pharmaceutical separations.

Steps in the Evolution of Chromatography

The subject of our book is the evolution of chromatography. It actually starts well before Tswett and we may even contribute one of Moses' miracles to "chromatographic separation," probably as a natural process. Although the use of adsorption-type (partial) separation had been reported in the second part of the 19th century, Tswett is without any question the true inventor of the technique. His life and activities are a fascinating subject and we try to capture as much of it as possible, particularly since earlier discussions and publications included some errors. We shall also report in detail on the activities of the five pioneering scientists — Kränzlin, Dhéré, Palmer, Lippmaa, and Coward — who, in the two decades following Tswett's work introduced chromatography in their investigations.

As already mentioned the situation suddenly changed in 1930–1931, after the first publication from Kuhn's laboratory: chromatography was reborn and within a decade it became widely used, first in laboratories dealing with the study of natural substances, but soon also in chemical and biochemical laboratories, where separation of various substances was desired.

In the 1940s the technique of chromatography was also further extended, adding partitioning and ion-exchange as a means of separation to adsorption–desorption, and demonstrating that chromatography can be performed not only in a column, but also on a planar surface (paper chromatography).

In the early 1950s chromatography underwent another quantum leap by the introduction of gas chromatography (GC). For many of us the next two decades represented the most interesting period of our lives, when the development accelerated and almost every day brought something new. At that time gas chromatography even eclipsed liquid chromatography (LC), and was on its way to dominate alone the field of analytical chemistry.[8] The evolution of gas chromatography was also accompanied by a detailed study of its theoretical background, and finally the technique was based on a sound theoretical foundation. Then, as the next step, this gain in the theory of GC was used to investigate the possibilities of improving classical LC. As a conclusion, while the 1950s and early 1960s saw the meteoric evolution of GC, the second part of the 1960s saw the introduction of a new, more sophisticated version of LC. Its principles remained the same as used by Tswett in the first decade of the 20th century, but the results were much improved by systematically applying the theoretical conclusions learned in GC to the separation process in LC. In fact, the difference was so striking that even a new term, high-performance liquid chromatography (HPLC) was adapted for its characterization. This new version of liquid chromatography resulted in the unparalleled rise of its application, within about two decades even surpassing GC. This evolution is still continuing.

In addition to gas chromatography the 1950s also saw the development of size-exclusion (gel filtration) chromatography, the extension of planar chromatography into thin-layer chromatography, and the development of the automated amino acid analyzer, which we may consider as the first sophisticated liquid chromatography instrument.

Before the advent of GC, chromatographers put together their "chromatograph" using simple laboratory hardware. This would not have been possible anymore with GC: construction of the needed instrument was beyond the capabilities of a standard laboratory. Fortunately the decade after the Second World War saw the establishment of the new scientific instrument industry which in turn became involved in the development, manufacturing, and marketing of the sophisticated and increasingly automated instruments, first making gas and then liquid chromatography everybody's tool.

The second half of the 20th century also had one interesting development that greatly accelerated the rapid spreading of the newest innovations, both in equipment and applications: the organization of frequently held symposia where the newest developments were reported to a truly international audience. At the end of these meetings the participants rushed home, to try and apply all the new innovations they have learned at the symposium, both from the formal presentations and during the intensive formal and informal discussions, characteristic of these gatherings. In the last chapters of this book we report on the most important early symposia both in GC and LC, setting the trend.

The 32 chapters of this book guide the reader through the fascinating evolution of chromatography in the 20th century. In addition to the development of the most important milestones of the technique, the background of the individual inventions is also provided and information is given on the scientists' life and activities.

References

1. L. Marchlewski, *Ber. Dtsch. Botan. Ges.* **25**, 225–228 (1907).
2. R. Willstätter and A. Stoll, *Untersuchungen über Chlorophyll. Methoden und Ergebnisse* (Springer Verlag, Berlin, 1913).
3. F. M. Schertz, *Plant Physiol.* **4**, 337–348 (1929).
4. P. Karrer, *Helv. Chim. Acta* **22**, 1149–1150 (1939).
5. L. Zechmeister and L. Cholnoky, *Die chromatographische Adsorptionsmethode. Grundlagen, Methodik und Anwendungen*, 2nd edn. (Springer Verlag, Vienna, 1938), pp. 298–329.
6. G. M. Schwab and K. Jockers, *Angew. Chem.* **50**, 546–553 (1937).
7. V. G. Berezkin, in *Chromatography, a Century of Discovery*, eds. C. W. Gehrke, R. L. Wixom and E. Bayer, (Elsevier, Amsterdam, 2001), p. 529.
8. Anon, *Chem. Eng. News* **39**, 76 (July 3, 1961).

Part One

The Precursors of Chromatography

Chapter

1

Chromatography in the Ancient World*

Science developed gradually and, usually, new developments are based on some observation of natural phenomena. At first the observer could not explain it and therefore interpreted it as the result of the intervention of some supernatural forces; often it took then centuries until finally the original empirical observation could be explained. If we search the history of mankind and its activities, we can find a number of such "predecessors" to new inventions.

The situation is the same in chromatography. We can find a number of publications preceding the actual invention of the technique that may be interpreted as using "chromatographic" principles: I only need to remind my readers of the activities of Runge, in the middle of the 19th century,[1] or D. T. Day and Engels in the decade just preceding Tswett's work[2]; we shall discuss these below, in Chapters 2 and 3. But we may even go back to ancient times, to the Romans or even to the Bible, and find description of some empirical procedures or

*Based on the article by L. S. Ettre, published in *LCGC* (*North America*) **24**, 1280–1283 (2006).

tests, which a superfluous observer may interpret as resembling "chromatography." For example, the general textbook of Heftmann, quite popular for some time,[3] traced chromatography back to the Moses-lead Exodus of the Jews from Egypt and to Pliny the Elder, the great savant of Imperial Rome, living two thousands of years ago.

While I would not consider the quoted works as "chromatography," it may be of interest to start the history of the evolution of chromatography by pointing out the capabilities of our distant ancestors to utilize natural phenomena.

1.1. Was Moses The First Chromatographer?

The Bible deals in details with the Exodus of the Jews from Egypt and their long wandering in the wilderness before finally reaching the land promised them by God. Among others we can find the following narrative[4]:

> So Moses brought Israel from the Red Sea, and they went out in the wilderness of Shur; and they went three days in the wilderness, and found no water.
>
> And when they came to Marah, they could not drink of the waters of Marah, for they were bitter; therefore the name of it was called Marah.
>
> And the people murmured against Moses, saying, What shall we drink?
>
> And he cried unto the Lord and the Lord shewed him a tree, which when he had cast into the waters, the waters were made sweet.

Figure 1.1 shows the most likely route of the wandering of the Jews, indicating the possible location of "Marah." It is a Hebrew word meaning bitterness, and the local people still refer to the bitter waters of this area as *Al Buhayrat al Murrah*. Evidently, the existence of certain shrubs in that area which could be used to sweeten bitter water had been known since time immortal: this is mentioned by J. H. Hertz in his commentaries to the Pentateuch.[5] Thus, Moses most likely used this observation to solve their critical problem.

Using our present knowledge we may interpret Moses' miracle as ion exchange, thus, we may conclude that Moses used a kind of

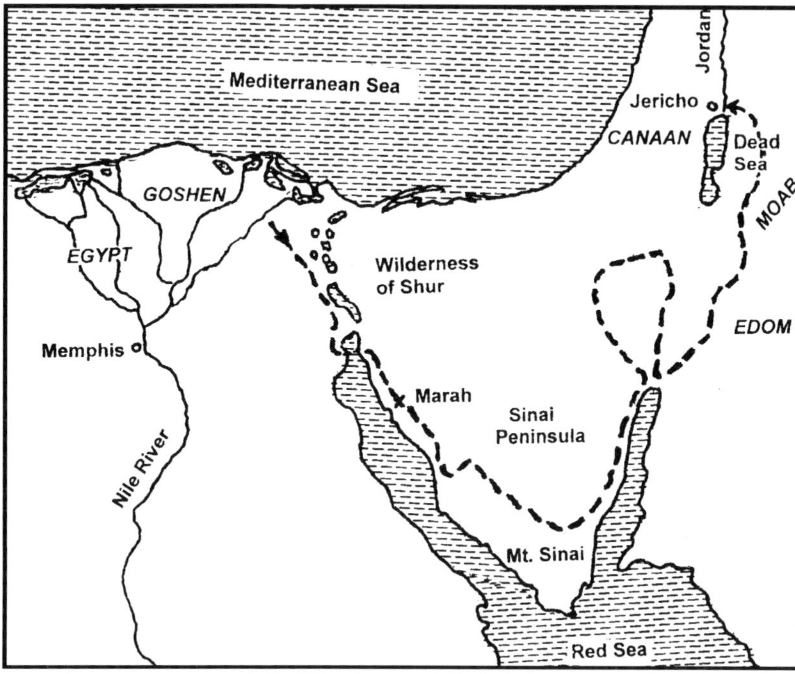

Fig. 1.1. The probable route of the Jews from Egypt to Canaan.

ion-exchange chromatography. It should, however, be mentioned that in "chromatography" we have a flowing stream, while the water of Marah was most likely stagnant. Thus, it is a matter of interpretation whether we consider Moses as the first chromatographer!

1.2. Did Pliny The Elder Use Planar Chromatography?

Gaius Plinius Secundus (23–79 AD) (Fig. 1.2) was the scion of a high-ranking Roman family. He had high government posts in Imperial Rome, has served in the military, and traveled widely in Roman-occupied Gallia, Spain, Germany, and North Africa. He was well educated, continuously studying nature and the foreign lands, and he wrote numerous books summarizing his findings. In 77 AD Pliny was appointed as the commander of the fleet in the Bay of Naples: He was there when in 79 AD Mount Vesuvius erupted, burying Pompeii

Fig. 1.2. Pliny the Elder. An imaginative portrait from the Middle Ages. (From *Botany Online*, an Internet text compiled by the late Professor Peter von Sengbusch (University of Hamburg).

and Herculaneum. Pliny went ashore to study closely the catastrophe and there he died through inhaling the poisonous gases emitted by the volcanic activity.

Pliny the Elder wrote a number of books; probably the greatest of his works is the *Historia Naturalis* ("Natural History"). This is an encyclopedic work encompassing all the knowledge of nature and science as known during Roman times: description of the world, geography, ethnography, anthropology, zoology, botany and agriculture, pharmacology, mineralogy, and art. In the Middle Ages Pliny's work was considered as the authoritative compilation of knowledge, regardless whether the discussion was based on observation and facts, or was simply pure fiction, reflecting even superstition.

The *Historia naturalis* is divided into 37 *libri* ("books"). Pliny the Elder published the first 10 books by himself in 77 AD, and was preparing the rest for publication when he died. Upon his sudden death, his nephew, Pliny the Younger (63–*ca.*113 AD), served as the executor of his will and the organizer of his literary remains, publishing also the rest of this work.

Two subjects in Pliny's work are sometimes related to what we call today planar chromatography: testing the purple dye used by the Roman upper class to dye the border of their toga, and checking whether *verdigris* is not adulterated. Pliny's full work is available in a modern bilingual (Latin–English) edition, translated and edited by H. Rakham.[6] Thus, I could study the pertinent passages.

The purple dye was extracted by the Romans from the fish *purpura* (sea purple) by a complex process well described by Pliny,[7] starting with the way to catch the fish, up to its testing. Historical discussions of chromatography sometimes mention this as spotting the dye on a piece of cloth and observing the development of the color, and interpret it as a forerunner of paper chromatography. However, the authors of these publications probably never read Pliny's original writing: what he actually suggested was to dip a piece of clean fleece in the aqueous dye extract and observe the intensity of the color. In other words, it was simply a dyer's test, and has nothing to do with chromatography.

The second test often referred to is related to *verdigris*. Verdigris (*vert de Grèce* or green of Greece; in Latin *aeruginis*) was prepared by the Romans by reacting copper and strong vinegar (forming copper acetate) and used as medicine. (Today the expression *verdigris* is also used for the greenish patina formed on copper, brass or bronze surfaces exposed to the atmosphere. However, that is chemically different: it is basic copper sulfate.)

Verdigris had been a popular remedy in Rome and was often adulterated; therefore, Pliny described the ways how it can be tested.[8] It is worthwhile to quote here the pertinent passage (my notes in Italics, in parenthesis):

> Rhodian verdigris is adulterated chiefly with pounded marble, though other use pumice-stone or gum. These adulterations can be detected by the teeth as they crackle when chewed. ... But the adulteration of verdigris that is the most difficult to detect is done with shoemakers' black (*this is iron sulfate*). ... Shoemakers' black can be detected by means of a papyrus previously steeped in an infusion

of plantgall (*containing tannic acid*): when smeared with genuine verdigris it turns at once black.

This is certainly a chemical test and is based on the reaction of copper with tannic acid. However, again, it is not chromatography: what happens is not separation, but reaction on a medium; interestingly, it is also somewhat similar to the work of Runge 1800 years later, who also carried out reactions on pieces of filter paper, and is close to the *Tüpfelanalyse* (spot tests) developed by Fritz Feigl in the 1930s. Thus, none of the tests mentioned by Pliny had anything to do with chromatography.

We can conclude that our ancestors were keen observers of natural phenomena and developed empirical tests based on such observations. However, we should not artificially make conclusions, with interpretations based on our present-day knowledge.

References

1. H. H. Bussemas and L. S. Ettre, *LCGC* (North America) **2**, 262–270 (2004).
2. L. S. Ettre, *LCGC* (North America) **23**, 1274–1280 (2005).
3. E. Heftmann, in *Chromatography — a Laboratory Handbook of Chromatographic and Electrophoretic Methods*, 3rd edn. ed. E. Heftmann (Van Nostrand Reinhold, New York, 1975), pp. 1–13.
4. *Exodus*, Chapter 15, §22–25 (King James Version).
5. J. H. Herz, ed., *The Pentateuch and Haftorahs*, 2nd edn. (Soncino Press, London, 1977), pp. 273–274.
6. H. Rackham, ed., and translator, *Pliny: Natural History, with an English Translation in Ten Volumes* (Harvard University Press, Cambridge, MA, 1938–1963).
7. Pliny the Elder, *Historia Naturalis*, Book IX, Chapters LXI–LXIII, §126–137.
8. Pliny the Elder, *Historia Naturalis*, Book XXXIV, Chapters XXV–XXVI, §108–113.

Chapter

2

Friedlieb Ferdinand Runge: "Self-Grown Pictures" as Precursors of Paper Chromatography*

Yale University's Beinecke Rare Book and Manuscript Library has many of the world's rarities. One could be of particular interest to chromatographers. It is a 43 cm × 26.5 cm book composed of black cardboard pages containing 32 multicolored pictures, each about 12 cm × 14 cm, and often shown in duplicate to illustrate that they can be reproducibly prepared. At first glance the pictures look like circular chromatograms on paper. They were made individually and glued on the cardboard pages: each has an explanation of how they were prepared. The title page has 22 color pictures around a centerpiece that gives the title of the book (in German), as the Od as the Driving Force of Formation of Substances, Visualized by Selfgrown

*Based on the article by H. H. Bussemas and L. S. Ettre, published in *LCGC* (North America) 22, 262–270 (2004).

Pictures.[1] The author is given as F. F. Runge, professor of technology; it is indicated that the book was privately published by him and its price is given as four Thaler. The book was published in Oranienburg (today a suburb of Berlin), in 1866.

Who was Professor Runge and why do we mention him as a milestone in chromatography? After all chromatography was invented almost 40 years after Runge's book. Still, due to his "self-grown pictures" Runge's work could be considered as the forerunner of paper chromatography.

2.1. Runge's Life and Activities

Friedlieb Ferdinand Runge is one of the important pioneers in medicinal and industrial chemistry of the first part of the 19th century, during the beginning of the industrial revolution. He was born on February 8, 1794, in Billwerder, a small village just southeast of the great German seaport Hamburg where his father was the town's pastor. When he was 16 years old Runge went to Lübeck, a town northeast of Hamburg, to start an apprenticeship in his uncle's pharmacy. After six years, however, he changed his mind and enrolled in the new University of Berlin (founded in 1809) to study medicine. Two years later, he moved to the University of Göttingen and then, within a year, to the University of Jena where he received his medical doctorate in 1819.

However, Runge never practiced medicine. In his last university years he turned more and more toward chemistry and the study of the poisonous compounds present in plants. Even his thesis for the MD degree was in this field and dealt with the study of atropine, a constituent of the plant belladonna. In his thesis Runge described a method he invented to prove the presence of atropine in solutions: inject a few drops of the poison-containing plant extract into a cat's eye, and significant dilatation of the pupil will immediately occur.

At that time Johann Wolfgang Goethe (1749–1832), the great German poet and natural philosopher, was living in Jena and he heard about Runge's finding; thus, he invited the young man to his house, to demonstrate his invention. The visit took place on October 3, 1819, and we have a detailed description of the visit by Runge himself.[2] His narrative explains how he, dressed in a borrowed tailcoat and a top hat, holding a cat under his arm, walked through the Market Square of the town to the great man's house, followed by a group of children.

While still in Jena, Runge started to write a book in which he summarized his investigations on the biologically active ingredients of plants. The first volume of this book was published in 1820, followed within a year by the second volume.[3] According to the explanation printed on the title page, the book is an "introduction to an improved analysis (*Zerlegungsweise*) of plants through theory and experiments." In addition to poisonous plants, Runge also investigated the extracts of a few other plants such as coffee beans and cinchona barks. The effect of these plants has been known for some time, but Runge was the first to discover the existence of the active substances, caffeine and quinine. (The coffee beans were actually given to him by Goethe during his visit, who suggested that he should investigate their active components.)

At the end of 1819, Runge went back to the University of Berlin to become a *Privat Dozent* in chemistry. There, in the spring of 1822, he finally obtained the PhD degree in chemistry (needed to become a *Privat Dozent*), with a thesis on the dye indigo. This represented the start of his lifelong interest in organic dyes and textile dyeing. As part of his thesis, Runge had to present two lectures, one in German to the faculty and the other in Latin to the public. The subject of the first lecture was the interaction of color and mass to the activity of plants, while the second was on the definition of poisons and nutrients. In this way, he finally had gained the right to lecture at the University on phytochemistry and technical chemistry. The latter subject — a brand new discipline — interested him more and more. Soon he would switch to it completely.

Runge was associated with the University of Berlin as a teacher for less than two years. In 1823 he moved to the University of Breslau, in Silesia (today: Wroclaw, in Poland), but soon embarked on a 2½-year long journey through Switzerland, France, the Netherlands, and England, as the companion of Carl Milde, the son of a rich manufacturer in Breslau, visiting laboratories and studying various industrial plants. When they returned in 1826, Runge became associated with the textile factory of Carl Milde's father. Finally, in November 1828, he was appointed an adjunct professor (*außerordentlicher Professor*) at the University of Breslau, to lecture on technical chemistry. He left the university in 1832 and joined the chemical factory of Dr Hempel in Oranienburg (a town just north of Berlin), as an industrial chemist. The factory was soon taken over by the Prussian State and renamed as the *Chemische Produkten Fabrik zu Oranienburg*. From 1832 until his retirement in 1852, Runge was associated with this factory, for many years as its technical director. However, in his books, he continued to call himself a professor of the University of Breslau.

The factory at Oranienburg was one of the very early German chemical factories; it was established in 1814. Until 1842 (when a fire destroyed most of the building, including Runge's apartment) it was partly located in the castle of Oranienburg (built in the 17th century), while production of chemicals in large volumes resulting in an unpleasant odor during manufacturing was carried out at the *Mühlenfeld*, an area just a few hundred feet outside city borders. After the fire this plant was enlarged to accommodate the whole company. Their products included sulfuric acid and various sulfate salts, potassium and sodium ferrocyanides (used in the production of "Prussian blue"), ammonium chloride, mixtures of unsaturated fatty acids (used by the textile industry), as well as soap and candles. A large part of their production was exported: data for 1845 show the export amounting to 46% of the total production, and of it, almost half went to New York City. At that time over 150 people worked at the factory.[4]

During his long association with the Oranienburg factory Runge contributed significantly to the improvement of its production methods and the widening of its product line. In addition, he also carried out major research, particularly related to coal tar: he was the first to isolate phenol and aniline, and prepare some colored dyes by their reaction with other compounds; he also isolated paraffin wax, which he used for the production of candles.

Runge retired from the factory by the end of 1852. From then on he devoted his time to his books, to his "chromatography" activities, and to some additional experimental work, among others making

Fig. 2.1. F. F. Runge with his home-made wine. Photography of F. W. Herms, Oranienburg, in the 1850s.

wine from various fruits. We have one photograph from this period showing Runge with a bottle of his favorite wine (Fig. 2.1).[a] During his retirement he received a pension from the King of Prussia. Runge died on March 25, 1867.

2.2. Runge's Chemistry Textbooks

Runge started to write chemistry textbooks while still in Breslau; the first was published in 1830[5] and he continued this activity for more than 15 years. His intention was to teach chemistry to everybody, and he apparently was successful in this. The popularity of his chemistry textbooks is best shown by the fact that his last book,[6] published in 1846–1847, was printed in 15,000 copies, an enormous number at that time.

[a]The figures used in this chapter are from the collection of H. H. Bussemas.

2.3. Investigation of Dyes

Runge's interest in colors was a consequence of his work at Milde's textile factory in Breslau, where he became involved in textile dyeing. He maintained his interest in this field even after moving to Oranienburg where he started to summarize his knowledge in a book series. The first volume of his *Farbenchemie* ("The Chemistry of Coloring") was published in 1834 and it discussed the dyeing of cotton;[7] the second volume dealing with textile printing was published eight years later;[8] and finally, the third volume handling the preparation of dyes was published in 1850.[9] All three books contain as illustrations small, original pieces of fabric, showing dye patterns. In the third volume he also dealt with the use of filter paper pieces to test dye solutions. To quote[9]:

> Due to its capillary force it separates a drop spotted on it into its components and ... creates a picture with a dark colored center part and lightly colored or even colorless rings or areas.

He also realized that filter papers can be used to study the reaction of dyes, and this is how he started to be involved in producing "chromatographic" patterns.

2.4. The Formation of Characteristic Patterns

Runge started to use filter strips to follow reactions fairly early. In his chemistry textbooks he illustrated chemical reactions in this way: when solutions of two reagents were spotted next to each other, the reaction products formed on the paper where the two different spots met. At the beginning of his work in this field, he definitely had some practical uses for such investigations. However, slowly he became more and more fascinated by the obtained pictures and started to "play" with the various possibilities, using a variety of reagents just to see what happened if they interacted on filter strips, forming various multicolored patterns.

By 1850 Runge collected a large number of such patterns and, simultaneously with the third volume of his *Farbenchemie*, he published a collection of these pictures. He announced its publication at the end of this volume, mentioning that he just published a book

Zur Farben-Chemie.

Musterbilder

für

Freunde des Schönen

und zum Gebrauch

für

Zeichner, Maler, Verzierer und Zeugdrucker.

1ste Lieferung.

Dargestellt

durch chemische Wechselwirkung

von

Dr F. F. Runge.

Professor an der Hochschule zu Breslau.

Berlin 1850.

Verlag von E. S. Mittler & Sohn

[Zimmerstraße Nº 84. 85]

Fig. 2.2. The title page of Runge's *Musterbilder* (1850).[10]

"about similar pictures that are of importance for the *Farbenkünstler* (color artist), and particularly for the pattern maker." It is a fairly large hardcover book[10] with a very long title, the English translation of which is To Color Chemistry. Pattern Pictures for the Friends of Beauty and for Use of Draftsmen, Painters, Decorators and Textile Printers, Prepared by Chemical Reactions (Fig. 2.2). In the literature this book is usually referred to by the main word of the German title as *Musterbilder*. The book was dedicated to King Frederic William IV, and Runge presented a copy to him; the king answered in a handwritten note, stating that he enjoyed the book very much.

A six-page long introduction explained the purpose of the book: to show that by mixing various substances on a filter paper strip, very special, multicolor pictures are obtained during the spreading of the spots. These pictures are highly specific to the starting substances and their mixing ratio, and even to the sequence of spotting. They are also highly reproducible. In Runge's opinion these pictures illustrate the reactions during mixing and separation of the various components much better than any experiment a teacher can carry out during a lecture. As he stated, "one can write a chemistry textbook with the help of these pictures and explain everything much easier." According to Runge, such pictures have a further advantage as compared with laboratory demonstration experiments: they are like a map, recording the whole reaction process, not just the end product of the reaction.

The introduction is followed by 21 pages with six 4 × 5 cm pictures on each page; the size of a page is 26 × 19.5 cm. In the text Runge already had begun to move away from the scientific view to an aesthetic — almost mystical — interpretation. He characterized the pictures as a "map true to nature" (*naturgetreite Landkarte*) of a certain area of chemistry, and described his own impressions when producing these pictures in the following way[10]:

> Suddenly a new world of formations, shapes and color mixtures displayed themselves here, which I could naturally never imagine and which were also contrary to all expectations, thus their reality surprised me even more.

Instead of further improvements in the analytical use of his "chromatograms," now Runge's main interest was directed toward making the "self-grown pictures" (*selbstständig gewachsene Bilder*) more colorful and aesthetic.

The pictures of the *Musterbilder* were produced by spotting the solution of one salt (or a salt mixture) onto the paper, which then was dried. Subsequently the solution of the second reagent (or a reagent mixture) was spotted on it. In some pictures he also added a third reagent. One example is shown in Fig. 2.3: it is picture No. 3 in *Musterbilder*. The reagents were potassium ferrocyanide ($K_4Fe(CN)_6$) and copper sulfate ($CuSO_4$): the paper was spotted with the cyanide solution and copper salt solution was then added.

Fig. 2.3. Picture No. 3 in Runge's *Musterbilder*.[10]

Five years after *Musterbilder*, Runge published its continuation entitled "The Driving Force for Formation of Substances, Visualized by Self-grown Pictures," this time by himself without the help of a publisher, although it still was sold by a bookstore. We refer to this book from the first German word of its title as the *Bildungstrieb*.[11] Now separation and analysis were practically forgotten: he characterized his goal as, "to exploit this art of painting and to obtain through various additives even more perfect and more expressive pictures."

The *Bildungstrieb* book looks more like a notebook compiled from large (43 × 26.5 cm) black cardboard pages with a total of 32 original pictures, each about 12 × 14 cm. Most of the pictures are included in duplicate to show that, by exactly reproducing the conditions, identical pictures are obtained. Each page has a listing of the reagents used, some explanation of their choice and a brief poetical description of the picture. Runge was using expressions such as "green sea with brown shores," "spear-like enclosures," "curled woman's collar," "a full flower with multileaved rays," or "two similar flowers fighting with each other to find the fortifications at their meeting boundary," etc. The book has an outside and an inside title page. The inside title

Fig. 2.4. The title page of Runge's *Bildungstrieb* (1855).[11]

page has a centerpiece giving the title, and other information, and is surrounded by 22 color pictures that have no explanation (Fig. 2.4). The outside title page is a slightly enlarged copy of the inside centerpiece without the surrounding pictures. At the end of the book, Runge explains that he engaged children to carry out the work of preparing the figures, by spotting the reagents on precut paper strips. (It should be emphasized that for both books, each figure had to be prepared individually for each copy!)

One example is shown here: it is picture No. 19 of *Bildungstrieb* (Fig. 2.5). The paper was spotted with solutions of manganese sulfate

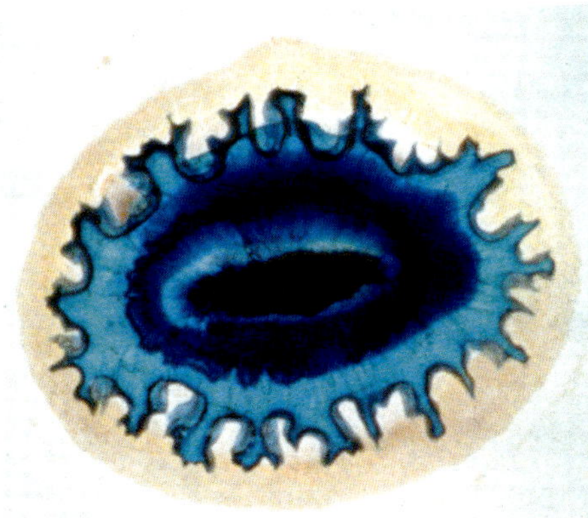

Fig. 2.5. Picture No. 19 in Runge's *Bildungstrieb*.[11]

($MnSO_4$) and ferrous sulfate ($FeSO_4$) to which, after drying, solutions of potassium ferrocyanide ($K_3Fe(CN)_6$) and oxalic acid were added.

2.5. Runge's Philosophy Concerning The "Self-Grown Pictures"

We already mentioned above that as early as 1850, in *Musterbilder* Runge moved toward an aesthetical, philosophical, and almost mystical interpretation of his pictures. This is even more evident in his *Bildungstrieb*.[11] Actually, this expression originated from Johann Friedrich Blumenbach (1752–1840), professor at the University of Göttingen, who considered *Bildungstrieb* ("driving force for formation") a special force existing within organisms that is "responsible for production, reproduction, and nutrition".[12] Similarly, Runge believed that color formation represents a new natural force, a special characteristic of life. To use Runge's own words[11]:

> I believe ... that in the formation of these pictures, a new, until now unknown, force is active. It has nothing in common with magnetism, electricity or galvanism. It is excited or attacked from the outside, but it is living within the substances, and becomes active when these equalize in their chemical opposition, that means bind and separate through attraction and repulsion. I name this force the driving force for formation (*Bildungstrieb*) and consider it as a sign of the vitality working in plants and animals.

With respect to Runge's belief that the formation of his colored pictures represents the manifestation of a special force of nature, it may be interesting to refer to the activities of Johann Wolfgang von Goethe. Goethe was also fascinated by the multitude of colors and spent decades investigating their interaction. His *Farbenlehre* ("The Science of Colors")[13] represents the summary of his work and his philosophical considerations. According to the summary in the book, Goethe's aim was

> ... to consider the chromatic appearances in connection with all usual physical phenomena, to place them in a sequence with the lessons of the magnet and tourmaline, with the revelations of electricity, galvanism and chemical processes, and in this way, prepare a unity of the physical knowledge through terminology and methodic.

It is known that after his visit in 1819, Runge became an admirer of Goethe, his writing and philosophy. Thus, it is not impossible that Runge's belief in the existence of a special force causing color formations has its origin in Goethe's *Farbenlehre*.

Today, we may consider these interpretations as somewhat outside the field of normal science and even philosophy. However, it is interesting to note that for *Bildungstrieb*, Runge received a special medal of the 1855 Paris World Exhibition, and the book was also honored at the World Industrial Exhibition held in London, in 1862. Thus, evidently Runge's contemporaries did not consider his interpretations as something "unscientific" or odd.

2.6. The "Od"

We cannot finish the discussion of Runge's work without mentioning this odd philosophical term and Runge's usage of it. It originated from Karl von Reichenbach (1788–1869), one of the important chemists of the first part of the 19th century. When studying different psycho-physical and psycho-physiological phenomena, Reichenbach introduced the idea of *das Od* ("the Od"). Under this term he meant[14]

> ... a hypothetical force to pervade all nature, manifesting itself in certain persons of sensitive temperament ... and exhibited especially by magnets, crystals, heat, light and chemical action.

Reichenbach published this concept in 1845 in the prestigious German periodical *Annalen der Chemie*. Although, eventually, leading scientists rejected his theory, the expression itself crept into the English language and remained there: it is, for example, still included in the 1961 edition of the *Oxford Dictionary of the English Language*[14] and in the 1976 edition of the *Webster*.[15]

Reichenbach's theory of the "Od" as a special force fitted well Runge's idea about the special color-forming force existing in nature. Thus, in 1866, 11 years after the publication of the original book and less than one year before his death, he republished *Bildungstrieb*, but now under a modified title: "The Od, as the Driving Force for Formation of Substances, Visualized by Self-grown Pictures".[1] The inner title page is identical to that of the 1855 edition, with the 22 pictures around the centerpiece. However, this centerpiece was now changed (Fig. 2.6): it was reset using different types of lettering indicating the new title, and a different ornament. There were also changes in the text of the title page: the 1855 edition mentioned that it is a continuation of *Musterbilder*; and this is now missing. The rest of the book — the pictures and the accompanying text — is identical to the 1855 edition.

We believe that Runge used the still-existing, unsold copies of *Bildungstrieb* for this "second edition." A close examination of the only existing copy in the Beinecke Rare Book & Manuscript Library of Yale University (New Haven, CT) revealed that the centerpiece of the title page is actually glued on another piece of paper of identical size. The most plausible explanation for this is that Runge changed the title but did not make an entirely new title page: he only covered the original centerpiece with a piece of paper that had the new title and information printed on it.

2.7. Runge's "Self-Grown Pictures" and Chromatography

In the 150 years that have passed since Runge's activities the evaluation of his work had been mixed. For example, Bechhold characterized

Das Od

als

BILDUNGSTRIEB

der

STOFFE.

Veranschaulicht

in

selbstständig gewachsenen

BILDERN

von

Dr. F. F. Runge,

Professor der Gewerbekunde.

ORANIENBURG, 1866.

(Selbstverlag.)

Fig. 2.6. Centerpiece of the title page of Runge's *Das Od als Bildungstrieb der Stoffe* (1866).[1]

it as "one of the most original scientific plays",[16] while Grüne called it "playing with colors and forms".[17] Weil and Williams used the expression "trick chromatography" when speaking about Runge's colored pictures; however, they also state that "one must recognize Runge's priority in the discovery of paper chromatography".[18] Thus, it is very difficult to make an objective judgment on Runge's place in the evolution of chromatography. There is no straight answer to this question.

As we have seen, at first Runge used his pictures on filter paper in a true analytical sense. Likewise the application to follow reactions can be considered primarily as analytical and not very far from the meaning of chromatography. Thus up to this point we are not incorrect if we term Runge's work as a precursor of chromatography. However, this cannot be said any longer about his mystical "self-grown pictures" where his only aim was to create complex, multicolor pictures by the reaction of various chemicals on filter paper, and particularly about his invocation of a special force of nature creating these pictures. Thus, although it would be unjustified to call Runge the originator of paper chromatography, we may still consider him as one of the precursors of the technique.

Independently of his possible role in the evolution of chromatography it is important to recognize that Runge has an important place in the history of chemistry: he was the first scientist who isolated a number of important substances from plants and coal tar and prepared synthetic organic dyes based on coal-tar chemicals. He also had noteworthy results in other fields, such as in the textile dyeing industry, and he contributed as a writer of textbooks to the popularization of chemistry as a scientific discipline.

References

1. F. F. Runge, *Das Od als Bildungstrieb der Stoffe, veranschaulicht in selbstständig gewachsenen Bildern* (Author's publication, Oranienburg, 1866).
2. F. F. Runge, *Hauswirthschaftliche Briefe: 1.–3. Dozen* (G. A. König's Verlag, Berlin, 1866; Reprint edition of the whole book: VCH Publisher, Weinheim, 1988). The visit to Goethe is described in the 36th Letter.
3. F. F. Runge, *Materialien zur Phytologie* (G. Reimer Verlag, Berlin, Vol. 1: 1820; Vol. 2: 1821).
4. M. Rehberg, *Friedlieb Ferdinand Runge, der Entdecker der Teerfarben* (Originally published in 1935; reprint edition: Museum of Oranienburg, 1993).
5. F. F. Runge, *Grundlehren der Chemie für Jedermann, besonders für Aerzte, Apotheker, Landwirte, Fabrikanten, Gewerbetreibende und alle Diejenigen, welche in dieser nützlichen Wissenschaft sich gründliche Kenntnisse erwerben wollen* (Grass, Barth & Co., Breslau, 1830; 2nd edn., 1833; 3rd edn., G. Reimer Verlag, Berlin, 1843).
6. F. F. Runge, *Grundriss der Chemie*, G. Franz, München; Part 1: 1846; Part 2: 1847.
7. F. F. Runge, *Farbenchemie, 1. Theil: Die Kunst zu färben, gegründet auf das chemische Verhalten der Baumwollenfaser zu den Salzen und*

Säuren. Lehrbuch der praktischen Baumwollfärberei (Mittler Verlag, Berlin, 1834).

8. F. F. Runge, *Farbenchemie, 2. Theil: Die Kunst zu drukken* (Mittler Verlag, Berlin, Posen and Bromberg, 1842).
9. F. F. Runge, *Farbenchemie, 3. Theil: Die Kunst der Farbenbereitung* (E. S. Mittler & Sohn, Berlin, 1850).
10. F. F. Runge, *Zur Farben-Chemie. Musterbilder für Freunde des Schönen und zum Gebrauch für Zeichner, Maler, Verzierer und Zeugdrucker. Dargestellt durch chemische Wechselwirkung* (E. S. Mittler & Sohn, Berlin, 1850).
11. F. F. Runge, *Der Bildungstrieb der Stoffe, veranschaulicht in selbstständig gewachsenen Bildern (Fortsetzung der Musterbilder)* (Author's publication, Oranienburg, 1855).
12. L. Kuhnert, Runge-Bilder und Liesegang-Ringe, in *Komplexität-Zeit-Methode. Gestalt und Selbstorganisation* ((ed. U. Niedersen) Wissenschaftliche Beiträge der Martin-Luther-Universität, Halle-Wittenberg, 1988), pp. 53–70.
13. J. W. v. Goethe, *Die Farbenlehre*, Vol. I–II. (J. G. Cottasche Buchhandlung, Tübingen, 1810). For the translation here the text of Vol. 10 of Johann Wolfgang Goethe's *Sämtliche Werke* was used. (Carl Hanser Verlag, München, 1989), p. 975.
14. *The Oxford Dictionary*, Vol. VII (Clarendon Press, Oxford, 1961), p. 58.
15. D. B. Guralnik (Editor-in-Chief), *Webster's New World Dictionary of the American Language*, 2nd College edn. (The World Publishing Co., Cleveland, OH, 1976), p. 985.
16. H. Bechhold, *Z. Phys. Chem.* **52**, 185–199 (1905).
17. A. Grüne, *Österr. Chem. Ztg.* **60**, 301–311 (1959).
18. H. Weil and T. Williams, *Naturwissenschaften* **40**, 1–7 (1953).

Chapter

3

Early Petroleum Chromatographers*

Toward the end of the 19th century the rapidly evolving petroleum production also initiated speculations concerning the origin of petroleum, and attempts have been made to explain the reason for the obvious difference in the characteristics of crude oils occurring at different locations. One prominent American chemist involved in such work was David T. Day, whose theory (the "filtration hypothesis") was based on the assumption of migration of crude oil in the earth through various strata and selective retardation of certain compounds or compound groups during this migration. In order to prove the validity of his assumption Day carried out some model experiments and demonstrated, that indeed some fractionation can be achieved in this way (whether it had anything to do with the origin of petroleum is another matter). His experiments were repeated in Germany by Carl Engler, and their publications also inspired some activities by other petroleum chemists. The technique of Day and Engler somewhat

*Based on the articles by L. S. Ettre published in *LCGC* (North America) **23**, 1274–1280 (2005) and **24**, 54–56 (2006).

resembled chromatography, however it was a kind of dead-end street, and there was no cross-influence between their work and Tswett's. Still, we may consider their approach as a forerunner of chromatography. Their activities are the subject of this chapter.

3.1. David T. Day

David Talbot Day (1889–1925), a graduate from the Johns Hopkins University in Baltimore, Maryland, had been associated from 1885 on with the US Geological Survey, in Washington, DC. First he pioneered in systematizing information on the mineral resources of the United States; then, in the 1890s, he turned his attention to various aspects of the newly emerging petroleum industry. Finally, in 1920 he established a private laboratory in California, investigating the possibilities of recovering oil from shale.

By the last decade of the 19th century oil production in Pennsylvania and Ohio was well established. Although chemical investigation of petroleum was still in its infancy, relying mainly on measuring its physical characteristics, it was noted that the characteristics of crude oils from different locations are different: one is heavier and contains more higher boiling components than the other. Various explanations have been suggested for this difference, among them the so-called filtration hypothesis of Day, first described in a paper presented in 1897.[1] He assumed that originally, primary oil existed at a central location; this oil then migrated through various rock formations with the help of diffusion. Various strata retarded more or less of the heavier fractions and thus, the oils which passed through them and then accumulated at their final locations had different compositions.

In 1900, a famous international exposition (forerunner of the present-day world's fairs) was held in Paris, France, and there, the United States had a representative pavilion, with a special display of the American petroleum industry: Day was in charge of this exhibit. In conjunction with the exposition, the First World Petroleum Congress was held in August and there, Day presented a paper on his hypothesis, "The variation of the characteristics of crude oils from Pennsylvania and Ohio".[2] After describing the characteristics of different samples, he mentioned that he also carried out some "filtration experiments" by conducting crude oil through fullers' earth, resulting in some fractionation: the lighter fractions moved faster, before the heavier fractions, and this process imitates the movement of oil in the earth. However, he apparently did not give any actual data, nor did he describe the experimental setup (at least it is not included in the published text of his paper). After referring to a "German scientist" (most likely to Engler, but no name was given) who stated that it would be important if somebody could develop a method enabling the separation of individual hydrocarbons present in the crude oil, Day made the often quoted (actually misquoted) statement that (translation of the text published in French)

> ... the filtration method offers good hope ... (and) I believe that before the next winter, I will be able to accomplish complete separation.

We do not know whether Day's statement reflected his optimism or an anticipation of future results. However, the realization of this hope was definitely beyond his capabilities and it took decades until his expectation could finally be accomplished by a different generation of petroleum chemists.

In the years following the Paris lecture, Day carried out some additional investigations but did not publish them; we only know of a lecture presented in 1903 at a meeting of the Geological Society of Washington, DC (briefly reported by Mendenhall[3]). In the following decade he had two more papers: one coauthored with Gilpin[4] and a more detailed paper in which he extended his theory to the oils of California, Louisiana, Mexico, Oklahoma, Texas, and Wyoming, but again, without actual data.[5] Finally, in 1922, Day published a handbook for the petroleum industry[6] and in it he briefly discussed adsorption processes; as an illustration he presented a table summarizing the results of Gilpin and Cram[7] (see below). Otherwise, however, he did not deal anymore with such investigations.

3.2. Joseph E. Gilpin

As mentioned earlier, Day was a graduate from the Johns Hopkins University and apparently, he maintained some contact with his

Table 3.2. Fractionation of an American crude oil sample according to Engler and Albrecht.[a,10]

Fraction	Fraction volume, cm^3	Specific weight (at 15°C)	Description of the fraction
1	70	0.7812	colorless clear
2	65	0.7870	clear with light fluorescence
3	120	0.7897	colorless with light fluorescence
4	70	0.7905	colorless with light fluorescence
5	50	0.7913	colorless with light fluorescence
6	110	0.7919	light yellow with green fluorescence
7	100	0.7920	somewhat more yellow
8	400	0.7965	yellow with green fluorescence
9	50	0.8020	yellow with green fluorescence
10	100	0.8032	yellow with green fluorescence
11	300	0.7976	orange with green fluorescence
12	500	0.7962	orange with green fluorescence

[a]The tube filled with fullers' earth was 86 cm (2.8 ft) long, with 4 cm diameter. The original oil sample was brown with green fluorescence; its specific weight was 0.7929. The rate of passage of the oil through the tube was about 100 cm^3 per 1–2 h. Viscosity was measured with an Engler viscosimeter. The fractions were collected after passing through the whole tube, at its top (see Fig. 3.2(A)). The selection of the fractions' volumes was arbitrary.

Most likely the reason for this was that Gilpin's primary aim was (just as that of Day) to investigate changes occurring during the passage of oil *in the strata* (modeled by the fullers' earth) through various geological formations, and not obtaining individual compounds. Another question one may ask is: since Gilpin's publications preceded Tswett's, why did he not cite him? Most likely Gilpin was not aware of Tswett's work (it would be very unlikely that he read the issues of the Bulletin of the German Botanical Journal), but even if he would have known about it, he would not consider any relationship to his investigations.

3.4. Other Scientists

We do not know of any further work of Engler along this line, however, a few petroleum chemists reported on some follow-up investigations. We should briefly mention here three: Ubbelohde and two Russian chemists.

3.4.1. *Leo Ubbelohde*

Leo Ubbelohde (1876–1964) studied at the University of Berlin and from 1907 on had been associated with the Technical University of Karlsruhe, from 1910 on as a professor. In 1933 he moved to the Technical University of Berlin–Charlottenburg. He had been very active in various associations in the field of petroleum research and production. Between 1907 and 1914 he served as the secretary general of the International Petroleum Commission and in 1933 he established the German Society for Mineral Oil Research. He was also active in the fields of textiles and natural fats. We are interested here in one of his papers from 1909 criticizing a paper of Rakusin (see below). In this paper he stated among others, that the theoretical basis of the "filtration experiments" of Day and Engler has nothing to do with capillary action but is based on selective adsorption.[12]

3.4.2. *Russian Petroleum Chemists*

Commercial exploitation of the oil deposits along the Caspian Sea, near Baku (in present-day Azerbaijan) began in 1872 and by the beginning of the 20th century it represented the largest oil field in the world. (Today it is practically unknown that these fields have been developed by brothers of Alfred Nobel and the income from oil production significantly contributed to the immense wealth of the family.) A number of important Russian chemists have been associated with the petroleum industry and they also carried out intensive research. However, although they published in well-known Russian and foreign scientific journals, a historical evaluation of their activities is still missing: essentially we know about them only from the review papers of Herbert Weil[13–15] and from the discussion of Camin and Raymond.[16]

With respect to our subject the work of two Russian scientists should be mentioned: they are M. A. Rakusin and V. E. Herr. Both followed the investigations of Day and Engler, studying the fractionation of crude oil by "filtration" through fullers' earth. M. A. Rakusin[17] also believed that capillary action serves as the basis for this process, and Ubbelohde's quoted paper[12] is a polemic discussion of this incorrect interpretation (and some other statements of Rakusin). The other

Russian chemist is V. E. Herr who, in 1909, published a paper in the German petroleum journal on "contribution to the filtration of crude oils from Baku through fullers' earth".[18] In this paper he concentrated on the possibility to eliminate the high-molecular-weight aromatic fraction of crude oil by selective adsorption. It is of interest to quote his statement that his investigations clearly showed that "fractionation in the usual sense (i.e. separation of individual components) does not take place."

3.5. Controversy

In 1927 the American Petroleum Institute started its Project No. 6, with the aim of separation and identification of the individual chemical constituents of petroleum and petroleum products, a project lasting for 40 years.[11,16] In this project fractionation on silica gel was first used, followed by increasingly sophisticated methods and the work of Day, Gilpin, and others achieving only limited fractionation was slowly forgotten. After some decades finally László Zechmeister, one of the pioneers of (classical) liquid chromatography (see Chapter 13), brought back their work from oblivion. In a lecture on the "History, Scope and Methods of Chromatography" presented at the Chromatography Conference of the New York Academy of Sciences held in November 1946, he discussed among others the origin of chromatography and the activities of Tswett as the inventor of the technique; he also mentioned the investigations of Day and Gilpin and correctly stated that their work "might have developed into systematic chromatography"; since, however, this did not happen, they can only be considered as forerunners of chromatography.[19] Zechmeister repeated this discussion in the introduction of his book published in 1950.[20]

Just a few months after the publication of this book, a polemic article was published in *Nature* by Herbert Weil and Trevor I. Williams,[21] criticizing Zechmeister that he was unjust to Day by naming Tswett as the inventor of chromatography: According to them this distinction belongs to Day. Soon after this brief article the first of Weil's series on the history of "industrial petroleum chromatography" was also published[18] and in it, he further elaborated on this question. Weil and Williams falsely interpreted Day's 1900 lecture in Paris, claiming that in it some form of chromatography had been described in detail (apparently they did not read Day's paper). Therefore, they stated that August 20, 1900 (the day of Day's lecture in Paris) should be considered as the "birthday of chromatography."

Zechmeister politely tried to answer Weil and Williams, emphasizing that instead of diminishing Day's merits, he actually resurrected it from oblivion.[22] However, Weil and Williams continued this polemics, repeating their (unjust) claim that Day "did in fact discover and clearly describe a system of adsorption chromatographic analysis".[23] Of course, we have seen in the discussion above that Day did *not* describe his system (this was done only eight years later by Gilpin) and that neither he nor his followers attempted to separate individual hydrocarbons. The papers of Weil and Williams clearly represented a misquotation, or misinterpretation, of Day's lecture.

Very wisely Zechmeister stopped this polemics and did not continue to argue with Weil and Williams. Today the role of Tswett as the true inventor of chromatography is universally recognized; at the same time, however, we should also recognize — just as suggested by Zechmeister — Day and Engler as *forerunners of chromatography,* first pointing out the possibility of fractionation of petroleum by selective adsorption.

3.6. Chromatography and the Cold War

As mentioned above, in 1950 Weil and Williams challenged Tswett's priority as the inventor of chromatography and claimed that this distinction should belong to D. T. Day, for his 1900-lecture at the First World Petroleum Congress, held in Paris. This polemics was not more than a tempest in a teapot and was soon forgotten. However, quite unexpectedly, a very harsh rebuttal was published from the other side of the Iron Curtain, accusing Weil and Williams with a sinister bourgeois-capitalist plot, trying to diminish the achievements of Russian science.

This polemics had been so typical for the Cold War period, that it is worthwhile to briefly deal with it. In order to understand the background of this ridiculous claim we have to go back 60 years to the end of the Second World War, to the victory of the Soviet Union over Germany, against great odds after immense sacrifices and losses. It is understandable that this resulted in a high pride among the people of the Soviet Union, and this feeling was exploited by their leaders, initiating an intensive propaganda using the victory as a sign of the superiority of their political system. This campaign was also extended to all other fields of life and the victory was considered as a proof of the superiority of the Russian people over "Western Capitalist-Imperialist" society. Suddenly, past accomplishments of the Tsarist Empire — until then condemned — were resurrected and considered as part of an unbroken tradition, and even minor advances were blown up to sound as at least equal or even superior to western achievements.

This propaganda also had a second aim. At the end of the Second World War the Soviet Union acquired a vast territory on its western borders, occupying a number of countries which, for centuries, belonged to the western world, were part of western culture and civilization. Now, after incorporating them into their empire, the Soviet leaders wanted to prove to them that the Communist system and culture are superior, and that past and present achievements of Russia and the Soviet Union surpassed those of the decadent west.

The Cold War further intensified this attitude. Besides the dangerous potential military confrontations the Soviet leaders became extremely sensitive to any statement that, in their interpretation, tried to belittle their achievements. Even harmless remarks were interpreted as deliberate provocation against the Soviet Union.

The polemics between Zechmeister, Weil, and Williams seemed to prove to them that their fear was valid: representatives of the bourgeois west again try to steal a Russian invention. Therefore, their polemic articles in *Nature*[21,23] initiated a quick answer from Moscow: two Russian scientists, Kh. S. Koshtoyants and K. F. Kalmikov, published a three-page long article in the Russian journal *Biokhimiya*[24] in which they defended Tswett's priority, but in a relatively objective summary of his work. But, of course, this alone would not satisfy the obligatory trend of demonstrating that their country was way ahead of everybody generations earlier. Therefore, they stated that actually, a Russian scientist in 1786 (more than a century before anybody!) already separated substances from solution by adsorption. In addition, they also cited S. K. Kvitka, a Russian petroleum engineer, who on June 17, 1900 (i.e. 2 weeks before Day's lecture at the First World Petroleum Congress, in Paris), was supposed to have submitted a report to the Technical Committee in Baku (the center of Russia's oil production) in which he "correctly interpreted" the process of separation by adsorption.

Of course, the references in the article of Koshtoyants and Kalmikov were untraceable; but still, the language of their article was within the norms of scientific publications. However, apparently, the editors of the journal *Biokhimiya* did not consider it politically correct enough, and therefore, as an introduction to the article they added a strong editorial entitled Questions of Priority of Native Biochemistry, signed by the Editorial Board.[25] In this, they referred to the existence of a "Western plot" aiming "to disparage the achievement of advanced science in the progressive countries" (i.e. the Soviet Union). Next, they stated that "the facts concerning the concealment of the research of Soviet scientists by the authors of bourgeois countries (particularly the United States) take on such a systematic nature that they leave no doubt of the premeditation of this phenomenon" and concluded that

> a demonstrative example of the nature of such an attempt is the rather cunning contrivance of certain "pseudo-historians" with regards to such an indisputable and extremely important achievement of Russian science as the creation of the basis of the chromatographic method by M. S. Tswett.

Almost certainly, the editors of the journal never read the questioned articles in *Nature:* they blame American authors although the questioned articles were published by British scientists, and obviously they did not know anything about the original articles of Zechmeister, an American scientist, who actually *defended* Tswett's priority. Mentioning this would have been contrary to the official anti-American policy.

This official Russian reaction made Mikhail Tswett — born in Italy and educated in Switzerland, a truly cosmopolitan scientist, who in his lifetime faced discrimination in Russia by the stratified science establishment because of his foreign background, and who actually carried out his work in (Russian-occupied) Poland — suddenly a native of Russia (what he was not) and a symbol of the superiority of Russian (and hence, also Soviet) science.

Note. I learned about this controversy about 30 years ago when I had the opportunity to go through the files of L. Zechmeister in the R. A. Millikan Library of California Institute of Technology, in Pasadena, preparing a study on Zechmeister's life and activities. I found there in a large box the typewritten translation of the two Russian papers.[24,25] Apparently Zechmeister learned about these papers and had them translated; however, according to my best knowledge, he never publicly discussed their content: probably he was just smiling when reading this manifestation of human stupidity...

References

1. D. T. Day, *Proc. Amer. Phil. Soc.* **36**, 112–115 (1897).
2. D. T. Day, La Variation des Caractères des Huils Brutes de Pennsylvania et de l'Ohio, *Congrès International du Pétrole, Notes, Memoires et Documents. Journal de Pétrole, Paris*, 1902; Vol. I, pp. 53–56.
3. W. C. Mendenhall, *Science* **17**, 1007–1008 (1903).
4. D. T. Day and J. E. Gilpin, *Ind. Eng. Chem.* **1**, 449–455 (1909).
5. D. T. Day, *Trans. Amer. Inst. Mining (Metal.) Engrs* **44**, 219–224 (1911).
6. D. T. Day, ed., *Handbook of Petroleum Industry* (Wiley, New York, 1922).
7. J. E. Gilpin and M. P. Cram, *Amer. Chem. J.* **40**, 495–537 (1908).
8. J. E. Gilpin and O. E. Bransky, *Amer. Chem. J.* **44**, 251–303 (1910).
9. J. E. Gilpin and P. Schneeberger, *Amer. Chem. J.* **50**, 59–100 (1913).
10. C. Engler and E. Albrecht, *Z. Angew. Chem.* **14**, 889–892 (1901).
11. L. S. Ettre, Evolution of Liquid Chromatography, in *HPLC — Advances and Perspectives*, Vol. 1, ed. Cs. Horváth (Academic Press, New York, 1980), pp. 1–74.
12. L. Ubbelohde, *Petroleum* (Berlin) **4**, 1394–1397 (1909).

13. H. Weil, *Petroleum* (London) **14**, 5–12, 16 (1951).
14. H. Weil, *Petroleum* (London) **14**, 205–210 (1951).
15. H. Weil, *Petroleum* (London) **15**, 9–12, 18 (1952).
16. D. L. Camin and A. J. Raymond, *J. Chromatogr. Sci.* **11**, 625–638 (1973).
17. M. A. Rakusin, *Petroleum* (Berlin) **5**, 760 (1910).
18. V. F. Herr, *Petroleum* (Berlin) **4**, 1284–1287 (1909).
19. L. Zechmeister, *Ann. N.Y. Acad. Sci.* **49**, 145–160 (1948).
20. L. Zechmeister, *Progress in Chromatography 1938–1947* (John Wiley & Sons, New York, 1950), p. 3.
21. H. Weil and T. I. Williams, *Nature* (London) **166**, 1000–1001 (1950).
22. L. Zechmeister, *Nature* (London) **167**, 405–406 (1951).
23. H. Weil and T. I. Williams, *Nature* (London) **167**, 906–907 (1951).
24. Kh. S. Koshtoyants and K. F. Kalmikov, *Biokhimiya* **16**, 479–481 (1951).
25. Editorial Board, *Biokhimiya* **16**, 478 (1951).

Part Two

M. S. Tswett and the Discovery of Chromatography

Chapter

4

M. S. Tswett, and the Invention of Chromatography Part I: Life and Early Work (1872–1903)*

On 8 March 1903 (old Russian calendar: it corresponds to March 21), M. S. Tswett, an assistant at Warsaw University, presented a lecture at the meeting of the Biological Section of the Warsaw Society of Natural Sciences, titled On a New Category of Adsorption Phenomena and Their Application to Biochemical Analysis. In this lecture, he discussed his wide-ranging investigations of leaf pigments performed during the previous couple of years. These investigations led to the development of a special adsorption technique that permitted the separation of the leaf pigments. In subsequent years, he further refined this technique, which eventually became known as chromatography.

*Based on the articles by L. S. Ettre and K. I. Sakodynskii, published in *Chromatographia* **35**, 223–231 (1993) and by L. S. Ettre, published in *LCGC* (*North America*) **21**, 458–467 (2003) and *LCGC Europe* **16**, 632–640 (2003). For information concerning the University of Warsaw the help of Prof. E. Soczewinski (University of Lublin) is gratefully acknowledged.

In this and the next chapter we shall investigate the stages of Tswett's thinking that led to the development of chromatography and consider the way chromatography eventually became the most widely used laboratory technique. We shall start with a brief summary of Tswett's life and his struggle with the Russian scientific establishment.

4.1. The Life of M. S. Tswett

Mikhail Semenovich Tswett was born on 14 May 1872 in the small Northern Italian town of Asti. His father, Semen Nikolaevich Tswett (1829–1900) was a Russian official, while his mother, Maria Dorozza (about 1846–1872) was a descendant of Venetian settlers in present-day Turkey, most likely from the 14th century. She was actually born in Turkey but grew up in Russia. They planned to spend an extended holiday in Arona, one of the resort towns on the *Lago di Maggiore*, the beautiful lake in Northern Italy. They arrived in Genoa, Italy, by ship from Russia, and continued traveling by train to their destination. However, they had to interrupt their journey in Asti: there Semen's wife gave birth prematurely to a son and died soon after childbirth. Mikhail's father took his infant son, with a wet nurse, to Lausanne, Switzerland, and during the next 24 years Mikhail lived in Switzerland, first in Lausanne and then in Geneva. His mother language was French and he learned Russian from his father only when he was a teenager. Actually, he had problems with this language even years after moving to Russia: according to recollections of his contemporaries he spoke Russian with a French accent and after he married in 1907, he preferred to speak French with his wife. He was also fluent in German, spoke some Italian, and understood English.

After finishing high school in Lausanne, Mikhail studied at the University of Geneva, majoring in botany. He received his Ph.D. in 1896 (Fig. 4.1). His thesis dealt with investigations of the structure of plant cells, the movement of the protoplasm, and the structure of chloroplasts, components of plant cells.[1] After finishing his studies, he suddenly decided to repatriate to Russia to join his father. He had high hopes that the Russian scientific establishment shall great him with open arms, with his Swiss doctorate. However, as they say, he fell flat

Fig. 4.1. M. S. Tswett in Geneva, circa 1896.

on his face: nobody was interested in him (according to his own words, he was "alien to everybody"). Apparently, he was unaware of the strict Russian system: even a junior academic position required a Russian magister's (master's) degree, while a senior position required a Russian doctor of science degree, and foreign degrees were not accepted. Thus, his Swiss Ph.D. was irrelevant.

Finally, in December 1896, Tswett found a temporary position in a laboratory in St. Petersburg. He first tried to resubmit his Swiss doctorate thesis to a Russian university, but this was not permitted. Thus, he had to start to work on a new thesis for his Russian master's degree, which he finally obtained in September 1901 from the University of Kazan'. At the end of the year, he accepted a junior position at the University of Warsaw, in the Russian-occupied part of Poland. He spent the next 14 years in Warsaw, first at the university and then, from 1908 on, at the Polytechnic Institute. In 1915 when German troops occupied Warsaw, Tswett had to flee with the

Polytechnic Institute which then settled in Nizhnii Novgorod where teaching started in August 1916. Finally in 1917, he obtained a full professorship at the University of Tartu (in present-day Estonia, then part of Russia), but soon he had to leave again because of the German occupation of the region. At the end of 1918, he started again as a professor at the new State University of Voronezh. By that time, he was already seriously ill, and he died on 26 June 1919.

4.2. Early Investigations

The subject of Tswett's Swiss Ph.D. thesis was the first demonstration of his interest in plant pigments. It was thus obvious that when he realized that he had to submit a new thesis for the Russian master's degree, he also selected its subject from this field. Tswett considered this process an unnecessary burden, only needed "to please protectionism in our country" (his own words), but he became more and more involved in these investigations. He selected the study of chlorophyll as the subject of his thesis and soon had new ideas, improving the existing, fairly scarce knowledge about this plant pigment. As we shall see, this work eventually led to the development of chromatography.

Naturally, the first step in his investigations was the extraction of the pigments from the leaves. He observed that different solvents behaved differently. For example, the pigments could be extracted easily from the leaves with ethanol or acetone; however, petroleum ether (a mixture of C_5–C_6 hydrocarbons) and ligroin (a mixture of higher paraffins with a boiling point range of 135–145°C), which easily dissolve chlorophyll and other associated pigments when they are available in isolated form, will extract only certain pigments (using our present-day nomenclature, the carotenoids) from the leaves, while chlorophyll will remain there. This observation was not new: however, past researchers attributed it to solubility problems or a chemical change of the structure of pigments rendering them soluble or insoluble. Not accepting this traditional thinking, Tswett assumed correctly that the reason for this behavior might be the interference of some molecular forces binding the pigments to the leaf substrate and that these forces depend upon the individual pigments; for some, such as chlorophyll, they are stronger than for others. Only solvents with a dissolving power stronger than that of the binding molecular forces can be used for the extraction of a particular pigment. On the other hand, after the pigment is extracted and these molecular forces no longer exist, even the weaker solvents can dissolve all the pigments easily. Tswett correctly identified adsorption as the basis of these molecular forces.

After drawing this conclusion the next logical step was to try to imitate the process by using a substrate that would behave similarly to the tissue of plant leaves. He selected filter paper, which also consists of cellulose. After extracting the pigments from the leaves with ethanol, he evaporated the solvent and redissolved the residue in ligroin; next, he impregnated the filter paper with this solution. The paper tainted with the pigments behaved exactly in the same way as the original green leaves: ligroin extracted only the carotenes, but after the addition of a small amount of ethanol, all pigments could be easily retrieved.

The title of his master's thesis submitted to the University of Kazan' and summarizing these experiments was The Physico-Chemical Structure of the Chlorophyll Particle: Experimental and Critical Study,[2,3] and it represented a detailed report of these studies. This degree finally qualified him for an appointment at a university, and he applied immediately for a position at the University of Kazan'. Meanwhile, however, an acquaintance of his from St. Petersburg, D. I. Ivanovskii (the discoverer of the tobacco mosaic virus) had just been appointed as a professor and the head of the Department of Plant Anatomy and Physiology at Warsaw University, and he invited Tswett to join him there. Tswett accepted his invitation and moved to Warsaw at the very end of 1901. However, first he participated at the Eleventh Congress of the Russian Society of Natural Scientists, held in St. Petersburg, on 20–30 December 1901. Senchenkova in his comprehensive biography of Tswett mentions that he presented three papers at the Congress of which one is of interest to us: he spoke on the last day on Methods and Objectives of Physiological Search of Chlorophyll.[4a] Senchenkova refers to the Proceedings of the Congress which was published in 1902; according to the summary in this publication, Tswett reported on the use of adsorption for the purpose of separating a mixture of plant pigments. However, we have no further information on this and probably it dealt with some investigations as part of his master's thesis.

4.3. In Warsaw (1901–1903)

Dmitrii I. Ivanovskii (1864–1920) was previously associated with the University of St. Petersburg. He and Tswett were on friendly terms and evidently he knew about Tswett's problems with acceptance by the Russian scientific establishment. Thus, it is logical to believe that he felt, Tswett will have a more tolerant atmosphere in Warsaw.

The University of Warsaw (Fig. 4.2) was founded in 1816 as a Polish school but was closed after the 1830 uprising against Russian occupation and reopened only in 1862. In 1869 it was transformed into a Russian Imperial University, now with Russian as the teaching language, and the Polish professors were forced to teach in Russian or resign. This is the period when great effort was made in the russification of the various ethnic groups within Imperial Russia: for example, Russian scholars joining Warsaw University as faculty

Fig. 4.2. The main building of Warsaw University in 1902. Tswett's laboratory was in this building. (Contemporary picture card; courtesy of Prof. E. Soczewinski.)

members even received some supplementary payment and the requirements for Russian faculty members was lowered. This forced trend is well reflected in the number of professors with Polish ethnic origin at Warsaw University: in 1870 there were 36; in 1880, 24; and in 1910 only one.

Tswett's starting position was only that of a laboratory assistant, and to supplement his meager salary he was also teaching at a secondary school. Finally, by the end of 1901 he was accepted as a *Privat-Dozent* at the University, enabling him to lecture to students.

In Warsaw Tswett was a lonely man and this remained so until his marriage in 1907. Every year starting in 1902 he spent a few months in Western Europe (mainly Germany), visiting universities and libraries, or just on holiday. It is interesting to note that all of his scientific publications (we know of a total of 58) were single-authored: he never had a co-author, not even an assistant.[4b]

As soon as he settled down in Warsaw, Tswett continued his investigations of plant pigments, as a follow-up of his master's thesis. His aim was to study them in their native state, separated from the leaves' substrate and from each other. As mentioned earlier, in his master's thesis he already explored the question of selective extraction, using solvents of different dissolution power. He strongly believed that in scientific investigations the proper methodology has a key role and he aimed to improve it. Later, in his 1910 book[5] he summarized his philosophy on this question:

> "Any scientific advance is an advance of the method." Regretfully, the method is not infrequently the weakest aspect of scientific research. Each generation inherits, as students do, techniques of the previous generation, and without subjecting them to serious criticism, being satisfied by the fact that they are generally accepted: they use them to obtain new results which win recognition by contemporaries, but have no lasting value.

(The first sentence of this quotation is actually from the writing of the French philosopher René Descartes (1596–1650). Tswett is citing it in his book.) Above we already mentioned that in his master's thesis Tswett already established that the pigments are bound to the leaves by adsorption forces and thus he believed that by proper combination of adsorption and extraction he would be able to accomplish to find the proper adsorbent and solvent. He investigated more than 100 inorganic and organic solid substances to study their adsorption characteristics toward the pigments. The powdered substance was packed into a small, narrow tube (Fig. 4.3) and the ligroin solution of the pigments added. Adsorption of the pigments could be observed by color changes of the adsorbent powder and of the pigment solution flowing out of the tube. After adsorption on the powder, the pigment could be desorbed (dissolved) by the selection of a suitable solvent. The best results were obtained by using inulin (a polysaccharide), calcium carbonate, and alumina.

From here on Tswett proceeded in two different directions. The first was stepwise selective adsorption and extraction, more or less following the principles of a procedure described in the 1870s by

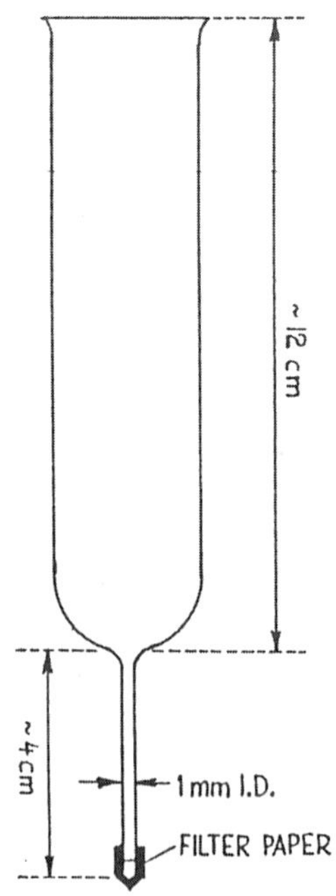

Fig. 4.3. The system used by Tswett in his first experiment examining the behavior of adsorbents toward pigments. After the description in Ref. 6.

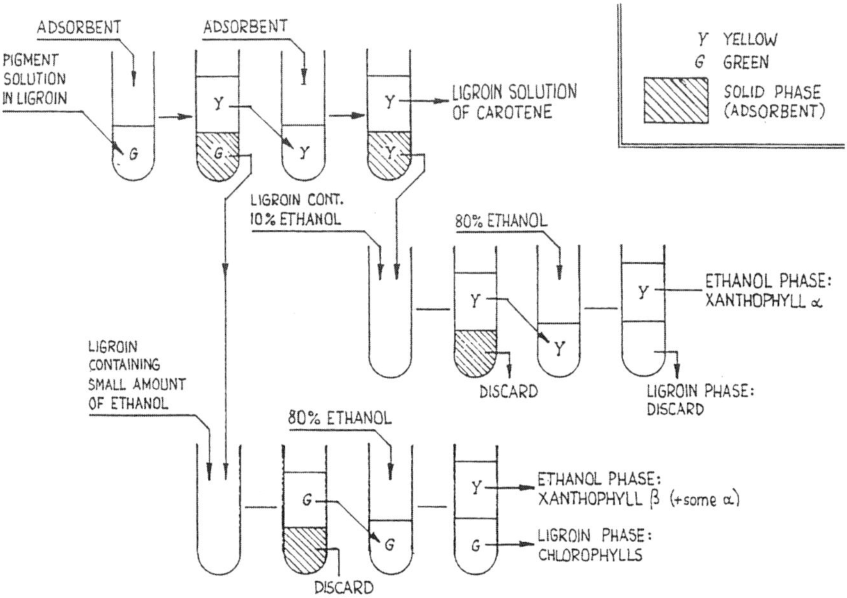

Fig. 4.4. Separation scheme used by Tswett in the stepwise differential adsorptive precipitation and extraction method. After the description in Ref. 6.

the German botanist G. Kraus (he called it adsorption precipitation). Tswett added the adsorbent powder to the pigment solution; it adsorbed the pigments which then could be selectively extracted from the filtered adsorbent. Repeating the process several times with different solvents resulted in separate solutions of the pure pigments, which could be identified by the color of their solution and by their UV-absorption spectra. Figure 4.4 (drawn based on his description) illustrates the multistage process used by Tswett in 1902–1903 for the separation of chlorophylls, carotenes, and xanthophylls.

The second method developed by Tswett was a dynamic method he called it as adsorption filtration: the adsorbent was packed into the narrow tube of a filter funnel and the pigment solution was filtrated through it by adding more solvent to facilitate the movement of the pigments in this "column" (to use our present nomenclature). Soon green and yellow rings started to form on the adsorbent, the rings became separated, widened, and moved down the column, and sometimes even separated into additional rings with different shades of color, indicating the presence of additional compounds. By the proper selection of the solvent, it also was possible to elute them successively from the column, which resulted in the solution of the separated individual pigments. This process, of course, is identical to chromatographic separation, although Tswett did not use this term as yet but called it adsorption filtration. At that time he did not decide which of the two techniques — adsorption precipitation or adsorption filtration — should be preferred.

At the beginning of 1903 Tswett had advanced well in his investigations and was able to summarize his results in the famous lecture on A New Category of Adsorption Phenomena and Their Application to Biochemical Analysis, he presented on the eighth of March (present calendar: March 21) at the meeting of the Warsaw Society of Natural Scientists. He considered this lecture to be only an interim report; therefore, he did not publish his results in any widely read German or French journal, not even in a Russian journal with nationwide distribution. The text of his lecture was printed only two years later in the periodical of the local society.[6–8] Thus, apart of his colleagues in Warsaw who attended the lecture (there were a total of 41 present, most likely also including students) most likely nobody else knew about it and about Tswett's results. Finally, in 1954, the chromatography supply house M. Woelm, in Eschwege, Germany, published the English (and also German) translation of this lecture as a company brochure.[7] In spite of this limited knowledge about Tswett's results we rightly consider the date of this lecture — 21 March 1903 (according to our present calendar) — as the birthday of chromatography.

References

1. M. Tswett, *Bulletin de Laboratoire de Botanique Générale de l'Université de Genève* **1**(3), 125–206 (1896).
2. M. S. Tswett, *Trudy Obshchestva Estestvoispytatelei pri Imperatorski Kazanskom Universitet* **35**(3), 1–268 (1901) (For a summary see Ref. 3).
3. M. S. Tswett, *Botanisches Centralblatt* **89**, 120–123 (1902) (Summary of the text published in Ref. 2).

4. E. M. Senchenkova, *Michael Tswett, the Creator of Chromatography* (Russian Academy of Sciences, Moscow, 2003) (Original Russian edition published by Mir Publisher, Moscow, 1997); (a) pp. 87–88; (b) pp. 308–312.
5. M. S. Tswett, *Khromofilly v Rastitel'nom i Zhivotnom Mire* (Chromophylls in the Plant and Animal Kingdom) (Karbasnikov Publishers, Warsaw, 1910).
6. M. S. Tswett, *Trudy Varshavskogo Obshchestva Estestvoispytatelei Otdelenie Biologii* **14**, 20–39 (1905) (For English translation, see Refs. 7 and 8).
7. G. Hesse and H. Weil, eds., *Michael Tswett's First Paper on Chromatography* (M. Woelm, Eschwege, 1954).
8. V. G. Berezkin, ed., *Chromatographic Adsorption Analysis: Selected Works of M. S. Tswett* (Ellis Horwood, New York, 1990).

Chapter

5

M. S. Tswett and the Invention of Chromatography Part II: Completion of the Development (1903–1910)*

After his lecture on 21 March 1903, Tswett continued his work on the new separation process. He slowly started to favor adsorption filtration (i.e., "chromatography") and gradually introduced it to his investigations of plant pigments. However, apart of the text of his 1903 lecture, he did not publish anything in 1902–1904.

Meanwhile, politics also interfered with the activities of the university. The start of the war between Russia and Japan, at the beginning of 1904, was followed by disturbances in Russia, and this was also felt at the universities. By the fall of 1904 the students disrupted classes and

---

*Based on the articles by L. S. Ettre and K. I. Sakodynskii published in *Chromatographia* 35, 329–338 (1993), and by L. S. Ettre, published in *LCGC* (*North America*) 24, 680–692 (2006).

by January 1905 an open revolt broke out in Russia, further disrupting the activities of the schools. Warsaw University was closed by the authorities for the school year of 1905–1906 and students were denied access to the university buildings. Professors and instructors still had opportunity to continue their research and work in the laboratories, but the continuous unrest certainly restricted their activities.

It is generally unknown but at this time Tswett also had a major family problem. His half-brother Aleksandr (1881–1912) was a junior naval officer on one of the torpedo boats part of the Russian fleet circumnavigating the world in 1904–1905 to join the hostilities in Asia. However, the Russian fleet commanded by Admiral Z. P. Rozhestvenskii was annihilated on 14 May 1905, in the Tsushima Strait between Korea and Japan by the Japanese fleet commanded by Admiral Togo. Unexpectedly Aleksandr's ship had a major role in the outcome of this battle because the severely wounded Admiral Rozhestvenskii was transferred to her after the sinking of his flagship, and the torpedo boat then surrounded without any resistance. After returning from Japanese captivity to Russia the admiral and all the officers of the boat were court martialled, but Alexandr and the other lower-rank officers were acquitted.[1]

Most likely Tswett used the closing of the school to further expand his foreign travels. We know that in 1903–1907 he spent each year a few months in Germany, going to libraries to check the newest periodicals and books unavailable in Warsaw, and was also active in conducting research there. For example, we know that he spent some time at the University of Kiel, in the laboratory of Professor J. Reinke, collecting algae in the harbor (Fig. 5.1), and also in Berlin, collecting samples of plants along the Spree River, and was investigating the pigment contents of these plants. But he still did not publish reports on his technique and results. Then, in 1905, the situation changed.

5.1. Controversy

At that time Hans Molisch (1856–1937) was one of the most respected botanists in Europe: between 1894 and 1908 he had served as professor and head of the Institute of Plant Physiology at the University

two consecutive issues of the journal. These are the two fundamental papers of Tswett describing the chromatographic separation method and summarizing his research on chlorophyll.

5.2. Tswett's Two Publications On Chromatography

Tswett's first 1906 paper[7] gives the background of the investigations, outlines the way how he reached his conclusions, presents his hypothesis on selective adsorption, and the effect of various solvents. The following quotation from this paper gives a wonderful summary of chromatographic separation:

> There is a definite adsorption sequence according to which the substances can displace each other. The following important application is based on this law. When a chlorophyll solution in petrol ether is filtered through the column of an adsorbent (I am mainly using calcium carbonate, tightly packed into a narrow glass tube), then the pigments will be separated from the top down in individual colored zones, based on this adsorption sequence, according to which the pigments which are adsorbed stronger will displace those which are retained more weakly. This separation will become practically complete if, after the pigment extract was passed through the adsorbent column, the latter is washed with pure solvent.

Continuing, Tswett then has his famous statement on the name of the new technique (Italics mine):

> Like light rays in the spectrum, the different components of a pigment mixture, obeying a law, are separated on the calcium carbonate column and can thus be qualitatively and quantitatively determined. *I call such a preparation a chromatogram and the corresponding method the chromatographic method.*

The second paper[8] contains the detailed description of chromatography, the apparatus used, and the results of separation when a pigment solution was analyzed. Tswett's chromatographic system was very simple: the adsorbent was packed in the narrow tube of the funnel (Fig. 5.3). The pigment solution and the necessary solvent were

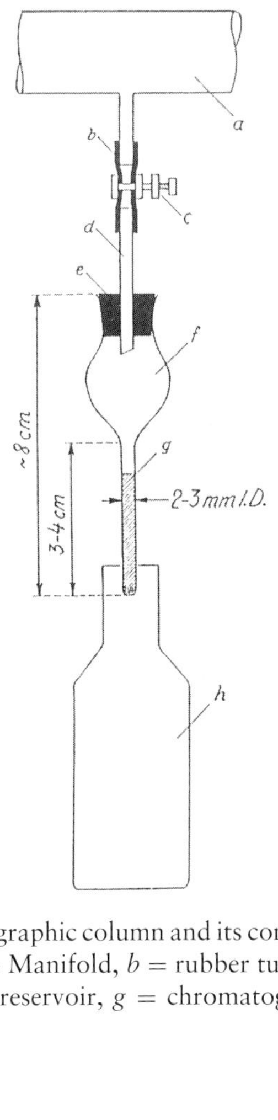

Fig. 5.3. Tswett's chromatographic column and its connection to the manifold of his chromatographic system. *a* = Manifold, *b* = rubber tube, *c* = pinch-cock, *d* = glass tube, *e* = cork, *f* = solvent reservoir, *g* = chromatographic column, *h* = flask to collect the eluting solvent.

added to the funnel which was then connected with other such funnels to a manifold through which some pressure could be applied to the columns with a small hand pump if needed (to speed up the movement of the solution in the column). The tube connecting the funnel to the manifold could be closed with help of a pinchcock, permitting the removal of one unit from the manifold without disturbing the others. When separation was completed, the units were removed, the adsorbent column — the packing with the separated multicolored bands — pushed out of the tube carefully with a wooden rod, and the individual bands cut with a scalpel. From these fractions, the adsorbed pure substances could be dissolved with a suitable solvent.

Tswett gave no explanation for the selection of the name "chromatography," and there have been some speculations on its origin. According to the most widely used explanation, "chromatography" is composed from two Greek words, *chroma (χρωμα)* meaning "color" and *graphein (μραφειν)* meaning "to write": thus, the word means "color writing." Historically, it had not been unusual to coin Greek words to define some scientific term, and since chromatography of a pigment mixture indeed produces individual color rings in the column (Tswett's second 1906 paper actually showed an illustration of a chromatographic column with the multicolored rings), the term seems to be logical. However, just two paragraphs later, Tswett emphasized that the technique is not restricted to the separation of colored pigments, it can also be used equally well to separate colorless chemical compounds; in other words, the "colored rings" are not a quintessential characteristics of the technique.

Some time ago, another explanation was also suggested for Tswett's selection of this name.[10] The interesting fact is, namely, that Tswett's surname is identical to the Russian word for "color" (ЦВЕТ, *tsvet)* (the spelling "Tswett" generally used by him corresponds to the German orthography); thus "chromatography" may also be interpreted as "Tswett's writing." Forty years ago Howard Purnell also mentioned this possibility, adding that it would be nice to think that Tswett, through the double meaning of *chroma* = *tsvet*, "took advantage of the opportunity to indulge his sense of humor."[11] It looks as Tswett wanted to revenge all those who did not appreciate his work, belittling it: he seemed to say that from now on, you will have to use my name when you use my separation method. But of course, we shall never know the real reasons for his selection.

5.3. Polemics

One would think that the logical and clear explanations in Tswett's papers convinced his opponents about the advantages of the new technique, particularly since the following year, he personally demonstrated the chromatographic technique twice, at the May 30 and June 28 meetings of the German Botanical Society, in Berlin, to a wide audience. This, however, did not calm down his critics; in fact, newer and even stronger voices joined the camp of his opponents.

The first was Leon Marchlewski (1869–1946), professor at the Jagiellonian University of Cracow, in the part of Poland belonging to the Austro-Hungarian Monarchy (Fig. 5.4). Marchlewski studied at the Federal Technical University in Zurich, Switzerland, and worked for years in England with E. Schunck, an early researcher of the chemistry of chlorophyll. In his papers Tswett (mildly) criticized the previous work of Marchlewski and Schunck and now, he obtained an immediate (and very acerbic) rebuttal. In it Marchlewski ridiculed the new technique, and warned Tswett that he should not believe that "a simple filtration experiment" would be enough "to swing himself to the height of a reformer of chlorophyll chemistry."[12] (It is an open question how much Marchlewski's antagonism was due to science,

Fig. 5.4. Leon Marchlewski (1869–1946). (Author's collection.)

vanity, feeling of personal insult, or patriotism. Although within the Austro-Hungarian Monarchy, the University in Cracow was a Polish school, with wide autonomy, and Marchlewski was a very patriotic Pole, naturally opposing the Russian Tswett, associated with a Russian Imperial University, in the Russian-occupied part of Poland, where the Polish language and national spirit was suppressed.)

In retrospect, Marchlewski's contribution to the development of science and plant pigments is negligible, and thus, we may dismiss his criticism. However, in the period discussed, he was a highly respected scientist, whose opinion counted. In fact, he was even nominated in 1913 for the Chemistry Nobel Prize, together with Willstätter[13] and this joint nomination indicates that at least some of their peers considered the contribution of Marchlewski and Willstätter to plant pigment research on an equal level.

An even more important — although more dignified — opponent of Tswett was Richard Willstätter (1872–1942), who, in 1915, received the Chemistry Nobel Prize for his chlorophyll research (Fig. 5.5). Willstätter called chromatography "an odd way" to carry out pigment research and stated that chromatographic separation is not reliable because chemical changes are occurring in the column.[14] Also, he expressed his opinion that chromatography cannot be used for preparative purposes, in other words to obtaining some amounts of the separated pigments for further investigations. With respect to the alleged chemical changes in the column, in his 1910 book (see below) Tswett always indicated whether a particular material was inert relative to the compounds to be separated or interacted chemically with them. It is worthwhile to note that 50 years later Richard Kuhn, one of Willstätter's students, admitted that they simply used the wrong adsorbent.[15] With respect to preparative use, Tswett already mentioned in his second 1906 paper,[8] that if needed, larger columns of 10–20 mm i.d. can be used; and in his 1910 book[16] he again addressed this question, describing the use of even larger columns.

In the first decades of the 20th century Willstätter was probably the most respected German organic chemist and the highest authority in chlorophyll research: therefore, his negative opinion about Tswett's results and chromatography as a valuable laboratory method certainly

Fig. 5.5. Richard Willstätter (1872–1942). (Author's collection.)

played an important role in the delay of the adaptation of chromatography by the international scientific community. Thus, it is strange that the chapter on Willstätter in a book about the chemistry Nobel laureates, published in 1993 by the American Chemical Society, stated that Willstätter's "use of chromatography made the technique more popular." [17] Obviously, the author of this chapter on Willstätter did not read his negative statements.

Besides his polemics with Marchlewski and Willstätter, Tswett was also fighting with other scientists of less importance. A typical case was J. Stoklasa at the University of Prague. In the first decade of the 20th century Stoklasa published a hypothesis that the chlorophyll molecules also contain phosphorus atom. Although Stocklasa did not mention him, Tswett immediately picked up this misinformation and hastened to rebuke it in a polemic paper submitted to the *Berichte* of the German Botanical Society.[18] Subsequently, Stoklasa repeated his investigations but now using chromatography, and was happy to answer Tswett, saying that his chromatographic experiments also showed the presence of a small amount of phosphorus in the chromatographically separated chlorophyll fractions[19]: thus, Tswett's own method shows that he (Tswett) is wrong! We do not know of any specific answer by Tswett to Stocklasa's answer, but it is obvious what happened: Stocklasa did poor separation and phosphorus-containing lipids co-eluted with chlorophyll in his chromatographic fraction.

5.4. Tswett's 1910 Book

Following the publication of his twin 1906 papers, Tswett continued to refine his technique and also to investigate various plant and animal pigments. He also started to summarize all the accumulated knowledge in a book, which was finally published in 1910 in Russian, by a Warsaw publisher. It was entitled Chromophylls in the Plant and Animal World[16] and it also served as his thesis for the Russian Doctor of Science degree (Fig. 5.6). The excellence of the book is best demonstrated by the fact that the Imperial Russian Academy of Science honored it in 1911 with its M. N. Akhmatov Prize, a major Russian scientific distinction with a fairly high (1000 Rubles) monetary award.

As its title shows, the primary subject of Tswett's book was the investigation of the various pigments occurring in nature. For him, chromatography was only a means for his studies, but he realized the importance of the new technique as a fundamental improvement in the ways of separation. This philosophy was in contrast to the opinion prevailing at that time, aiming the isolation of a *single substance* for further study: in Tswett's opinion a scientist always must consider the *whole sample* and separate *all* the substances present. He also emphasized that a chromatographically separated substance is at least as pure as one obtained by traditional means such as chemical reactions, distillation, and crystallization. Today this is self-evident; however, this was questioned for a long time after Tswett. As late as in 1929, we can still find papers stating that chromatography is an inferior technique, because without crystallization one could never produce a pure plant pigment.[20] Only in the 1930s, after the work of

M. TSVETT Les Chromophylles dans les Mondes Végétal et Animal

М. С. ЦВѢТЪ

ХРОМОФИЛЛЫ

ВЪ РАСТИТЕЛЬНОМЪ И ЖИВОТНОМЪ МІРѢ

Neue Tatsachen sind an sich nur eine geringfügige Bereicherung der Wissenschaft. Ihren Wert erhalten sie erst durch den Zusammenhang in welchem sie mit bereits bekannten und mit künftig bekannt werdenden stehen.

Ostwald.

ВАРШАВА
ТИПОГРАФІЯ ВАРШАВСКАГО УЧЕБНАГО ОКРУГА.
1910.

Fig. 5.6. The title page of Tswett's 1910 book.[16]

Kuhn, Karrer and their associates, was the superiority of chromatography finally accepted (see Chapters 12 and 13).

In his book, Tswett discussed chromatography in a systematic way, including all the information discussed in his previous publications and also adding significant new material. For example, he further extended the list of suitable solvents and emphasized that each has advantages, disadvantages, and particular fields of applications. He also illustrated the use of solvent mixtures and the possibility of gradually changing the solvent during the chromatographic process (a forerunner of gradient-elution chromatography). When dealing with the mechanism of separation, Tswett treated the process theoretically and divided the adsorbent into consecutive segments, considering the adsorption equilibrium in each segment. This concept is not far from the theoretical plate concept. He also showed that the rate at which each sample component travels along the adsorbent bed depends upon its adsorption coefficient and is independent of the other components present, and treated the separation process from the point of adsorption, quoting also the publications of J. Willard Gibbs. At that time very few scientists were aware of Gibbs' revolutionary theories.

In addition to the standard method of stopping the chromatographic process after the separated rings are formed on the column, Tswett also mentioned the possibility of washing out (eluting) the separated compounds from the column, collecting the individual fractions. This is in fact the same way as we carry out chromatography today: it was actually introduced in the second part of the 1930s under the name *Durchflusschromatogramm* (flow-through chromatogram, see Chapter 13). Here again, Tswett was way ahead of his time.

Unfortunately, Tswett's book was published only in Russian, at a local publisher, and thus its direct influence outside Russia was negligible. However, it is interesting to note that it had been included among the references of a few early western publications (Dhéré, Palmer), giving its title *in French* (as indicated in the upper part of the title page; see Fig. 5.6). We also know that Willstätter personally ordered a German translation for his own use, and as we shall see in Chapter 12, this copy had a major role in the rebirth of chromatography, in 1930–1931.

5.5. Postwords

When finishing his book, Tswett was still very active and full of plans. In 1911 he published seven papers in German and French journals; one of these included a very detailed discussion on the various carotenoids: in fact this term was proposed by him in this paper and soon was generally accepted.[21] He also went on an extended study trip to Germany, the Netherlands, Belgium, and France.

Tswett married in 1907; in that year he left Warsaw University, becoming associated with the Veterinary Institute and then the following year, with the Warsaw Polytechnic Institute (Fig. 5.7).

Fig. 5.7. The main building of Warsaw Polytechnic Institute. (Contemporary picture card; courtesy of Prof. E. Soczewinski.)

Fig. 5.8. The author in front of a plaque commemorating Tswett at the University of Tartu (1981).

However, after 1910 his health started to decline and he was forced to reduce his activities. In 1915 he had to leave Warsaw before the advancing German troops, moving temporary to Moscow and Nizhnii Novgorod. Finally, in 1917, he obtained an appointment as a full professor at the highly respected University of Tartu (in present-day Estonia; then part of Russia). He moved to Tartu (Fig. 5.8), but within a year he was again on the road because of German occupation of the Baltic area. He was appointed as a professor at the newly organized state university of Voronezh, but by then he was seriously ill, and he died on 26 June 1919.

References

1. C. Pleshakov, *The Tsar's Last Armada: the Epic Voyage to the Battle of Tsushima* (Basic Books, New York, NY, 2002).
2. H. Molisch, *Botan. Z.* **64**(I), 131–162 (1905).
3. M. Tswett, *Botan. Z.* **64**(II), 273–278 (1905).
4. H. Molisch, *Botan. Z.* **63**(II), 369–372 (1905).
5. F. G. Kohl, *Ber. Dtsch. Botan. Ges.* **24**, 124–134 (1906).
6. M. Tswett, *Ber. Dtsch. Botan. Ges.* **24**, 235–244 (1906).
7. M. Tswett, *Ber. Dtsch. Botan. Ges.* **24**, 316–326 (1906). For English translation see Ref. 9.
8. M. Tswett, *Ber. Dtsch. Botan. Ges.* **24**, 384–392 (1906). For English translation see Ref. 9.
9. V. G. Berezkin, ed., *Chromatographic Adsorption Analysis: Selected Works of M.S. Tswett* (Ellis Horwood, New York, 1990).
10. D. J. Campbell-Gamble, *Chem. Ind.* **59**, 598 (1940).
11. H. Purnell, *Gas Chromatography* (Wiley, New York, 1962), p. 1.
12. L. Marchlewski, *Ber. Dtsch. Botan. Ges.* **25**, 225–228 (1907).
13. E. Crawford, J. L. Heilbron and R. Wrich, *The Nobel Population 1901–1937*. (Office of the History of Science & Technology, University of California at Berkeley, 1987).
14. R. Willstätter and A. Stoll, *Untersuchung über Chlorophyll: Methoden und Ergebnisse* (Springer Verlag, Berlin, 1913).
15. R. Kuhn, in *Gas Chromatography 1962* (Hamburg Symposium), ed. M. Van Swaay (Butterworths, London, 1962), pp. xvii–xxvi.
16. M. S. Tswett, *Khromofilli v Rastitel'nom i Zhivotnom Mire* (Chromophylls in the Plant and Animal World) (Karbasnikov Publishers, Warsaw, 1910).

17. Zexia Barner, Richard Martin Willlstätter, in *Nobel Laureates in Chemistry 1901–1992*, ed. L. K. James (American Chemical Society, Washington, DC, 1993), pp. 108–113.
18. M. Tswett, *Ber. Dtsch. Botan. Ges.* **26A**, 214–220 (1908).
19. J. Stoklasa, V. Bradlick and A. Ernest, *Ber. Dtsch. Botan. Ges.* **27**, 10–20 (1909).
20. F. M. Schertz, *Plant Physiol.* **4**, 337–348 (1929).
21. M. Tswett, *Ber. Dtsch. Botan. Ges.* **29**, 630–636 (1911).

Chapter

6

M. S. Tswett and the 1918 Nobel Prize in Chemistry*

Almost every time the activities of M. S. Tswett (1872–1919) are discussed, one can find a remark noting that his achievements were not recognized in his lifetime. This question is usually based on two facts: that until 1917 he could not obtain a senior university appointment in Russia, and that the general acceptance of chromatography was delayed. However, as mentioned earlier, in Imperial Russia foreign degrees were not sufficient to obtain any senior position at a university or other scientific establishment: for that, degrees earned at Russian universities were needed. Tswett obtained his Russian "magister" (masters) degree in 1901 and his Russian doctor of science degree only

*Based on an article by L. S. Ettre, published in *Chromatographia* **42**, 343–351 (1996). The permission of the Nobel Archive of the Center for History of Science of the Royal Swedish Academy of Sciences to study the proceedings of the Chemistry Nobel Prize Committee and providing copies of the pertinent parts is gratefully acknowledged. Personal information about Van Wisselingh was obtained through the generosity of Ms. Denise Tjallome of the University of Technology, Eindhoven, The Netherlands.

in 1910. It is true that in the 1910s, there was an unfortunate delay for him finding a senior university position and for this, the stratified social structure of Imperial Russia can partly be blamed. However, we should not forget that Russia had only nine universities, and also that from August 1914 on, the Great War upset the regular civilian life. Finally, on March 24, 1917, Tswett was appointed to the chair of botany and director of the botanical gardens of the University of Tartu, certainly one of the highly respected universities within Imperial Russia. Unfortunately, he could not enjoy this position anymore because of the political events.

Soon after his Russian doctorate Tswett obtained in 1911 a serious recognition in Russia: he was awarded the M. N. Akhmatov Prize of the Imperial Russian Academy of Sciences, for his book on chromatography. This was a major science award, carrying with it a cash premium of 1000 Rubles, an extremely high amount at that time. This shows that by then, his activities were well known within Russia, and favorably accepted by at least some influential Russian scientists.

It is true that his scientific results were accepted only by a very few western scientists (see Chapters 7–10). However, we should not forget that his main field of research was the investigation of plant pigments, and in this — as discussed in the previous chapter — he had a few major antagonists, and their negative influence certainly influenced the acceptance of his results. However, he was not unknown in the western scientific world, particularly among botanists; he traveled widely in Western Europe on extended study trips, visiting universities and botanical institutes, and he has even worked for shorter times in such places.

That he was well known is best demonstrated by the fact that in 1918 he was proposed for his achievements in the field of chlorophyll and other plant pigments (not for chromatography!), for the highest award a scientist can receive: the Nobel Prize. This fact previously has been unknown and is nowhere mentioned in the literature. The circumstances of this proposal and its handling are the subjects of this chapter. However, we first must understand the background: the rules regulating the Nobel awards, and the situation in the European scientific world at the end of the First World War.

6.1. The Nobel Prizes

The Nobel Foundation, with the five annual prizes (chemistry, physics, physiology and medicine, literature, and peace), was established by Alfred Nobel (1833–1896) in 1895, one year before his death. Its Statutes were approved by the King of Sweden on June 29, 1900 and the first prizes were awarded in 1901. Apart from a few amendments the rules of the original Statutes are valid even today. The chemistry prize — our present concern — is awarded by the Royal Swedish Academy of Sciences.

According to the Statutes, nomination for the awards can be made by selected members of the international scientific community, and the nomination must reach the Nobel Prize Committee by February 1 of each year. These proposals are carefully considered in each field by the respective Committees, evaluating the activities of the nominees. The minutes of their deliberation and their final proposal is forwarded to the main Nobel Prize Committee which then makes the final decision late September to early October; they may change or disregard the proposals. The yearly awards are formally presented by the King of Sweden on December 10, the day of Nobel's death.[1]

The Academy may also decide that in a particular year no prize should be awarded, or defer a prize for one year. Such deferred prizes were not unusual in the first part of the 20th century: actually the 1918 Chemistry Prize (the subject of our chapter) was also awarded and announced only in 1919, and the same happened with the 1918 Physics Prize.[1]

It should be noted that the nominations, the deliberation of the Nobel Prize Committees, and their proposals are kept confidential for 50 years: after that period they may be made available to qualified researchers. This rule gave me the opportunity to study Tswett's nomination.

With respect to the period discussed here, we should also realize that during the First World War, even if announced, prizes were not presented: the ceremonies were postponed until the end of the war, and the laureates received their awards in a special ceremony only in 1920.

Rarely does somebody receive the Nobel Prize at the first nomination, and the same person is often nominated independently by a number of people. For example, Wilstätter, the great German organic chemist (and Tswett's antagonist) who received the 1915 Nobel Prize in Chemistry, was first nominated in 1911 and then repeatedly in 1912, 1913, 1914 and 1915: in 1914, he received eight, and in 1915 nine nominations.[1] The Committees usually wait for a few years until it is proven that the achievements of a person pass the test of time. The best example — well known to chromatographers — is the Chemistry Prize to A. J. P. Martin and R. L. M. Synge: their seminal paper on partition chromatography was actually published in 1941 but the Prize was awarded to them only in 1952 (see Chapter 14).

6.2. The Nominations for the 1918 Chemistry Prize

A large number of nominations were received for 1918; of these, 15 were disallowed (because they were not specific enough, or arrived late) while nine nominations — among them Tswett's — were considered. The Nobel Committee in Chemistry at that time had five members: Åke Gerhard Ekstrand, Peter Klason, Henrik Söderbaum, Oskar Widman, and Olof Hammersten, the chairman of the Committee.[1] They have served on the Committee for a number of years and most of them were of advanced age: Hammersten (professor of medicinal and physiological chemistry at Uppsala University since 1893) was 77, Ekstrand 72, Klason 70, Widman 66, and Söderbaum 56 years old.

As usual, each member was responsible for the preparation of a detailed report on certain nominees, and these reports were then discussed by the whole Committee. The protocol of their final session was dated September 18, 1918: after a detailed discussion of the nominees and their relative merits, the Committee concluded to propose awarding the 1918 Nobel Prize in Chemistry to Fritz Haber, director of the Kaiser Wilhelm Institute for Physical Chemistry and Electrochemistry in Berlin-Dahlem, for the development of the synthesis of ammonia from hydrogen and nitrogen, which opened a new era in the production of fertilizers. As stated in the Committee's report, Haber had already been recommended for the Chemistry Prize several times, in 1912, 1913, 1915, and 1916, and the only reason that prevented the Committee from recommending him for the Prize in the previous years was that, due to the ongoing war, they could not obtain "fully reliable accounts...of the practical usefulness, as well as of the scientific, industrial and politico–economical consequences of the ammonia synthesis method." However, in 1918 they finally could obtain all the necessary data and it was clear to them that "no one has, for at least a generation, made a discovery of such great importance to agriculture and to the nutrition of the population as Haber's ammonia synthesis."

The decision was not unanimous: Hammersten, the Committee chairman, dissociated himself from the decision but the other four members voted for Haber, and thus this proposal was forwarded to the Academy of Sciences for final decision, and it was accepted. However, for some reason which is not indicated in the Committee's protocol, the Prize was deferred for one year and announced only in November 1919. This announcement caused an uproar in the international community, because of Haber's leading role during the War in the development of the poison gases used by the German Army in France.[2] As a protest some of the laureates of Nobel prizes awarded during the war decided not to be present in Stockholm in June 1920, when finally the awards of the last six years were presented, because they did not want to share the ceremonies with Haber.

6.3. Tswett's Nomination

Let us now investigate the circumstances of Tswett's nomination. He was nominated for the Nobel Prize in Chemistry by Cornelis Van Wisselingh, of the University of Groningen, in the Netherlands, one of the foreign scientists who were asked by the Swedish Academy of Sciences to submit nominations for the 1918 Nobel Prize in Chemistry. Cornelis Van Wisselingh, who was born on July 30, 1859 in Utrecht, was trained as a pharmacist. He graduated from the University of Utrecht in 1882. In the next 24 years he worked as a pharmacist, but he also became involved in botanical studies. In 1899 he received an honorary doctorate in botany from the University of Groningen in recognition of his achievements. In 1906 the university appointed him professor in pharmaceutics and toxicology, and during 1916–1917 he served as the Rector Magnificus of the University. During his 19 years of professorship at Groningen, he found time for his botanical studies and became nationally and internationally recognized in this discipline. His main field of interest was cytology, but he also published on the chemistry of the cell wall, on carotenoids, osmosis, and the role of the nucleus of the cell. Van Wisselingh resigned from the professorship because of illness during September 1925 and died two months later, on November 30. His eulogy particularly emphasized that he was a modest and very honest man.[3,4]

We do not know whether Van Wisselingh met Tswett during his trip to the Netherlands in 1911, but it is most likely that at least they corresponded with each other. As we shall see below, he was aware of Tswett's whereabouts until 1916.

Van Wisselingh's letter to the Royal Swedish Academy of Sciences is dated January 8, 1918, and the Academy received it on January 14. The letter is written in German and is fairly brief. It starts by saying that

> In answering your letter of September 17, 1917, I have the honor to inform you that among the researchers who are involved in phytochemical investigations, I would select Professor M. Tswett of Nizhnii Novgorod, formerly of Warsaw, as worthy of considerations for the Chemistry Nobel Prize, on the basis of his investigations on chlorophyll and other pigments.

Van Wisselingh indicates in the letter that Tswett's results had been disclosed in various publications, and he quotes 12 papers plus Tswett's 1910 book, published in Russian.[5] It is interesting to note that Van Wisselingh gave the title of this book in French as *Les Chromophylles dans les Mondes Végétal et Animal*, the French title printed on top of the title page of the Russian edition. This, and the fact that Van Wisselingh characterized this book as the major treatise of Tswett's results, would indicate that he knew it. (However, it is clear from the report prepared by Hammersten that he could not have considered this book in his evaluation, because it was unavailable to him.) Van Wisselingh finished the letter by saying that "the adsorption analysis and chromatographic method invented by Tswett is very ingenious and is also praised by Willstätter," giving reference to a 1912 paper by Willstätter.[6] (This was probably the only paper by Willstätter in which he had a positive evaluation of some of Tswett's results.)

Van Wisselingh's letter indicating Tswett as "of Nizhnii Novgorod, formerly of Warsaw" reveals some interesting information. As mentioned in the previous chapter, the Warsaw Polytechnic Institute with which Tswett had been affiliated since 1908 was evacuated before the German occupation of the city (in August 1915) and eventually settled in Nizhnii Novgorod in August 1916. Then, in March 1917 Tswett was appointed as professor of botany at the University of Tartu and assumed his duties there in September 1917. This would clearly indicate that Van Wisselingh still had contact with Tswett in the period between August 1916 and early 1917, but did not know anything about him since then. This is a very interesting information because it would indicate that even after the evacuation from Warsaw, living in temporary locations, Tswett still maintained contact with western scientists. This is nowhere documented in the extensive literature about Tswett.

Tswett's nomination was assigned to Olof Hammersten, the chairman of the Chemistry Committee who, on April 1, 1918, produced a long report entitled Account on M. Tswett's Investigations on Chlorophyll and Other Pigments. In this report Hammersten summarizes Tswett's investigations on plant pigments, characterizing them as mainly dealing with "their optical properties, their behavior in different solvents, their detection, and their reactions with different chemical reagents." According to the report, in Tswett's papers "a more in-depth chemical investigation of the isolated, pure pigments has not been presented"; in contrast to Willstätter who studied the chemical nature of the chlorophyll pigments and the composition of the compounds formed by its degradation, Tswett's work was really "limited to the investigation of solubility matters and absorption spectra, apart from some polemic essays and questions concerning priorities."

This citation immediately shows the fundamental handicap Tswett had: his work was judged in comparison with Willstätter, the most important German organic chemist of that period who received the Nobel Prize just a few years earlier for his achievements in chlorophyll research. It is important to emphasize that this was also the subject of Tswett's nomination by Van Wisselingh and *not* chromatography. Chromatography was a *technique*, the use of which was not yet proven by others, and obviously Van Wisselingh realized that its development alone would not be sufficient for a nomination. He mentioned the chromatographic method developed by Tswett only in one sentence, almost like a postscript. Thus, it is particularly interesting that in his evaluation report, Hammersten dealt in detail with it and he definitely considered the chromatographic method as Tswett's "most original and most revered contribution." The report summarized the way in which pigments were separated by adsorption on "a $CaCO_3$ cylinder," which then reminds one of a spectrum, with multicolored bands. Hammersten also gave examples of some of the separations obtained by Tswett, but he stated that similar separations (by other methods) were also carried out by Marchlewski and Willstätter, and although Tswett might have priority in the demonstration that both the amorphous and crystalline form of chlorophyll are actually mixtures, his work cannot be compared to the meticulous investigations of Willstätter. According to Hammersten, while it is true that Tswett was the first to demonstrate that the so-called crystallized chlorophyll, described first by I. P. Borodin (professor at St Petersburg University) in 1882, is not a natural form but an artifact, Tswett "left the question of the nature and origin of the crystallized chlorophyll unresolved." Hammersten's report concludes with the statement that

> ... when comparing the research of Tswett with Willstätter's investigations of plant pigments, especially chlorophyll, for which he was awarded the 1915 Chemistry Nobel Prize, it should be obvious that Tswett's investigations on chlorophyll and other pigments cannot be considered for the Nobel Prize in Chemistry.

After discussing this report, the Committee agreed with its conclusion.

6.4. Evaluation

For us chromatographers who now can fully judge Tswett's epochal contribution to science — the introduction of differential migration techniques in separation, the realization that instead of isolation of a single substance, the separation of all the components present is needed, and that chromatographic separation provides a purer compound than the classical techniques — the 1918 decision of the

Chemistry Nobel Committee obviously represents a great disappointment. However, we should not be surprised by this decision, for a number of reasons. To understand these, we must look into our subject in the proper context, considering the status of European science in 1918.

First of all, we must emphasize that we are dealing with the Nobel Prize in *Chemistry*. What Willstätter did in his chlorophyll research was primary chemistry, however, Tswett did not do "chemistry" in the classical sense: he studied the pigments in the plants and their role, and his achievement was primarily the simultaneous *separation* of many components. However, for scientists of that period, separation was meaningless: what was important to them was *isolation* of a single substance from a large amount of raw material. But even isolation in itself was not considered science: it was done by junior associates in the basement of the laboratory building (as mentioned by Willstätter in his autobiography[7]), the reactions carried out with the isolated substance (upstairs, in the real laboratory) were considered as "science."

We should also realize that the time was not yet ripe to award a Nobel Prize for a *technique*; in fact even major engineering achievements were not considered scientific enough. The best proof of this is that Carl Bosch (1874–1940), who was responsible for "translating" Haber's laboratory results into an industrial scale, developing high temperature and pressure systems that up to then were considered technically impossible, was not even mentioned in Haber's nominations and in the elaboration of the Nobel Committee, even though the ammonia synthesis method was rightly named as the Haber–Bosch process. Bosch had to wait 13 additional years until finally he was considered in 1931 for the Chemistry Nobel Prize, then together with Friedrich Bergius (1884–1949), for the development of processes to produce gasoline by high-pressure hydrogenation of coal. By that time Bosch was one of the directors of I. G. Farben, the largest and most influential chemical concern of Europe, becoming its chairman of the board of directors. How much his position helped him to receive the Nobel Prize, is an open question.

As we have seen, the report of Hammersten also criticized Tswett's publications for containing too much of "polemic assays and questions concerning priorities." This criticism is true: Tswett was continuously fighting with his peers, particularly with Marchlewski (see the previous chapter). The question that concerns us is not who was right — in general, Tswett was — but the language of the polemics and the personal attacks, which today would be unacceptable in any scientific publication, and no editor would permit the publication of any paper using such language. With full objectivity we have to say that Tswett was at least partly responsible for allowing this dialogue to get out of hand.

As a conclusion we can state that it is explainable why in his lifetime, Tswett's achievements could not get the full international recognition he deserved: the field was simply not yet ripe for it. His greatness was to realize the importance of *separation* almost 30 years before its general recognition. By the introduction of chromatography, a new, universal method based on differential migration, his contribution to science can be considered at the same level as those of Lavoisier and Bunsen.[8]

References

1. E. Crawford, J. L. Heilbron and R. Wrich, *The Nobel Population 1901–1937*. (Office of History of Science & Technology, University of California at Berkeley, 1987.)
2. M. Goran, *The Story of Fritz Haber* (University of Oklahoma Press, Norman, OK, 1967), pp. 85–86.
3. "Biographical Notes of Professors Appointed in 1905–1906," *Yearbook of the University of Groningen 1905–1906*, pp. 42–43 (in Dutch).
4. "Notes on Emeriti Professors Who Passed Away in 1925–1926," *Yearbook of the University of Groningen, 1925–1926*, pp. 38–39 (in Dutch).
5. M. S. Tswett, *Khromofilly v Rastitel'nom i Zhivotnom Mire* (Chromophylls in the Plant and Animal World) (Karbasnikov Publishers, Warsaw, 1910).
6. R. Willstätter and M. Iser, *Ann. der Chemie* **390**, 269–339 (1912).
7. R. Willstätter, *Aus meinem Leben* (My Life), 2nd edn. (Verlag Chemie, Weinheim, 1973), p. 156.
8. Cs. Horváth, Differential Migration Progress: Milestones in Separation Science Over the Last 40 Years, *The Chromatography Yearbook, 1994*, eds. M. B. Evans and A. E. Fell (The Chromatographic Society, Nottingham, UK, 1994), pp. 31–35.

Part 3

The First Pioneers in the Use of Chromatography

Chapter

7

Gottfried Kränzlin, the First Follower of Tswett*

It is interesting that the first person who successfully used chromatography in his investigations was a young graduate student at the Botanical Institute of the University of Berlin. He started his thesis work around the middle of 1906 and, as all graduate students do, he first studied the available literature, among others naturally the issues of the *Berichte* of the *Deutsche Botanische Gesellschaft*, the German Botanical Society. Tswett's two seminal papers were just published at this time[1,2] and our graduate student liked them so much, that he immediately decided to use Tswett's methodology for his own work.

The graduate student was Gottfried Kränzlin. This chapter discusses his investigations and doctoral thesis representing the first use of chromatography immediately after Tswett.

7.1. G. Kränzlin and his Work

Most of the information we have about Kränzlin is from his autobiography, included in his doctorate thesis. Gottfried Ernst Richard

*Based on the paper by H. H. Bussemas and L. S. Ettre, published in *Chromatographia* 39, 369–374 (1994).

Kränzlin was born on 6 September 1882, in Lichterfelde, near Berlin (today part of the city). His father was Dr. Fritz Kränzlin (1847–1934), teacher at *das Graue Kloster* (the gray cloister). This was the most famous, very exclusive *gymnasium* (high school) in the Berlin area, maintained by the Evangelical (Lutheran) Church to educate the elite of Germany. Its most famous student was Otto von Bismarck (1815–1898), the great German statesman and the founder of the German Empire, who studied there in 1830–1832. Fritz Kränzlin was a botanist, member of the German Botanical Society, and the author of a number of books about orchids; thus, it was logical that — after studying at his father's school and finishing it in 1903 — Gottfried also majored in botany at the University of Berlin. Gottfried Kränzlin finished his undergraduate studies in 1905 when he started his graduate work in the Botanical Institute of the University, the director of which was Professor S. Schwendener; Dr. E. Baur, *privat-dozent* at the Institute was his thesis advisor. Kränzlin finished his work toward the end of 1907 (his thesis is dated 23 November 1907), and finally received the Ph.D. degree on 29 February 1908.

7.2. Kränzlin's Thesis

The title of Gottfried Kränzlin's thesis is Anatomical and Pigment-Analytical Investigations of Variegated Plants[3] (Fig. 7.1). The aim of his work was twofold. The first was related to the origin of plant variegation which can be natural or due to infection. In this respect, the question raised was whether there is any characteristic data in the composition of the leaves that would distinguish between the two origins. The second aim of the investigations was to see whether there are similarities in the anatomy of variegated leaves and in their pigment composition: In other words, whether they represent mutations of a basic form or are completely different. The study of plants becoming variegated by infection was the specialty of Dr. E. Baur at the Botanical Institute: Kränzlin cites three papers by Baur from 1906 to 1907 on this subject. Thus, Kränzlin's investigations represented a continuation of Baur's work.

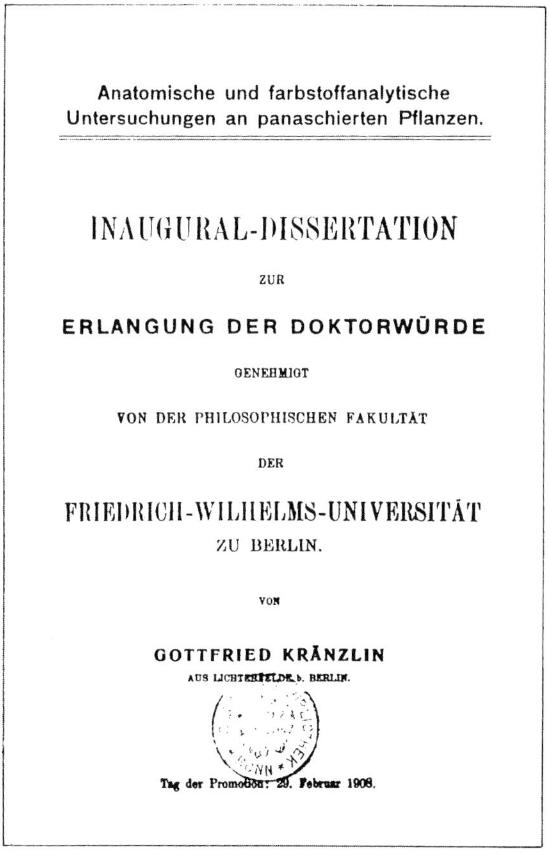
Anatomische und farbstoffanalytische
Untersuchungen an panaschierten Pflanzen.

INAUGURAL-DISSERTATION

ZUR

ERLANGUNG DER DOKTORWÜRDE

GENEHMIGT

VON DER PHILOSOPHISCHEN FAKULTÄT

DER

FRIEDRICH-WILHELMS-UNIVERSITÄT
ZU BERLIN.

VON

GOTTFRIED KRÄNZLIN
AUS LICHTERFELDE b. BERLIN.

Tag der Promotion: 29. Februar 1908.

Fig. 7.1. Title page of the Doctorate Thesis of Gottfried Kränzlin submitted to the University of Berlin, at the end of 1907.

Kränzlin studied 17 different plants. In addition to anatomical investigations he also measured the amount of some inorganic and organic substances (e.g., nitrates, starch, proteins) and the relative amounts of various plant pigments. We are interested in the latter part of his work.

At Kränzlin's time investigation of the individual plant pigments was in its infancy, and Tswett's two basic 1906 publications[1,2] were the first that characterized a number of pigments by using chromatography for separation. As already mentioned Kränzlin adopted the technique of chromatography and Tswett's terminology almost immediately after the publication of these papers, when he had just started his thesis work. He continued to follow the literature and also refer to a paper of Tswett from 1907.[4] It is interesting to note that although Tswett did not yet specifically deal with carotene in his 1906 papers, Kränzlin extended his investigations to it.

We have already mentioned (see Chapter 5) that in 1907, when Tswett was on an extended trip to countries outside Russia, he participated twice — on May 31 and June 28 — at the meetings of the German Botanical Society, in Berlin, and there he demonstrated the chromatographic technique.[5,6] Kränzlin was present at the second meeting, and he mentions in his thesis that Tswett discussed methodology not only for chlorophyll determination, but also for the investigation of carotene (based on adsorptive precipitation and extraction: see Fig. 4.4 in Chapter 4). This was more advanced than the method used by Kränzlin, and he realized that its use would have given better results. However, he was already well advanced in his investigations and therefore he did not change his methodology for carotenes.

7.3. Chromatography in Kränzlin's Thesis

Kränzlin used the following methodology in his thesis. Leaves (5 g) were cut to small pieces, ground, and mixed with a small amount of $CaCO_3$ in order to neutralize the acidic substances present. Then the ground leaves were extracted with absolute alcohol and the clear solution was mixed with carbon disulfide. After repeated extraction the pigments were present in the CS_2 solution, which was then freed from the remaining alcohol by washing with water. As emphasized by Kränzlin, this double extraction procedure — extraction of the leaves by ethanol and then transfer of the pigments into CS_2 solution — was necessary to assure a quantitative extraction of all pigments from the leaves, and it represented an improvement to Tswett's original extraction technique, which used only CS_2. According to Kränzlin the fresh leaves contain water: a separate water phase is formed when CS_2 is the primary solvent, and part of the pigments may stay in this water phase. On the other hand, when using ethanol as the solvent, it will form a single phase with water, and the pigments can then be quantitatively extracted from it with CS_2.

The pigment solution in CS_2 was chromatographed in a 100-mm long, 15-mm i.d. column containing $CaCO_3$ as the adsorbent to a height of about 50 mm, using suction with a water-jet pump. Carotene runs through the column with the solvent while the other pigments form the known colored bands. After recording the lengths of the individual bands and their relative position, Kränzlin added benzene to the column as a second mobile phase and observed the changes in the relative position and separation of the individual bands on the column. Finally, the packing was carefully removed from the column and the colored bands were separated from each other and from the rest of the packing. After dissolution and partitioning between two solvents for further purification, the solutions of the individual pigments were characterized by measuring their UV absorption maxima.

Based on the measurements of the lengths of the individual bands, Kränzlin characterized each chromatogram by four values. Three of these represented the sum of the lengths of (a) all colored bands, (b) the green bands and (c) the yellow bands. The fourth value represented the relative concentration of carotene in the column eluent (see below). These values were tabulated in the thesis and the 2×17 chromatograms were also presented as schematic drawings. Figure 7.2 shows typical chromatograms of three samples, obtained (a) from the original CS_2 extract and (b) after further development with benzene The symbols are identified in Table 7.1.

Based on these and other investigations Kränzlin made a number of conclusions with regard to the comparison of the investigated variegated plants. Here, we only want to mention some of his observations concerning chromatography and Tswett's data about the individual pigments.

Although Kränzlin essentially copied Tswett's methodology, he also further improved it in two aspects. The first was the double extraction procedure which provided better quantitative extraction of the pigments from the leaves, while the second was the additional development of the original CS_2 chromatogram with benzene. Tswett already had discussed the different elution characteristics of

Dhéré's student at the University of Fribourg (see Chapter 8), Gottfried Kränzlin's involvement in chromatography was a single-work affair. Just as Rogowski, he pioneered in his Ph.D. thesis in the use of the new technique, but after that he never again was involved in its use.

References

1. M. Tswett, *Ber. Dtsch. Botan. Ges.* **24**, 316–323 (1906).
2. M. Tswett, *Ber. Dtsch. Botan. Ges.* **24**, 384–393 (1906).
3. G. Kränzlin, *Anatomische und Farbstoffanalytische Untersuchungen an panaschierten Pflanzen.* Inaugural Dissertation (Friedrich-Wilhelms University, Berlin, 1908), 63 pp. + 17 plates.
4. M. Tswett, *Ber. Dtsch. Botan. Ges.* **25**, 137 (1907).
5. *Ber. Dtsch. Botan. Ges.* **25**, 217–219 (1907).
6. *Ber. Dtsch. Botan. Ges.* **25**, 267 (1907).
7. G. Kränzlin, *Zeitschrift für Pflanzenkrankheiten* **18**, 193–203 (1908).
8. M. S. Tswett, *Khromofilly v Rastitel'nom i Zhivotnom Mire* (Chlorophylls in the Plant and Animal World) (Karbasnikov Publishers, Warsaw, 1910).
9. L. Zechmeister and L. Cholnoky, *Die Chromatographische Adsorptionsmethode* (Springer, Vienna, 1937) (2nd ed, 1938), p. 12.
10. G. Kränzlin and A. Marcus, *Baumwolle*, Wohltmann Bücher No. 9, Deutsche Auslandverlag Walter Bangert, Berlin-Charlottenburg, 1931.
11. Listed in: F. Dietrich, ed., *Bibliographie der deutschen Zeitschriften Literature* (Felix Dietrich Verlag, Gautsch bei Leipzig).

Chapter

8

Charles Dhéré – Pioneer and Tswett Biographer*

Tswett's invention, chromatography, and his results in the field of plant pigments, did not receive the recognition they deserved: they were greeted with skepticism and even ridiculing them. This so-called dormant period finally ended in 1931, when the group of Richard Kuhn, in Heidelberg, demonstrated the superiority of chromatography (see Chapter 12). In the 25 years which passed between Tswett's two fundamental papers of 1906 and the start of the work at Heidelberg, only a very few scientists recognized the importance of chromatography as a separation technique and utilized it in their research. One of these was Charles Dhéré, in Switzerland. In this chapter we shall discuss his activities.

8.1. Dhéré's Life; His Field of Interest

Charles Dhéré (Fig. 8.1) was born on 5 March 1876, in Paris, and first studied medicine, receiving his *Docteur en Médecine* degree in

*Based on the article by L. S. Ettre and V. R. Meyer published in the *Journal of Chromatography* 600, 3–15 (1992).

Fig. 8.1. Charles Dhéré, in the 1930s. (Courtesy of Mrs. H. Urgi, Chancellery of the University of Fribourg.)

1898 with a thesis dealing with the variation of nerve centers as a function of size.[1] However, he never practiced as a physician. First he became an assistant at the Sorbonne in Paris, in the Department of Natural Sciences; then, in 1900, he joined the University of Fribourg, in Switzerland, as an associate professor (*professeur extraordinaire*) of physiology, biological chemistry and microbiology. In 1908 he became a full professor and, in 1909, he received the *Docteur ès Sciences* degree from the Sorbonne, with a thesis on the investigation using ultraviolet spectroscopy.[2] During his long tenure at the University of Fribourg, Dhéré served twice (in 1916–1917 and 1933–1934) as the dean of the Faculty of Science. He retired in 1938, when the title of a *professeur honoraire* was bestowed on him. After his retirement Dhéré moved to

Geneva where an office and laboratory space was provided for him in the Institute of Zoology of the University. He continued his scientific studies until 1951. He died on 18 January 1955.[3]

Dhéré's main interest was the investigation of biologically important substances, particularly by ultraviolet and fluorescence spectroscopy. He even coined two special terms to express his field of interest: *optochimie* and *optochimie biologique*. As a summary of his life's activities in 1937 Dhéré published a major book on the use of fluorescence in biochemistry,[4] which was considered at that time so important that in 1938 he received the Swiss Marcel Benoit Prize for it. This prize was originally established to honor annually important scientific inventions which particularly help to improve human life; it was first presented in 1919 and the winner was always selected by the Swiss Federal Department of Interior. In 1938 the Prize carried an amount of 30,000 Swiss Francs, a very large sum at that time.

Naturally, a prerequisite of any spectroscopic investigation is the ability to prepare pure substances. For this reason Dhéré became interested already at the beginning of his professional career in fundamental laboratory techniques which can be utilized for this purpose. One of these was electrodialysis and Dhéré's contributions to the advancement of this technique were important. His interest in methods permitting the separation of biologically important substances and their preparation in pure form led him to chromatography.

Dhéré started to use chromatography around 1911, and the first work in which this is documented is the doctorate thesis of Wladyslaw de Rogowski, a student from Warsaw. A few years later, the chromatographic technique was further improved in the work of another graduate student, Guglielmo Vegezzi. These activities represent the first major use of chromatography since its invention by Tswett.

It is interesting to note that Dhéré never met personally Tswett, although in the years before the First World War Tswett visited a number of times in Geneva his friend Edouard Claparède (1873–1940), a former classmate and now professor of physiology at Geneva University. Most likely Dhéré's knowledge of Tswett's work came through his graduate student, Wladyslaw de Rogowski.

8.2. Rogowski and His Chromatography Work

8.2.1. *Rogowski's Life*

Wladyslaw Franciszek de Rogowski was born on 3 December 1886, in Warsaw, in the Russian-occupied part of Poland. ("Wladyslaw" is the correct Polish spelling of his first name; one can also find it written as *Wladislas* or *Ladislas.*) His grandfather was involved in the 1831 Polish uprising against Russia and this rebellious nature was evidently inherited by his grandson: in 1905, as a senior in high school in Warsaw, he organized a strike, induced by the disturbances in conjunction with the 1905 (short-lived) Russian revolution. Therefore, he had to transfer to a private school (the Jezewski School of Commerce) in order to finish his secondary education. Subsequently, he went to Switzerland to start his university studies. Again, this was probably connected with the political events in Russian-occupied Poland where, for a period, the universities were closed to students.

We have found Rogowski's police registration in the Swiss city of Bern dated 12 October 1906, indicating him as a student at the University. However, he evidently left within one year: an entry at the police dated 14 September 1907 indicates that he departed to Russia which undoubtedly means his return to Warsaw. He also studied at the Jagiellonian University of Cracow, in the Austrian-occupied part of Poland, where large number of Polish students from territories under Russian and German rule (and not only those under Habsburg rule) studied; in addition a Polish biography[5] indicates further studies by Rogowski at the universities of München and Leipzig, in Germany, but his own autobiography attached to his doctorate thesis (see below) does not mention any of these.

The next definite information we have about Rogowski is an entry in the student registration book of the University of Fribourg dated 19 October 1911 (Fig. 8.2). There, he carried out his thesis work under Professor Dhéré and received his doctorate on 21 December 1912. According to an entry in Dhéré's paper on Tswett written 30 years later[6] "my student and collaborator W. de Rogowski left my laboratory (and most likely, also Switzerland) immediately after finishing his doctorate examination in Fribourg, in 1912." It is interesting to

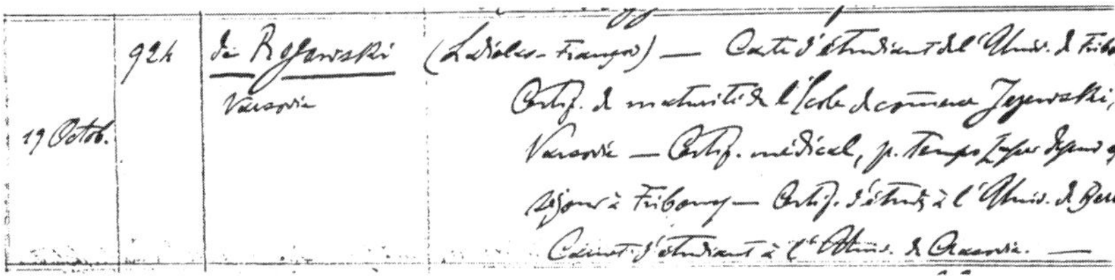

Fig. 8.2. Entry in the Student Registration Book of the University of Fribourg, for the 1911–1912 Winter Semester. The transcript of the hand-written text is: *de Rogowski, Varsovie (Ladislas–Franzen) — Carte d'étudient de l'Univ(ersité) de Fribourg — Certif(icat) de maturité de l'Ecole de commerce Jezewski, à Varsovie — Certif(icat) medical, p.Temps. (?), séjour à Fribourg — certif(icat) d'étud(iant) à l'Univ(ersité) de Berne. Carnet d'étudiant à l'Univ(ersité) de Cracovie.*

note that contrary to the customs of that time, Rogowski's name is not included in the Annual Reports of the University of Fribourg, listing all the persons who received a doctorate: he is missing from the Report of 1912 as well as from the Reports of the following years. We could not find his thesis either in the Swiss National Library, where all doctorate theses submitted to Swiss universities are deposited. Finally, we found at the University of Fribourg a poor copy of a typewritten French text[7] of Rogowski's thesis. This typed text is unusual in a number of aspects. Although the title page specifies that it is presented by Rogowski to the Faculty of Science of Fribourg University in order to obtain the degree of a *Docteur ès Sciences*, the date (year) is given as 1914, and not as 1912, and the city as Warsaw, and not Fribourg. The brief autobiography of the author included with the thesis also specifies that he received the doctor degree in Fribourg on 21 December 1912, and not in 1914 what is the date on the typed copy of the thesis: "*A la suite de mes examens à la Faculté des Sciences de Fribourg j'ai obtenu le title de docteur le 21 decembre 1912.*" This would indicate that this thesis is not the one submitted originally to Fribourg University; or it may mean that he actually did not have a formal written thesis at the time of his final examinations mentioned by Dhéré. Finally, upon investigating this typed copy one can immediately make a very interesting observation: the accented vowels used in French (e.g., *à, è, é,* etc.) were not in the keyboard of the typewriter used, and the accents were added by hand. I cannot believe that a typewriter in the French-speaking

Fribourg University would not have had such characters. In other words, the thesis had to be typed in Warsaw (the Polish language does not use such accents).

Zechmeister and Cholnoky, in the bibliography section of their fundamental chromatography book published in 1937[8] list Rogowski's thesis as submitted to *two* universities: to Fribourg in 1912, and to Warsaw in 1914. Thus, this seems to indicate that after his doctorate examination in Fribourg Rogowski also submitted a thesis (most likely identical to his work in Fribourg) to Warsaw University, for a second doctorate. He most likely wrote this thesis in Polish for the University of Warsaw, but also translated it into French, and sent the French copy from Warsaw to Dhéré.

This "second doctorate" theory opens the interesting possibility that in Warsaw, Rogowski actually met Tswett. From 1908 on, Tswett was affiliated with the polytechnic institute and not with the university; however, undoubtedly, he maintained connections with the university where he had been active between 1901 and 1908. In his thesis Rogowski frequently cited Tswett, among other publications also his 1910 book,[9] and it is given by its Russian title, written with Cyrillic letters. In other words, Rogowski definitely had known Tswett's book. At that time Warsaw was not a large city and scientists formed a fairly close-knit society. Thus, it would not be unusual if his colleagues gave Tswett this thesis for review or at least, to read it. We should also not forget that in Russia, defending a doctorate thesis had always been a well-publicized public affair. It would have been unusual for Tswett not to go to this open session: after all, the candidate used *his* method and the main subject of the thesis was to prove that *his* (Tswett's) results disputed by others were correct. The only reason which could have prevented this was Tswett's illness: we know that in 1914 he was absent of the university between the end of March and the middle of November. Unfortunately, we have no information about the exact date when Rogowski submitted his thesis to Warsaw University and of his examination.

Although it is not our subject, we should briefly summarize the information about Rogowski's further activities and life (Fig. 8.3).

Fig. 8.3. Wladyslaw Franciszek de Rogowski, in 1935. (Courtesy of Mrs. Barbara Rogowska-Pietrowska, Rogowski's daughter.)

After his doctorate, he was not engaged anymore in original scientific work: he was an educator, both in Poland and Brazil (setting up schools for the children of Polish immigrants); a poet and a writer; a pioneer in modern agricultural methods and a patriotic Pole who, when needed, served his country against those who wanted to destroy it. He was in Warsaw during German occupation in 1939–1945 and was a member of the Home Army, the illegal Polish resistance army; he participated in the 1944 uprising but, on 20 January 1945, a few days after the liberation of Warsaw, he was arrested by the Red Army, together with many other members of the Home Army. He died on April 1945, in the Soviet concentration camp Baskoje, in the Ural Mountains. His wife survived her husband and died on 4 January 1973; they had two daughters, Barbara (born in 1923) and Kalina (born in 1927).

8.2.2. Rogowski's Thesis Work

In 1912 Dhéré submitted a paper coauthored with Rogowski to the journal of the French Academy of Sciences (where most of his papers were published); according to the French customs, it was presented at the 12 October 1912, session of the Academy and subsequently published.[10] This short paper reported on the UV spectra of chlorophylls and the so-called crystallizable chlorophyll, and chromatography is mentioned only in a single sentence. However, in Rogowski's thesis[7] the technique, the system used by him, and his results are described in more detail.

In Rogowski's system (Fig. 8.4) the glass column was not tapered at its lower end (as originally proposed by Tswett) but closed by a

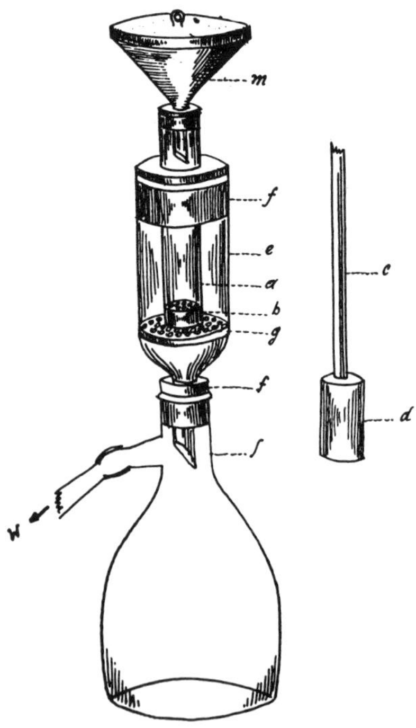

Fig. 8.4. Rogowski's chromatography system.[7] a = Chromatographic column (250 mm × 20 mm i.d. glass tube); b = cork (10 mm high) with multiple perforations; c, d = wooden pestle; e = outer glass sleeve; f = rubber stoppers; g = perforated porcelain disk; l = glass flask; m = funnel; w = water pump.

cork having multiple perforations; it was standing on a perforated porcelain disk held by a tapered outer glass sleeve. Calcium carbonate, pre-baked at 150°C for 10 h, was used as the adsorbent; the height of the chromatographic packing in the column was 60–80 mm. After separation and formation of the colored rings, the column packing (the "chromatogram") was slowly pushed out of the tube and the colored rings were carefully separated from each other and from the rest of the packing.

Rogowski's technique represented an improvement as compared to Tswett's methodology: how to isolate and collect the separated colored rings, without loss and contamination by components present in the other rings, had always been a problem. Let us not forget that elution of the separated compounds from the column by a continuous solvent flow (obtaining a "liquid chromatogram", an expression used at that time to distinguish it from the column serving as the "chromatogram") started only to be used in the second half of the 1930s.[11] In his later paper on the evolution of chromatography published in 1943[6] (discussed below) Dhéré emphasized this special feature of Rogowski's system, that it permits one to obtain the individual fractions without rupture of the column packing.

The subject of Rogowski's thesis was the spectroscopic investigation of chlorophyll *a* (α) and *b* (β), of the so-called "crystallizable chlorophyll," and of certain carotenoids. Here a brief explanation is in order. Chlorophyll *a* and *b* are identical to chlorophyll α and β: Tswett used Greek letters while later literature adopted the use of Roman symbols. The so-called "crystallizable chlorophyll" was a substance isolated by Willstätter who believed it to be a single, native pigment. Tswett demonstrated in a number of publications[12–14] that it was not a native pigment but an artifact, formed during the long alcoholic extraction of the living tissue, and it was actually a mixture of two substances. This debate added to the controversy between Tswett and Willstätter (see Chapter 5).

The main subject of Rogowski's work was the UV adsorption of chlorophyll *a* (α) and *b* (β), and an important point in the investigation was to check the purity of the isolated fractions, comparing data of Tswett vs. Willstätter and his associates. Twenty-five years later Oscar

Biermacher, an American graduate student of Dhéré, who obtained his Ph.D. in Fribourg in 1936, summarized in his thesis[15] the situation in the following way:

> By the classical method of Willstätter and Stoll of 1913, fractionation between immiscible solvents was employed to separate chlorophyll components *a* and *b* from the yellow pigments and from each other. The spectrograms inserted in the thesis of W. de Rogowski show, however, that the chlorophyll *b* previously prepared by Dhéré and de Rogowski by the chromatographic adsorption method was notably more pure than the chlorophyll *b* prepared in the same period by Willstätter and Stoll.

The same fact was emphasized in 1940 by Hans Fischer and Adolf Stern[16] who stated that

> M. Tswett was able to establish exactly the existence of two green components with help of the adsorption analysis introduced by him and to describe in detail the spectra of these pigments. These investigations were, however, later violently contradicted and they were forgotten, although, as we know it today, M. Tswett was the first who actually obtained — be it in solution — really pure chlorophyll. Only Ch. Dhéré and W. de Rogowski (1912) could again prepare pure chlorophyll solutions according to the method of M. Tswett and describe the fluorescent spectra of both components.

Although this is not the place to discuss and evaluate Rogowski's spectra, it is interesting to compare his visible wavelength spectra of chlorophyll *a* and *b* with the ones published by Willstätter. Twenty years later Alfred Winterstein analyzed in detail Willstätter's spectra[17,18] and proved that his chlorophyll *b* was not pure but had a 15% impurity of chlorophyll *a*. This is evident from the spectra published by Willstätter and coworkers from "pure" chlorophylls in the early part of the 1910s[19,20] where the 644 nm absorption band of chlorophyll *b* was accompanied by a weak band at 663 nm (being the most prominent but not the only extra band in the spectrum): the 663 nm absorption was actually caused by chlorophyll *a* present in the solution. Today, it is amusing to read that Rogowski, in his thesis, worried about this missing band and apologized that he only obtained *une légère ombre* (a weak shadow) at this wavelength when increasing the optical path length to 26 mm.[7] Obviously, the chlorophylls prepared by him were really pure and indeed confirmed Tswett's results. However, in those days, Willstätter was the big authority in this field and people believed him more than the little-known Russian botanist. It took 25 more years until doubt in Tswett's methodology was confuted by P. Karrer who stated[21] that

> ... it would be a mistake to believe that a preparation purified by crystallization should be purer than one obtained from chromatographic analysis. In all recent investigations chromatographic purification widely surpassed that of crystallization.

These facts emphasize even more the farsightedness of Dhéré who realized at such an early stage of the evolution of science that chromatography can give purer compounds than the classical methods.

Concerning the leaf carotenoids investigated by Rogowski in his thesis, Dhéré emphasized later[6] that this work was the first after Tswett preparing pure substances and demonstrating (independently of Tswett) that pure xanthophylls do not have a red fluorescence in alcoholic solution, as described by Escher — an assistant of Willstätter — in his Ph.D. thesis of 1909.[22] As stated by Dhéré, "if Escher had used purification by the chromatographic method, he would not have committed this error."

8.2.3. *Dhéré and Tswett*

As already mentioned, Dhéré never met Tswett. However, we can safely assume that he already knew about his activities not only through Tswett's numerous papers published in German and French journals, but was almost certainly aware of Tswett's major 1910 book published in Warsaw,[9] most likely through Rogowski, his graduate student. We have already mentioned that Rogowski's thesis cites Tswett's book with the original Russian title. Obviously, Rogowski spoke Russian (after all, he grew up in Russian-occupied Warsaw) and we would not be surprised if he brought in 1911 a copy of the book with him to Fribourg. Here, we have a direct connection between Tswett's major publication and the work in Dhéré's laboratory.

8.3. Vegezzi and His Thesis Work

In 1913–1914 Dhéré had seven publications on animal and plant pigments and their spectroscopic investigations. Among these only one, coauthored with L. Ryncki and reporting on the UV spectroscopic investigations of carotenoid pigments,[23] mentioned that pure carotene was prepared according to the method of Tswett, using the technique described by Rogowski, and giving reference to the paper of Dhéré and Rogowski.[10] The next major work in Dhéré's laboratory in which the chromatographic technique was further improved was the thesis of Vegezzi.[24]

Guglielmo Vegezzi was born on 1 August 1890, in Ticino, the Italian-speaking area of Switzerland (Fig. 8.5). He did his undergraduate studies at the Universities of Zürich and Fribourg; then he started his thesis work under Dhéré in 1912, and mostly finished it by the summer of 1914. However, his military service in the Swiss army in 1914–1915 prevented him from finishing it until the spring of 1916

Fig. 8.5. Guglielmo Vegezzi, in his middle age. (Courtesy of Dr. G. Vegezzi, Jr.)

when he received his *Docteur ès Sciences* degree on July 25. Subsequently, he joined the Swiss Federal Administration of Alcohol, in Bern, as a chemist, advancing later to the position of the vice-director of the Agency. Vegezzi died on 5 September 1955.

The subject of Vegezzi's thesis was the spectroscopic investigation of various pigments present in invertebrates, such as in the bile and liver of the escargot *Helix pomatia*, and in the eggs of the spider crab *Maja squinado*. These investigations were very important because they represented the first application of chromatography in the preparation of such pigments in pure form, permitting the measurements of their UV spectra and fluorescence. In addition to the printed thesis of Vegezzi[24] the results were reported in six papers coauthored by Dhéré and Vegezzi.[25–30]

The chromatographic system of Vegezzi (Fig. 8.6) differed only slightly of that of Rogowski. Now, the outside glass sleeve extended to the whole length of the column. In his 1943 paper[6] Dhéré mentioned a shortcoming of Rogowski's design: that the upper part of the "chromatogram" could not be seen well because the upper rubber stopper blocked the view of it; therefore, now they changed this. In his later book on fluorescence[4] Dhéré recommend the Vegezzi design of a chromatographic system.

8.4. Later Work of Dhéré

It is interesting to note that after Vegezzi left his laboratory, Dhéré evidently did not carry out any more chromatography work, and he also slowed down in having graduate students. In the list of Laszt[3] there is one thesis from 1906, six from 1910–1912, one each from 1916 (Vegezzi's), 1924, 1927, and 1928, two each from 1932 and 1936, and one from 1941 (well after his retirement!). In other words, there is an eight-year gap after Vegezzi, and only eight thesis work from the remaining 17 years; none of these theses reported on chromatography investigations although a great number of naturally occurring pigments were prepared by other methods. The only thesis referring to chromatography is the one by Oscar Biermacher (1936) who mentioned that he was first thinking to use chromatography for the isolation and purification of chlorophylls, but finally decided to use liquid–liquid partition for

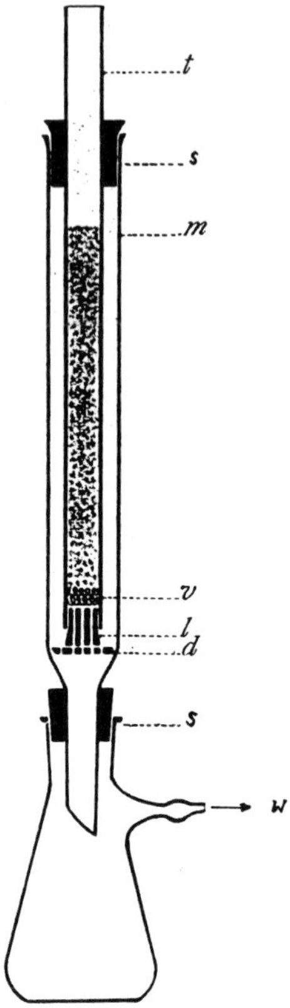

Fig. 8.6. The chromatographic system of Vegezzi.[24] t = Chromatographic column (350 mm × 16 mm i.d. glass tube), m = outer glass sleeve, v = glass wool layer (about 5 mm high), l = cork with multiple perforations, d = perforated porcelain disk, s = rubber stoppers, w = water pump.

this purpose. This is somewhat strange because by 1936, chromatography was in use in Switzerland, in a number of laboratories.

8.5. Dhéré's Paper on Tswett

In 1937 Zechmeister and Cholnoky published a bestseller, their book on chromatography,[8] and one year later a second, greatly enlarged edition, was published (see Chapter 13). In the preface of the second edition, the authors stated that "it was intended to begin this volume with a biography of Tswett, but reliable data about the life of this pioneer were not available up to this time". Indeed this was true: the first brief biography of Tswett (a total of 28 lines, with a bibliography listing 56 publications) was published only in 1940, in a collection of the biographies of botanists from Geneva[31] written by John Briquet (1870–1931), professor of botany at the University of Geneva, a former friend of Tswett.

Evidently, this publication induced Dhéré to start collecting data on Tswett. Dhéré was now retired, living in Geneva with strong connection to the University. Briquet, Tswett's friend from college days with whom he remained in touch until almost his death, was dead, and his other Geneva friend Edouard Claparède, with whom he studied at the University and whom he visited frequently from Russia, had just died. However, Dhéré could still contact P. G. Hochreutiner, professor at the University, who was a graduate student together with Tswett in the laboratory of Professor Marc Thury.

It took about two years for Dhéré to compile his 50-page long paper which he finally submitted on 23 March 1943, to the journal *Condollea*.[6] This paper not only dealt with the life of Tswett but it also discussed in detail his scientific work, presenting a critical evaluation of the controversies related to the early years of chromatography and to Tswett's chromatographic investigations of plant pigments. In addition, Dhéré also presented a summary of the evolution of chromatographic analysis up to 1940, and finished the article with a brief discussion of the influence of Tswett's method on the results of "the princes of contemporary science", the Laureates of the Nobel Prize in Chemistry. A detailed bibliography listing the most important papers related to chromatography, including 36 publications by Tswett, was also given.

Dhéré did an excellent job in collecting all the information although he made two errors: he was giving the time of Tswett's death as May 1920, while he actually died on 26 June 1919 (the error was taken over from the obituary in the *Berichte der deutschen botanischen Gesellschaft*), and stated that Claparède died in 1939 (instead of 1940). For 30 years, until the start of the publications of Sakodynskii, Senchenkova and others, this was the most comprehensive discussion of Tswett's life and activities.

Fig. 9.3. Leroy Sheldon Palmer, in 1940–1942. (Courtesy of the Archives of the University of Minnesota.)

research on this subject, continued in his PhD thesis work.[2–4] This latter research followed from Eckles' interest in relating the composition of the pigments present in milk to the pigments present in the food intake of the cows. It was an ambitious project on a subject which, at that time, was barely explored. The problem tackled by Palmer can be summarized as follows. It had been observed that in the spring and summer, when the cows are in pasture eating fresh grass or green alfalfa hay, the butter produced from their milk has a deep yellow color. However, in winter, when eating stored foodstuff, the color of the butter is usually very light. Thus, two obvious questions were raised: (a) Is the change in the color of the butter (and other milk products) related to the change in the food intake of the animal? (b) Is there any link between the constituents of milk and milk products, and the food intake of the cow?

When Palmer started his research in 1909 nobody knew the answer to these questions; also almost nothing was known about the compounds causing the yellow color which today, we call the carotenoids. Even the name "carotenoids" describing this particular group of compounds was proposed by Tswett only in 1911, when Palmer was already halfway through his research, but the final nomenclature of these compounds was not established until the late 1930s. Without any question Palmer's PhD thesis was the first to establish the direct relationship between the carotenoids present in the food intake of the cow and the composition of milk and butter of the animal. It represented the first direct proof that the carotenoids present in milk, milk products, and certain animal tissues are the same pigments as present in the food intake of the animal: in other words, the animals do not produce them. Palmer was fully aware of the importance of his work: he emphasized at the end of the first part of his thesis[3] that

> the investigations which will be reported in the succeeding papers are the first to show that there is a definite relation other than chemical between the yellow plant and animal pigments.

Palmer's PhD thesis[2–4] consisted of four main parts. The first reviewed the literature concerning the yellow pigments present in plants and animals. This was followed by a detailed discussion of the chemical and physiological relations between the pigments present in milk fat and the carotenes and xanthophylls of green plants. The laboratory methods used for their isolation and separation (including chromatography) were also elaborated here. The third part dealt with the pigments of body fat, corpus luteum, and skin secretions of the cow. Finally, the fourth part consisted of three sections dealing with the yellow pigments of blood serum, the fate of carotenes and xanthophylls during digestion, and the pigments present in human milk fat.

In the second part of the 1910s Palmer extended his investigations to other animals such as chicken,[5–8] sheep, goat, swine, and horse.[9] He clearly demonstrated that not every animal utilizes the same carotenoids, while some apparently do not store them at all. For example, the hen has no apparent use for carotene, but preserves the xanthophylls present in the food intake and carries it into the egg yolk and the fatty tissue. At the same time other animals — sheep, swine, rabbit, etc. — have no carotenoids present in their fatty tissue. His chemical/chromatographic analyses documented the differences within the different animals in the deposition and metabolism of these pigments.

Through these studies Palmer became one of the top experts in the field of carotenoids. These activities culminated in his book on carotenoids, published in 1922[10]; this, the first modern summary of this subject (Fig. 9.4), consists of 316 pages with a 17-page bibliography. It represents an exhaustive discussion of the subject and also reports on the possibility of the separation of carotenoids present in various plants and animals. We shall discuss the importance of this book below.

After his move to Minnesota Palmer's interest extended into nutrition studies in general and also dealt with the influence of minerals and vitamins present in food on the growth and health of the animals. However, these questions are beyond the scope of this book.

9.3. Chromatography in Palmer's Work

When Palmer started his research project in 1909, the obvious path for him to follow would have been the assiduous method of Willstätter and others[11]: to carefully isolate the individual pigments utilizing their selective solubility and purify them in a multistep process ending with crystallization. Instead of this, however, he immediately realized the superiority of chromatography, and extensively utilized it in his work. This happened just three years after Tswett's basic publications, at the time when practically nobody in Europe paid any attention to them.

Palmer's decision to rely on chromatography was a courageous act, because it was not obvious. Let us not forget that even as late as in 1929, some people were still convinced that chromatography, without crystallization, could never produce a pure pigment.[12]

Fortunately, Palmer could get the information on chromatography directly from the primary source, the Bulletins of the German Botanical Society (*Berichte der deutschen botanischen Gesellschaft*) in which Tswett's major papers were published: records indicate that

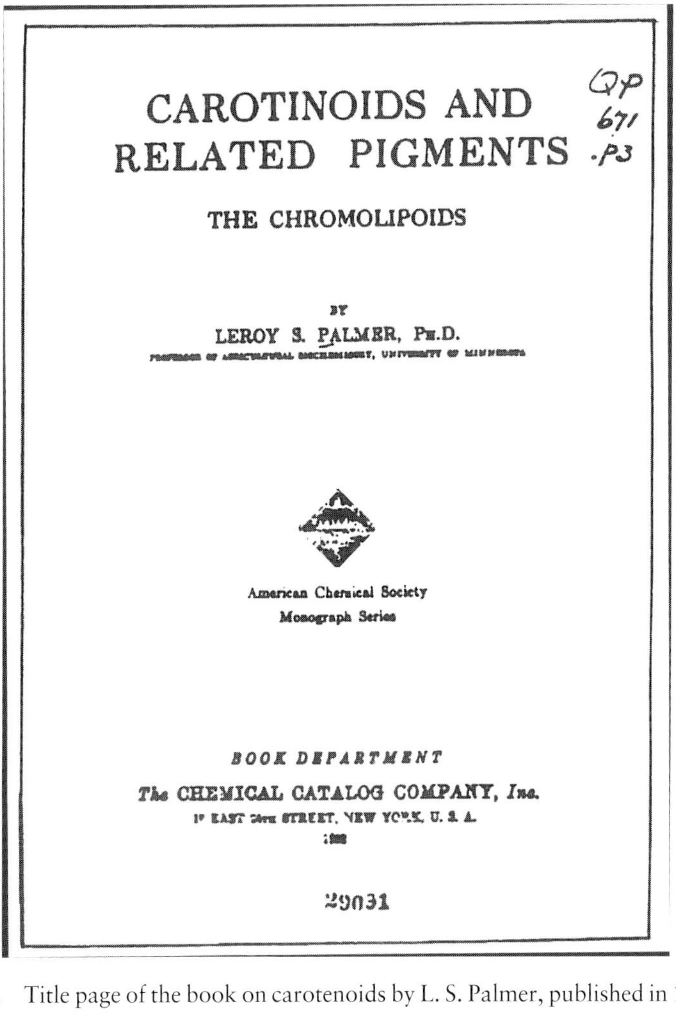
CAROTINOIDS AND RELATED PIGMENTS

THE CHROMOLIPOIDS

BY

LEROY S. PALMER, PH.D.

American Chemical Society
Monograph Series

BOOK DEPARTMENT

The CHEMICAL CATALOG COMPANY, Inc.

Fig. 9.4. Title page of the book on carotenoids by L. S. Palmer, published in 1922.[10]

the library of the University of Missouri regularly received the issues of this journal. Palmer's methodology was based mainly on the two classical papers of Tswett published in 1906[13,14]; he also utilized the results of Tswett's 1911 paper[15] that was published when he was halfway through his thesis work, and — as we shall mention below — he might even have known Tswett's book published in Russian in 1910.

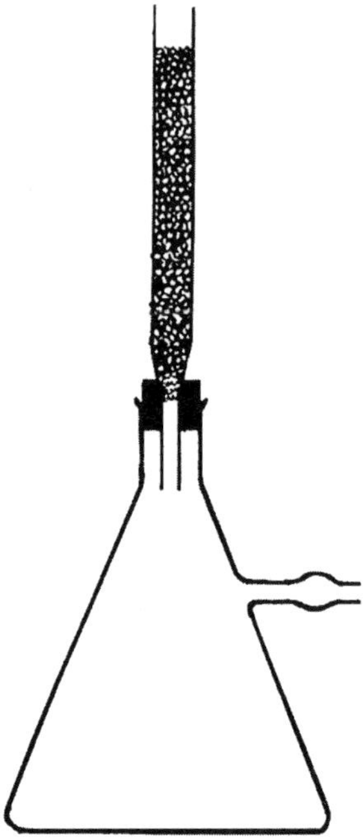

Fig. 9.5. Typical chromatographic arrangement used by Palmer.[2–4]

Palmer used a simple setup to chromatograph the pigment solutions; it is shown in Fig. 9.5. It consisted of a 15–20 cm long, 1–2 cm i.d. glass tube, tapered at one end; a small piece of cotton was placed in the small end of the tube and the tube was filled with a dry adsorbent. The process could be speeded up by suction with a water-jet pump. This setup is identical to the system recommended by Tswett for larger sample volumes.[14] It is particularly interesting that Palmer correctly listed inulin and sugar as suitable alternative adsorbents to calcium carbonate. Their use was specifically proposed by Tswett for the separation of compounds that can easily undergo chemical or structural transformation. Willstätter's objection against chromatography was based mainly on his assistants' findings that some of the pigments underwent changes with calcium carbonate as the adsorbent.[11] Today we know that Willstätter's assistants did not read carefully Tswett's papers and neglected his warnings.[16]

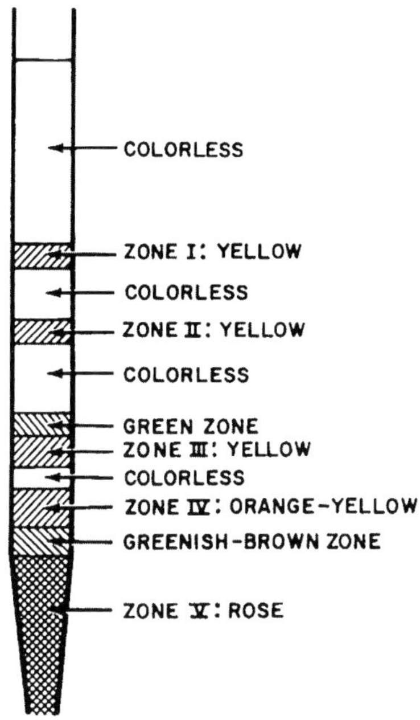

Fig. 9.6. Chromatogram of the CS_2 extract of dry alfalfa hay.[2–4]

In Palmer's work — as in the work of Tswett and Dhéré — the sample solution was introduced at the top of the column and enough solvent was added to wash it down the column. During this process the separated fractions appeared as colored rings. Figures 9.6 and 9.7 show two "chromatograms" from Palmer's thesis.

After the separation on the column was finished, Palmer again added the solvent to the column and carefully eluted the colored fractions one by one, separately collecting each fraction (the solution of a pure pigment). In this way, Palmer's method significantly differed from Dhéré's (Chapter 8): In Dhéré's lab, the solid column packing was slowly pushed out of the tube; the colored rings were carefully separated (cut) from each other and from the rest of the packing; and only then were the individual pigments extracted from the packing for further investigation. In retrospect, Palmer's adaptation of Tswett's method may be considered as the embryo of the continuous

knowledge of the distribution of these pigments in all forms of plant and animal matter. Tswett's important observations are accessible to us in a series of papers[13–15,18] from 1906 to 1911. The last paper is more of the nature of a summary, but by reason of its clear-cut statements, it may well serve today as our best laboratory outline for working with the class of pigments with which this monograph deals. It was in this paper that Tswett proposed the nomenclature for the carotinoids which has been adopted in this monograph.

9.5. Palmer as the Transition Between Tswett and The "Rebirth" of Chromatography

In spite of the wide distribution of Palmer's book, only a few researchers utilized chromatography in the 1920s. As we shall see later chromatography was "reborn" only at the beginning of the 1930s; this coincided with a breakthrough in the carotenoid field, with the clarification of the individual compounds, the establishment of their structure, and the realization of the existence of isomeric compounds. In these activities, Palmer's book served as the primary reference material about the chromatographic technique of Tswett. Thus, Palmer and his activities represent the transition between Tswett's work and modern chromatography.

References

1. L. S. Palmer, The Coloring Matter in Fat From Cow's Milk, Thesis for the M. A. Degree in Chemistry, College of Agriculture, University of Missouri, Columbia, MO, 1919, 50 pp.
2. L. S. Palmer, A Study of the Natural Pigment of the Fat of Cow's Milk, Thesis for the PhD Degree in Dairy Husbandry, College of Agriculture, University of Missouri, Columbia, MO, 1913, 205 pp.
3. L. S. Palmer and C. H. Eckles, Carotin: The Principal Natural Yellow Pigment of Milk Fat. Chemical and Physiological Relations of Pigments of Milk Fat to the Carotin and Xanthophylls of Green Plants, *Bulletins of the University of Missouri Agricultural Experiment Station* (Columbia, MO, 1913), No. 9, pp. 313–336; No. 10, pp. 339–387; No. 11, pp. 391–411; and No. 12, pp. 414–451.

4. L. S. Palmer and C. H. Eckles, *J. Biol. Chem.* **17**, 191–210, 211–221, 223–236, 237–243, 245–249 (1914).
5. L. S. Palmer, *J. Biol. Chem.* **23**, 261–279 (1915).
6. L. S. Palmer and H. L. Kempster, *J. Biol. Chem.* **39**, 299–312 (1919).
7. L. S. Palmer and H. L. Kempster, *J. Biol. Chem.* **39**, 313–330 (1919).
8. L. S. Palmer and H. L. Kempster, *J. Biol. Chem.* **39**, 331–337 (1919).
9. L. S. Palmer, *J. Biol. Chem.* **27**, 27–32 (1916).
10. L. S. Palmer, *Carotinoids and Related Pigments. The Chromolipids* (Chemical Catalog. Co., New York, NY, 1922), pp. 42–44.
11. R. Willstätter and A. Stoll, *Untersuchungen über Chlorophyll: Methoden and Ergebnisse* (Springer Verlag Berlin, 1913); English edition: F. M. Schertz and A. R. Merz (Translators), *Investigations on Chlorophyll: Methods and Results* (Science Press, Lancaster, PA, 1928).
12. F. M. Schertz, *Plant Physiology* **4**, 337–348 (1929).
13. M. Tswett, *Ber. Dtsch. Botan. Ges.* **24**, 316–323 (1906).
14. M. Tswett, *Ber. Dtsch. Botan. Ges.* **24**, 384–393 (1906).
15. M. Tswett, *Ber. Dtsch. Botan. Ges.* **29**, 630–636(1911).
16. R. Kuhn, in *Gas Chromatography 1962* (*Hamburg Symposium*), ed. M. Van Swaay (Butterworths, London, 1962), pp. xvii–xxvi.
17. M. S. Tswett, *Khromofilly v Rastitel'nom i Zhivotnom Mire* (Chromophylls in the Plant and Animal World) (Karbasnikov Publishers, Warsaw, 1910).
18. M. Tswett, *Ber. Dtsch. Botan. Ges.* **24**, 235–244 (1906).

Chapter

10

Katharine Hope Coward: A Pioneering User of Chromatography*

Besides Gottfried Kränzlin in Berlin, Charles Dhéré and his students at the University of Fribourg, in Switzerland, Leroy S. Palmer at the University of Missouri, in Columbia, MO, USA, and Theodor Lippmaa at the University of Tartu, in Estonia (see next chapter), the fifth pioneer in the so-called dormant period of chromatography was Katharine Hope Coward, a British scientist. It is however interesting to note that except for the inclusion of one of her papers from 1924 (L, Table 10.1) in the extensive bibliography of the book of Zechmeister and Cholnoky,[1] nothing is known about her activities.

Like the other four scientists mentioned earlier, Coward was also a *user* of chromatography, who realized the advantages of Tswett's method and utilized it in her work. This chapter summarizes her activities in the 1920s, in the context of the status of science at that time.

---

*Based on the article by L. S. Ettre and P. J. T. Morris, published in *Chromatographia* 60, 613–617 (2004).

10.1. K. H. Coward — Her Life

We know about the life and activities of Katharine Hope Coward (Fig. 10.1) from the entries in *The Pharmaceutical Journal*[2–4] and in *Poggendorff's Handwörterbuch*.[5] She was born in 1885 in Blackburn, Lancashire, studied at the University of Manchester gaining an honors degree in botany in 1906, and an MSc in botany in 1908. We know nothing about her activities in the next 12 years; then in January 1920, she entered University College in London, to study biochemistry and began research under Professor J. C. Drummond. She received a personal grant from the Medical Research Council and was appointed an assistant in the Department; then, in December 1920, she was awarded a Beit Fellowship for Medical Research. After receiving her DSc degree in biochemistry and the Schafer Prize in 1924, she remained in Dr Drummond's laboratory; then, in 1925, she was granted a Rockefeller Fellowship to the University of Wisconsin, in Madison WS, United States, to work in the laboratory of Dr Harry Steenbock, Professor of Biochemistry and also a pioneer in nutrition and vitamin research.

Fig. 10.1. Katharine Hope Coward, around 1937.[2] (Courtesy: *Pharmaceutical Journal.*)

After returning to England in 1927, Dr Coward joined the newly organized laboratories of the Pharmaceutical Society (today: the Royal Pharmaceutical Society) as the head of the Nutrition Department. In 1936 she was appointed a Reader in biochemistry at the School of Pharmacy of the University of London, and in 1937 she was elected an honorary member of the Pharmaceutical Society. Dr Coward died in 1978, at the advanced age of 93.

10.2. The State of Science in Coward's Time

In order to understand the subject of Katharine Coward's work we have to consider it in the context of knowledge in the fields of nutrition and vitamins and naturally occurring pigments in the first two decades of the 20th century. Also a brief summary of the scientists' activities in whose laboratories she had worked is useful.

10.2.1. *Nutrition and Vitamins*

Until the end of the 19th century nutrition was considered as a function of the protein, carbohydrate, fat, and mineral content of the food. It was assumed that if sufficient amounts of these substances are consumed, then the growth and well-being of the person or animal are assured. However, slowly this prevailing opinion was questioned by some observations, suggesting that some unknown, vital factor(s) must also be present in foods to support normal growth and life. In 1911 Casimir Funk demonstrated the effect of one such compound, present in the shavings from polished rice and found that this compound was an amine. He suggested that various diseases could be prevented by the addition of such "vital amines" to the food. This expression was the origin of the collective noun "vitamins" used to characterize these micronutrients. In the subsequent years it had been found that not all of these substances are amines; therefore, in order to avoid the possibility of any misunderstanding, the term was shortened to "vitamins."

By the beginning of the 20th century the State of Wisconsin, in the United States of America, already had a key role in the production of butter, and the Agricultural Experiment Station at the University of

Wisconsin, in Madison, represented an important place where agricultural research was carried out.[6] One of their important experiments had to do with the diet of dairy cows: animals were fed different mixtures having identical compositions concerning their protein, carbohydrate, fat, and salt contents, but using different basic materials. It was found that cows which had corn in their food were healthy, produced milk in large amounts and reproduced normally; however, if wheat was substituted for corn, the cows were unable to reproduce and lactate. The conclusion was that corn had to contain some unknown nutrients not present in wheat. This result initiated a search for these nutrients. The leading scientist in this work was E. C. McCollum.[7,8]

After receiving his PhD at Yale University, in New Haven, Connecticut, Elmer Verner McCollum (1879–1967) joined the University of Wisconsin in 1907. His early work concerned chemical analyses of the feed of dairy cattle. Frustrated by the long procedures entailed in the use of such large animals he instituted the use of rats as the experimental animals. By 1913 McCollum was able to report that deficiency in the diet of rats could be compensated by adding extracts of eggs or butter to their feed. He concluded that a "fat-soluble growth factor" must be present in eggs or butter. Almost simultaneously Osborne and Mendel, at Yale University, also demonstrated the existence of this micronutrient. Further studies by McCollum indicated that the presence of other substances is also necessary for normal health: a "water-soluble factor", the absence of which can cause severe neuritis and anemia, and another fat-soluble factor, the absence of which is responsible for *rachitis*. These factors were identified as vitamins and were later assigned the letters A, B, and D, respectively.

In 1917 McCollum accepted the directorship of the newly created School of Public Health at Johns Hopkins University, in Baltimore, Maryland. At Wisconsin his activities were continued by Harry Steenbock (1886–1967), a graduate of the University, who became an Associate Professor in 1917 and Professor of Biochemistry in 1920.[9] Steenbock developed a method to produce vitamin D by the UV irradiation of sterols present in foodstuffs. He also postulated in 1919 a possible correlation between the presence of carotene in food and the occurrence of vitamin A; however, for a decade the nature of this relationship was not clear and the chemical composition of these compounds was not known.

The fact that the chemical composition of the vitamins and their true relationship to other naturally occurring substances — carotenoids, sterols — were unknown resulted in a number of misconceptions. These were aggrieved by the lack of proper analytical methods to determine the individual vitamins. Rather, their presence or absence was established by nutritional experiments, administering certain foods to test animals (rats) and observing the resulting effects. This method was particularly misleading in the case of vitamin A. The fact that the animal's liver produces this vitamin from carotene was not known; thus, if normal growth of the animal was observed, this was attributed to the presence of vitamin A in the food. One can find a number of publications from this period indicating the presence of this vitamin in various plants, e.g., in dried peas[10] and in yellow corn seeds (S, Table 10.1). These misconceptions were only eliminated by the end of the 1920s and early 1930s, after V. Euler, at the University of Stockholm,[11] and Moore, at Cambridge University,[12] have clearly shown the *in vivo* transformation of carotene into vitamin A; and Karrer, at the University of Zurich, described the chemical structures of these compounds,[13,14] establishing their chemical link.

To be objective we should mention that relying on animal experiments was not only due to the lack of absolute chemical analytical methods: it followed from the main aim of the investigations. Their primary aim was to carry out *nutrition studies*, in order to establish the usefulness of various human and animal foods and find ways to prevent or cure certain illnesses. The conclusions resulting from such experiments were fundamental and very useful: the shortcoming was in the interpretation of the results.

Up to now we have concentrated on the activities of the researchers associated with the University of Wisconsin. Naturally, intensive research has also been carried out in Europe. One of the principal European scientists involved in early research in the field of nutrition and vitamins was Jack Cecil Drummond (1891–1952), a graduate of Queen Mary College, London. After graduation in 1914 he joined Casimir Funk at the Cancer Hospital in London and became his successor in 1918. The following year he was appointed a Reader in physiological chemistry at University College, London, and in 1922 he was promoted to Professor of Biochemistry.[15] As mentioned in the introduction, Katharine Coward joined Dr Drummond's group at the beginning of 1920.

10.2.2. *Carotenoids*

In the first decades of the 20th century, the interest in substances occurring in the nature grew. One widely investigated group of compounds was the so-called lipochromes, colored pigments present in the fatty tissues of plants and animals. It was generally recognized that these may actually be divided into two groups, based on their solubility in light petroleum fractions (petroleum ether or ligroin) and in a polar solvent such as ethanol. Willstätter and Mieg showed in 1907 that the compounds of the first group are hydrocarbons, with the general formula of $C_{40}H_{56}$, while those of the second group also contain some oxygen, corresponding to the formula of $C_{40}H_{56}O_2$.[16] The main compound of the first group was carotene (carotin), the yellow pigment of carrots, while the name xanthophyll was used for the oxygenated compound(s). Besides this basic information, however, very little was known about these substances and there was a general confusion concerning the identity of the various compounds isolated from different plants. There was a constant debate whether substance A isolated by researcher X was identical to substance B found by researcher Y, or whether they were different substances. Tswett's investigations clearly indicated the presence of at least four different xanthophylls,[17] however, his peers questioned his results, particularly since he conducted no chemical tests and also did not show the isolated substances in pure crystalline form. Let us not forget that at that time the validity and the meaning of chromatographic separation were questioned by some of the internationally recognized chemists (Willstätter, Marchlewski). Recognizing the apparent similarity of the various lipochromes, Tswett proposed in 1911 to use the collective name *carotenoids (carotinoids)* for them,[18] however, it took 10 years until this name started to be accepted. In fact, the chemical structure of the individual pigments and hence, the existence of a large number of carotenoids with similar structure was established only in the 1930s.

The confusion in the identification of the individual carotenoids and the ignorance of Tswett's results also reflected in the inadequacy of the methods used in the investigation of the chromolipids. Instead of relying on chromatography permitting the separation of the individual compounds, Willstätter's "phase test" was used generally. After suitable preparation of the sample the carotenoids were dissolved in petroleum ether and then this solution was extracted with alcohol: the carotenes remain in the petroleum ether while the xanthophylls will be present in the alcoholic solution. The amount of the two groups in the respective solutions was estimated by colorimetric measurement, comparing color intensity with that of a standard 0.2% solution of potassium dichromate ($K_2Cr_2O_7$), using pre-established calibration plots. The individual samples were characterized by the carotin : xanthophyll ratio, and by their total lipochrome content.[19] Naturally this fairly primitive method could only give a rough estimate of the actual composition of the chromolipids present in the various samples and did not permit the establishment of any meaningful connection between vitamin A and the various carotenoids. Therefore, research conducted in this field was essentially restricted to data collection concerning the lipochromes present in various plant and animal tissues and their nutritional properties.

10.3. The Scope of Coward's Work in the 1920s

When Katharine Coward joined Professor Drummond's group at University College, London, his group had already been engaged in intensive nutritional studies, concentrating on questions associated with vitamin A. Thus, it was obvious that Coward will also be involved in these studies and this can be seen from the titles of their joint papers (Table 10.1). Her doctoral thesis was also in this field, concerning the study of lipochromes present in plant tissues and the possible formation of vitamin A. We have found three papers representing parts of her thesis (E, J, K, Table 10.1). These papers reported on a large amount

Table 10.1. Publications of Katharine Hope Coward during 1920–1927.

[A]	K. H. Coward and J. C. Drummond, Researches on the fat-soluble accessory substance. Nuts as a source of vitamin A, *Biochem. J.* **14**, 665–667 (1920).
[B]	J. C. Drummond and K. H. Coward, Researches on the fat-soluble accessory substance. The nutritive value of animal and vegetable oils and fats considered in relation to their color, *Biochem. J.* **14**, 668–677 (1920).
[C]	J. C. Drummond and K. H. Coward, Researches on the fat-soluble accessory factor. Effect of heat and oxygen on the nutritive value of butter, *Biochem. J.* **14**, 734–739 (1920).
[D]	J. C. Drummond and K. H. Coward, Nutrition and growth on diets devoid of true fats, *Lancet* **2**, 698–700 (1921).
[E]	K. H. Coward and J. C. Drummond, The formation of vitamin A in living plant tissues, *Biochem. J.* **15**, 530–539 (1921).
[F]	J. C. Drummond and K. H. Coward, Researches on vitamin A. Notes on the factors influencing the value of milk and butter as sources of vitamin A, *Biochem. J.* **15**, 540–552 (1921).
[G]	H. L. Jameson, J. C. Drummond and K. H. Coward, Synthesis of vitamin A by a marine diatom growing in pure culture, *Biochem. J.* **16**, 482–485 (1922).
[H]	K. H. Coward and J. C. Drummond, On the significance of vitamin A in the nutrition of fish, *Biochem. J.* **16**, 631–636 (1922).
[I]	K. H. Coward and A. J. Clark, The vitamin content of certain proprietary preparations, *Brit Med. J.* **1**, 13–15 (1923).
[J]	K. H. Coward, The formation of vitamin A in plant tissues, II, *Biochem. J.* **17**, 134–144 (1923).
[K]	K. H. Coward, The association of vitamin A with the lipochromes of plant tissues, *Biochem. J.* **17**, 145–155 (1923).
[L]	K. H. Coward, Some observations on the extraction and estimation of lipochromes from animal and plant tissues, *Biochem. J.* **18**, 1114–1122 (1924).
[M]	K. H. Coward, The lipochromes of etiolated wheat seedlings, *Biochem. J.* **18**, 1123–1126 (1924).
[N]	J. C. Drummond, O. Rosenheim and K. H. Coward, The relations of steroids to vitamin A, *J. Soc. Chem. Ind.* **44**, 123–124 (1925).
[O]	K. H. Coward, Synthesis of vitamin A by a freshwater alga, *Biochem. J.* **19**, 240–241 (1925).
[P]	K. H. Coward, The persistence of vitamin A in plant tissues, *Biochem. J.* **19**, 500–506 (1925).
[Q]	J. C. Drummond, H. J. Channon and K. H. Coward, Studies on the chemical nature for the presence of vitamin A, *Biochem. J.* **19**, 1047–1067 (1925).

(*Continued*)

Table 10.1. (*Continued*)

[R]	J. C. Drummond, K. H. Coward and J. Hardy, On the technique of testing for the presence of vitamin A. *Biochem. J.* **19**, 1068–1974 (1925).
[S]	H. Steenbock and K. H. Coward, Fat-soluble vitamins. The quantitative estimation of vitamin A, *J. Biol. Chem.* **72**, 765–779 (1927).
[T]	K. H. Coward, The influence of light and heat on the formation of vitamin A in plant tissues, *J. Biol. Chem.* **72**, 781–799 (1927).

of investigations; however, they were handicapped by the general misconceptions prevailing at that time, discussed earlier in this chapter.

It should be noted that both in London and during her one-year stay in Wisconsin, Dr Coward was also involved in some investigations which touched a new field: the possible interaction of vitamins A and D (R, S, Table 10.1). This is, however, outside our present field of interest.

10.3.1. *Coward and Chromatography*

Our interest in Coward's activities is related to her utilization of chromatography in some of her investigations. In one of her 1923 papers (K, Table 10.1) we can find a footnote noting that "a full account of the lipochromes (carotinoids) and their occurrence in animal and plant tissues has just been written by Palmer." This is with reference to the comprehensive book of Palmer published in 1922[20]: In it he summarized knowledge on the carotenoids and also described in detail Tswett's results and the chromatographic technique (see Chapter 9). Evidently Coward studied the book and from it, learned about Tswett's results and the potentialities of chromatography the first time. As a conclusion she reopened her investigations and carefully re-examined the relative merits of the by then generally accepted methods of extraction, separation, and estimation. These new investigations were reported by her in two papers (L, M, Table 10.1).

What interests us is Coward's use of chromatography to see whether she can also find four different xanthophylls in the plant extracts, as reported by Tswett.[17,18] She followed Tswett's methodology without any change, using calcium carbonate as the adsorbent, adding the lipochrome solution to the top of the column and eluting the substances with the addition of further volumes of the solvent. The separated fractions (colored rings on the column packing) were then carefully removed and physically separated from each other and from the rest of the packing; the individual pigments were extracted from the packing and their amounts were estimated by colorimetry; the fractions were also characterized by the bands of their absorption spectra. Coward's conclusion was the confirmation of the existence of the four xanthophylls in all the samples investigated.

Coward went even one step further. When chromatographing various plant extracts, carotene usually passed already through the column with the solvent while the separated rings of the xanthophylls were still formed. She noted that the color of the eluate changed slightly with time: when collecting the effluent into small fractions, the absorption bands of the very first fraction were also different than that of the rest. She carried out these experiments with the light petroleum ether extract of both the corollas of African marigold and tomatoes and observed in both cases such differences. Her conclusion from these experiments was that apparently they indicate — in addition to carotene — the presence of some other substances that were closely related. This again showed the correctness of Tswett's assumption about the possible existence of a number of closely related carotene-like substances.

As seen Coward was close to developing a method for the chromatographic separation of carotenoids and vitamin A. However, we must not forget that her main interest was nutrition and the chromatographic investigations only served to better characterize the food samples used in the experiments. Thus, the demonstration of this possibility had to wait one more decade until shown by Karrer and Schöpp.[21]

10.4. Postscript

Although it is not our task to discuss the post-1927 activities of the scientists mentioned here, a brief summary might be useful. All three continued their involvement in nutrition and vitamin research. E. V. McCollum remained at Johns Hopkins University until his retirement, in 1945, when he became an emeritus professor. He received numerous awards and was elected a member of the US National Academy of Sciences and a fellow of the Royal Society of London. In 1957 he published a comprehensive book about nutrition.[22]

After patenting his irradiation process to produce vitamin D, Harry Steenbock assigned the incomes from licenses to the Wisconsin Alumni Research Foundation, established for this purpose. He became an emeritus professor in 1956. After his death, in 1967, an Endowment has been established at the University of Wisconsin sponsoring a yearly Steenbock Symposium and Lecture in Biochemistry.

J. C. Drummond continued his activities at the University of London in the field of nutrition and vitamins and as the head of a key center of biochemical teaching. Chromatography had been widely used in his laboratory for the analysis of carotenoids and vitamins A and E, and he also improved the technique: a 1932-paper from his laboratory described the first time a column chromatography system permitting to carry out separation in an inert atmosphere, with the exclusion of air.[23] In 1939 Professor Drummond co-authored with his wife a book on *The Englishman's Food*.[24] During World War II he has served as a key adviser to the Allied Armies' Supreme Command on the questions of nutrition; for his activities he was knighted in 1944 and elected a fellow of the Royal Society. In 1946 he became director of research at Boots Pure Drug Inc. In 1952, he was murdered with his wife and daughter, while on a vacation in France. In the 1970s the Drummond Memorial Fund was established at the University of London to promote education and research in nutrition which is organizing yearly lectures in the memory of Professor Drummond.

Finally a few words about the later activities of Dr Coward. As mentioned in the introduction, after returning from Wisconsin she joined the newly organized laboratory of the British Pharmaceutical Society as the head of its Nutrition Department and she remained in that position until her retirement. During this period she played a prominent role in devising standards and standard analytical methods — utilizing also chromatography — for vitamins and other related compounds and published widely in various journals. Dr Coward served in a number of committees of the British Medical Research

Council, the British Pharmacopoeia, and the Biochemical Society, and also served as an adviser to the committees of the League of Nations and the World Health Organization.

References[a]

1. L. Zechmeister and L. Cholnoky, *Die chromatographische Adsorptionsmethode*, 2nd edn. (Springer Verlag, Vienna, 1937), 1938.
2. anon., *Pharm. J.* **138**, 484 (1937).
3. anon., *Pharm. J.* **165**, 261 (1950).
4. anon., *Pharm. J.* **221**, 134 (1978).
5. *Poggendorff's Biographisch-literarisches Handwörterbuch*, Vol. VI, Part I (A–E) (Verlag Chemie, Berlin, 1923–1931) p. 486.
6. P. L. Nelson, B. C. Soltved, eds., *One Hundred Years of Agricultural Chemistry and Biochemistry at Wisconsin* (Science Tech Publishers, Madison, WI, 1984).
7. C. C. Gillispie, ed., *Dictionary of Scientific Biography*, Vol. 7 (Charles Scribner's Sons, New York, 1981), pp. 590–591.
8. B. Narius, ed., *Notable Scientists from 1900 to Present*, Vol. 3 (Gale Group, Farmington Hills, MI, 2001), pp. 1502–1503.
9. F. L. Holmes, ed., *Dictionary of Scientific Biography*, Vol. 18 (Charles Scribner's Sons, New York, 1990), pp. 849–851.
10. H. Steenbock, M. Y. Sell and P. W. Boutwell, *J. Biol. Chem.* **47**, 303–308 (1921).
11. B. V. Euler, H. V. Euler and H. Hellström, *Biochem. Z.* **203**, 370–384 (1928).
12. T. Moore, *Biochem. J.* **24**, 692–702 (1930).
13. P. Karrer, A. Helfenstein, H. Wehrli and A. Wettstein, *Helv. Chim. Acta* **13**, 1084–1099 (1930).
14. P. Karrer, R. Morf and K. Schöpp, *Helv. Chim. Acta* **14**, 1036–1040, 1441–1446 (1931).
15. P. Campbell, *A Brief History of Biochemistry at UCL Before the Founding of the Department in 1946*, University College London, URL: http://www.biochem.ucl.uk/about-the-department/detailedhistory.doc. (Viewed August 3, 2004).
16. R. Willstätter and W. Mieg, *Ann. Chem.* **355**, 1–28 (1907).

[a]We express our gratitude to Ms Brioni Hudson of the Royal Pharmaceutical Society, London, for providing copies of the Pharmaceutical Journal articles.

17. M. Tswett, *Ber. Dtsch. Botan. Ges.* **24**, 384–393 (1906).
18. M. Tswett, *Ber. Dtsch. Botan. Ges.* **29**, 630–636 (1911).
19. R. Willstätter and A. Stoll, *Untersuchungen über Chlorophyll. Methoden und Ergebnisse* (Springer Verlag, Berlin, 1913).
20. L. S. Palmer, *Carotinoids and Related Pigments. The Chromolipids* (Chemical Catalog Co., New York, 1922).
21. P. Karrer and K. Schöpp, *Helv. Chim. Acta* **15**, 745–746 (1932).
22. E. V. McCollum, *A History of Nutrition* (Hughton-Mifflin, Boston, 1957).
23. I. M. Heilbron, R. N. Heslop, R. A. Morton, E. T. Webster, J. L. Rea, and J. C. Drummond, *Biochem. J.* **26**, 1178–1193 (1932).
24. J. C. Drummond and A. Wibraham, *The Englishman's Food* (Jonathan Cape, London, 1939).

Chapter

11

Theodor Lippmaa, A Forgotten Chromatographer*

We have seen that in 1917, Tswett finally was appointed a full professor at the University of Tartu, in present-day Estonia, then part of the Russian Empire, and as the head of the University's Botanical Gardens. Tartu is the Estonian name of the town; in Tswett's time, it was known both by its German (Dorpat) and Russian (Yure'ev) name. The university was founded in 1632 by the Swedish king, Gustavus II Adolphus (at that time, the Baltic area belonged to Sweden), and in Tsarist Russia, it was considered as one of the best universities. However, due to the events of the First World War, Tswett only stayed at Tartu for a few months and was evacuated with the other Russian professors when the area was occupied by German troops. Eventually, the situation consolidated in the Baltic area: for a period of 20 years Estonia became an independent republic, and the university was reorganized, now as an Estonian school of higher learning. At that time

*Based on the article by L. S. Ettre, published in *Chromatographia* 20, 399–402 (1985).

Fig. 11.1. Theodor Lippmaa (1892–1943). (Courtesy: Prof. Endel Küllik.)

Theodor Lippmaa was appointed as the successor of Tswett, occupying the Chair of Botany and serving as the Head of the Botanical Gardens. Lippmaa is one of the few pioneers who in the 1920s utilized chromatography in his work; thus we can consider him as a member of the chain of scientists connecting Tswett with the rebirth of chromatography in the early 1930s.

Theodor Lippmaa (Fig. 11.1) was born in 1892 in Riga, presently the capital of the Republic of Latvia. The Baltic area was then a part of Russia, and this area including Estonia was called Livonia. I could not find any information about his life prior to the 1920s, but obviously it was interrupted by the First World War: let us not forget that he was 22 when the war broke out. The first information we have about Lippmaa is that he studied at Tartu University majoring in botany and graduated in 1924. In that year he presented a paper at the meeting of the *Naturforscher Gesellschaft* (the Scientific Society of the University) (see A in Table 11.1). The 1926 Yearbook of the

Deutsche Botanische Gesellschaft (the German Botanical Society) lists him as a member, indicating his degree as "Magister der Botanik".[1] However, this membership list was obviously compiled earlier in that year because he actually obtained his doctorate (Dr phil. nat.) in 1926. This was followed by his habilitation in 1927, and appointment as Professor of Botany and head of the Botanical Gardens at Tartu University in 1930. The Second World War tragically ended his life: on January 27, 1944, a stray bomb fell on the house next to the Botanical Gardens in which he was living and killed him along with his wife and daughter. (His son, Endel, happened to go to the bakery for bread, and thus survived.)

11.1. The Separation of Carotenoids

Lippmaa's main research interest at the University concerned botany and not chemistry. He was the first to compile a catalog of Estonian plants in 1933. He also studied the flora of several parts of the world: Northern Africa (Algeria, Morocco, and Tunesia), the Mediterranean area of France, northern Finland, the northern part of the United States, and the Labrador peninsula of Canada. He also pioneered in what we call today environmental protection.

Chromatography represented a relatively short period of Lippmaa's activities at the beginning of his professional career. Evidently, the study of red plant pigments was the subject of his doctoral thesis at the University of Tartu and he summarized his investigations in a number of papers during 1924 to 1926 (see Table 11.1.). I found two papers in the publications of Tartu University [A, B]; three papers in the *Comptes Rendus*, the biweekly scientific journal of the French Academy of Sciences [C, D, E]; and one paper in the journal of the German Botanical Society [F], which (according to a note below its title) Lippmaa personally presented at the December 1926 meeting of the Society in Berlin. Of these six papers, two [B, C] specifically mention the use of chromatography for the separation of rhodoxanthin (the red pigment) and other carotenoids (Fig. 11.2 and Table 11.2). In the literature compilation of Zechmeister and Cholnoky[2] only one of these papers [C] is cited. It is an irony of fate that in the heading of this particular paper, Lippmaa's name was misspelled (it was written

Table 11.1. Theodor Lippmaa's papers on rhodoxanthin and chromatography.

[A]	Theodor Lippmaa, Über den Parallelismus im Auftreten der Karotine and Anthocyane in vegetativen Pflanzenorganen, *Sitzungsberichte der Naturforscher Gesellschaft der Universität Dorpat* 30(3/4), 58 (1924).
[B]	Theodor Lippmaa, Das Rhodoxanthin. seine Eigenschaften, Bildungsbedingungen und seine Funktion in der Pflanze, *Schriften herausgegeben von der Naturforscher Gesellschaft bei der Universität Tartu (Dorpat)*, **24** (1925).
[C]	Theodor Lipmaa,[a] Sur les Proprietés Physiques et Chimiques de la Rhodoxanthine, *Comptes Rendus Hebdomadaires des Séances de l'Academie des Sciences* **182**, 867–868 (1926).
[D]	Theodor Lippmaa, Sur la Formation des Chromoplastes chez les Phanérogames, *Comptes Rendus Hebdomadaires des Séances de l'Academie des Sciences* **182**, 1040–1042 (1926).
[E]	Theodor Lippmaa, Sur les Hématocarotinoides et les Xanthocarotinoides, *Comptes Rendus Hebdomadaires des Séances de l'Academie des Sciences* **182**, 1350–1352 (1926).
[F]	Theodor Lippmaa, Über den vermuteten Rhodoxantingehalt der Chloroplasten, *Berichte der deutschen botanischen Gesellschaft* **44**, 643–648 (1926).

[a]The name is misspelled, with one "p".

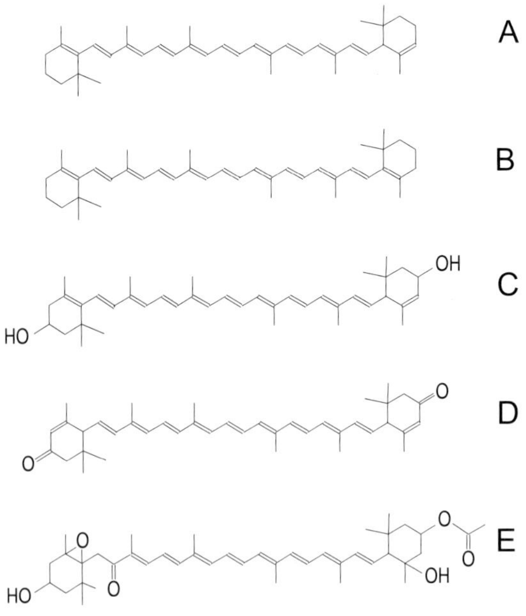

Fig. 11.2. Molecular structure of some carotenoids. For identification of the compounds see Table 11.2.

Table 11.2. Identification of the carotenoids in Fig. 11.2.

(A)	α-Carotene	$C_{40}H_{56}$
(B)	β-Carotene	$C_{40}H_{56}$
(C)	Lutein (xanthophyll)	$C_{40}H_{56}O_2$
(D)	Rhodoxanthin	$C_{40}H_{50}O_2$
(E)	Fucoxanthin	$C_{42}H_{58}O_6$

with one "p"); therefore, this is the way his name is now spelled in the literature.

In order to understand Lippmaa's work and its importance, we have to place it in the proper historical context. In the period discussed here, the structures of the substances we call carotenoids had not yet been established. Tswett, in the second of his two basic papers,[3,4] had already demonstrated that chromatographic adsorption analysis provides a clear proof for the existence of individual carotenoids and the presence of four "xanthophylls" in leaves. However, Willstätter (at that time the most respected scientist in the field of chlorophyll and carotenoid research) did not accept Tswett's results and considered the four xanthophylls as artifacts, formed by isomerization from a single, naturally occurring xanthophyll during adsorption. Due to the prestige of Willstätter, his opinion was generally accepted in the field. Thus, since chromatography was not yet considered a reliable identification (and separation) method, researchers had to rely on other data, for example, spectroscopic measurements from solutions. However, one could never be sure whether the solution contained a single compound or was contaminated by some other compounds.

Owing to this uncertainty in the identification of the individual compounds, there was always the problem of whether a newly-isolated compound from a plant was really a new compound, or identical to a compound already described by somebody else. In fact, as we shall see in the next chapter, the rebirth of chromatography in 1931 was actually related to such a question: The goal of Lederer's work in Heidelberg was to find out whether the "lutein" of egg yolk is a separate, pure substance, or a mixture of other carotenoids already found in leaves.

Another consequence of the rejection (or disregard) of chromatography as *the* superior separation technique was the confusion in trying to combine the plant pigments into various groups and trying to derive one of these groups from the others. Not being able to disregard the great differences among these groups and the different plant extracts, but denying the existence of a number of individual carotenoids (e.g., the four xanthophylls of Tswett), researchers tried to present a simplistic explanation of the obvious differences among the groups by saying they were simply due to differences in the relative amounts of the components present. As Lippmaa summarized very correctly, they believed in quantitative, but not in qualitative differences.

Lippmaa was one of the few scientists of this period who specifically stated his belief that the substances present in different plants are indeed different, with different chemical structures, and who used Tswett's chromatographic technique to prove his point.

As already mentioned, the subject of Lippmaa's research was the red pigment present in certain plants. Tswett was the first, who in 1911,[5] described and isolated the red pigment in the leaves of *Thuja orientalis* and other plants. He called it "thuyorhodine." A few years later, Lubimenko, another Russian scientist, also isolated this substance from other plants. He called it rhodoxanthin (the name we use today), and considered it an isomer of xanthophyll. Some scientists, however, felt that the red color of plants is not due to carotenoids but due to anthocyan pigments.

Lippmaa used *Reseda odorata* as the raw material. He extracted the pigments from its leaves and then purified them in the classical way, by partitioning between two immiscible solvents, for example between petrol ether and methyl alcohol, or between carbon disulfide and methyl alcohol. However, full separation of rhodoxanthin from the other carotenoids was accomplished by chromatography. Let us quote (in English translation) Lippmaa's report about his own work [B]:

> As it is known, Tswett (1906) invented a particular method for the analysis of dye mixtures, with which it was possible for him to separate e.g., the xanthophyll of Willstätter and Mieg into four xanthophylls: $\alpha, \alpha', \alpha''$, and β. The chromatographic method I used is identical to that of Tswett. Freshly precipitated calcium carbonate, dried at 150°C, was used as the adsorbent.

> The chromatogram will look completely different, depending whether one is using carbon disulfide or petrol ether as the solvent. In the $CaCO_3$–petrol ether system, rhodoxanthin is adsorbed (together with xanthophyll) by the $CaCO_3$ and thus can be separated easily from carotene. In the $CaCO_3$–carbon disulfide system, xanthophyll and rhodoxanthin behave completely differently: while the first carotenoid is also adsorbed here, rhodoxanthin can be washed out of the $CaCO_3$ column with carbon disulfide, and thus separated from xanthophyll.

These quotations clearly show that Lippmaa was well aware of Tswett's work and was not influenced by the general opinion mentioned earlier.

Lippmaa carried out detailed investigations on the separated xanthophyll and rhodoxanthin fractions. In conclusion ([C], English translation of the French original) he stated that

> These characteristics clearly show that the opinion on the isomeric character of rhodoxanthin and xanthophyll is unfounded. It is quite possible that rhodoxanthin, just as fucoxanthin, represents a carotenoid substance in a higher oxidation state than xanthophyll.

This is really a very clear statement, well ahead of its time; just a look at the structure of these carotenoids (see Fig. 11.2) shows how right Lippmaa was. This is even more laudable if we consider that at that time very little was known about the structure of these substances.

In addition to these investigations on the rhodoxanthin from *Reseda*, Lippmaa also helped to correct some misconceptions published by others, notably by Professor Harald Kylin of the University of Lund, in Sweden. Kylin claimed to have found rhodoxanthin in the extracts of green leaves, using the capillary analysis of Goppelsröder.[6] Repeating the experiments by using leaves in which rhodoxanthin was definitely present, Lippmaa demonstrated that the narrow band found by Kylin was not rhodoxanthin as it forms a band at a different place.

Lippmaa's investigations, which demonstrated the superiority of Tswett's method, were carried out just a few years before the decisive work of the Heidelberg group that is generally considered as the rebirth of chromatography. It is a pity that, being a botanist, Lippmaa did not further pursue these investigations. Even so, he deserves more than a passing reference in the history of chromatography.

11.2. Postscript

I visited Tallin and Tartu in 1979 and 1982 and wanted to use this opportunity to collect as much information about Lippmaa as possible. However, the situation was not simple. At that time Estonia was still part of the Soviet Union, and tension was already obvious between Estonians and the official Soviet authorities as well as the newcomer Russians living in the area. I was warned to be very careful when searching information about Lippmaa and writing about him, because the first independent period of Estonia between the two world wars ("the bourgeois republic") was a non-subject. Because of this, while his existence could not be denied, the study of Lippmaa's life and activities was strongly discouraged. I was not permitted to request material about Lippmaa from the library at Tartu University, however, privately I received a lot of help from Estonian scientists; but I was told that I must not acknowledge their cooperation. My host, Professor Endel Küllik (1929–1990) of the Chemistry Institute of the Estonian Academy of Sciences obtained for me the pertinent copies of Lippmaa's papers; however, I was again warned not to mention his help, but rather state that literature search was done by myself here, in the United States. A special subject that could not be discussed was the period of the Second World War when Estonia was under German occupation: I was particularly forbidden to mention that the bomb that killed Theodor Lippmaa along with his wife and daughter was actually dropped by a *Russian* airplane.

Even though I followed these warnings, my paper about Lippmaa published in 1985[7] created a sensation in Estonia: it broke a taboo, providing the first published discussion of a scientist from the "bourgeois republic" period. However, within five years, the situation changed: Estonia was the first Baltic Republic declaring its independence. In this struggle, Professor Endel Lippmaa, the son of Theodor Lippma, played a key role.

References

Lippmaa's papers listed in Table 11.1 are referenced in the text by using the capital letters A through F.

1. *Ber. Dtsch. Botan. Ges.*, p. 22 of the yearly report at the end of Vol. **44** (1962).
2. L. Zechmeister and L. Cholnoky, *Die chromatographische Adsorptionsmethode*, 2nd edn. (Springer Verlag, Vienna, 1938).
3. M. Tswett, *Ber. Dtsch. Botan. Ges.* **24**, 316–323 (1906).
4. M. Tswett, *Ber. Dtsch. Botan. Ges.* **24**, 384–393 (1906).
5. M. Tswett, *C. R. Acad. Sci. Paris* **152**, 788–789 (1911).
6. L. S. Ettre, Evolution of Liquid Chromatography: A Historical Overview, in *High-Performance Liquid Chromatography — Advances and Perspectives*, Vol. I, ed. Cs. Horváth (Academic Press, New York, 1980), pp. 1–74 (pp. 13–15).
7. L. S. Ettre, *Chromatographia* **20**, 399–402 (1985).

Part Four

The Rebirth of Chromatography

Chapter

12

The Rebirth of Chromatography

In the previous chapters we have already mentioned that Tswett's contemporaries were skeptical about the advantages of chromatography and in the first 25 years, only a very few scientists — Kränzlin, Dhéré, Palmer, Lippmaa, and Coward — utilized the technique in their work. Some, like e.g., Stocklasa,[1] used it with the specific aim to discredit it, and the major researchers investigating chlorophyll pigments (Marchlewski, Willstätter) expressed their reservations about its usefulness. In their fundamental monograph, the first textbook providing detailed information about chromatography, Zechmeister and Cholnoky called this 25-year period the *Latenzzeit* (dormant period) of chromatography.[2]

Then, in 1931, a change occurred: increased interest in the study of complex natural substances necessitated the adaptation of new methods for their separation, and chromatography was found to ideally suit this purpose. As a contrast to the Latenzzeit, Zechmeister and Cholnoky called the period from 1931 on the *Blütezeit* (flowering time) of chromatography, and we can consider 1931 as the year of the rebirth of the technique.

The rebirth of chromatography started in Heidelberg, Germany, and was initiated by the activities of Edgar Lederer in the laboratories of Richard Kuhn. Their work was followed by a fantastic explosion in the use of the technique, and within less than a decade, chromatography became a generally accepted technique in both Europe and the United States.

This chapter is devoted to the activities of Richard Kuhn and Edgar Lederer, in Heidelberg; the next chapter will then discuss the rapid expansion of the technique in Europe and the United States.

12.1. Richard Kuhn

A native of Vienna (Austria) Richard Kuhn (1900–1967) started to study chemistry at the University of Vienna but after one year, transferred to the University of Munich, Germany, where he graduated in 1921. He was then accepted by Richard Willstätter (1872–1942), the most famous German chemist of that time, as a PhD student. We have already mentioned Willstätter earlier (see Chapter 4) when speaking about Tswett's struggle for acceptance. Willstätter graduated at Munich and has served for some years as a special assistant to Adolf von Baeyer; in 1905–1912 he occupied the chemistry chair at the Swiss Federal Technical University (ETH) in Zurich (where most of his chlorophyll work was carried out), and in 1912–1915 served as the first director of the Kaiser Wilhelm Institute for Chemistry, in Berlin–Dahlem (succeeded in Zurich by Hermann Staudinger). Finally, in 1915, upon the retirement of Baeyer, Willstätter took over Baeyer's chair in Munich. By then Willstätter turned his attention from the chlorophylls to the exploration of enzyme chemistry and naturally, Kuhn's doctorate thesis was also in this field. He finished it within one year — almost unheard of at that time — with the highest honors, and remained at the University carrying out both teaching and research. In 1926 upon the recommendation of Willstätter he was offered full professorship at the E.T.H., in Zurich, as the successor of Staudinger who moved to the University of Freiburg, in Germany. Kuhn was staying in Zurich for less than three years: in 1929 he was appointed as the director of the Chemistry Institute within the Kaiser Wilhelm Institute for Medical Research, in Heidelberg,

Germany. (This Institute actually consisted of four sub-institutes, the Chemistry Institute being one of these.) From Zurich he took with him several assistants, among them Alfred Winterstein (1889–1960); within a short time he further expanded the scientific staff of his institute in Heidelberg.

While still in Zurich, Kuhn's interest turned toward the complex plant pigments called carotenoids, compounds with long chains of carbon atoms with alternating single and double bonds. He continued this work in Heidelberg. The rebirth of chromatography is associated with these activities.

12.2. The Field of Carotenoids

Although their existence has been known for a long time, very little was known about these compounds. As mentioned in the previous chapters, in 1907 Willstätter and Mieg[3] clearly established the existence of two such compounds (as later realized, compound groups): "carotene," a hydrocarbon with the elementary composition of $C_{40}H_{56}$ and "xanthophyll," with the composition of $C_{40}H_{56}O_2$; and then in 1912, Willstätter and Escher found another compound with the same composition as carotene, which they called lycopene.[4] Tswett also studied these compounds and already in his 1910 book, he had expressed his opinion that carotene is not a single compound but a mixture of two or more homologues which could be separated by chromatography.[5] Then in a paper published in 1911 Tswett also stated that xanthophyll is not a pure compound, but a mixture of some isomeric compounds.[6] It is interesting to note that Willstätter disputed this opinion, saying that "it is unlikely that this assumption of the respected botanist (i.e. Tswett) is true."[7]

This confusion about the various carotenoid compounds existed until the early 1930s: such compounds were observed in various plants, but it was not clear whether compound A found in source X is identical or not to compound B found in source Y, and whether an isolated substance is a pure compound of a mixture of some isomeric compounds. This is clear from the monograph of L. S. Palmer published in 1922, representing the first relatively modern summary of the information available about carotenoids[8]: This book reported on the compounds found in various plants and animals, but the information presented was mainly empirical. It is important to realize that none of these researchers had any information about the actual structures of these compounds; these were only gradually established in the 1930s by Kuhn, Karrer, and others (Fig. 12.1 and Table 12.1).

A special problem in the investigation of carotenoids was that these compounds are present in nature only in very small amounts; therefore, when using classical methods in their study, very large amounts

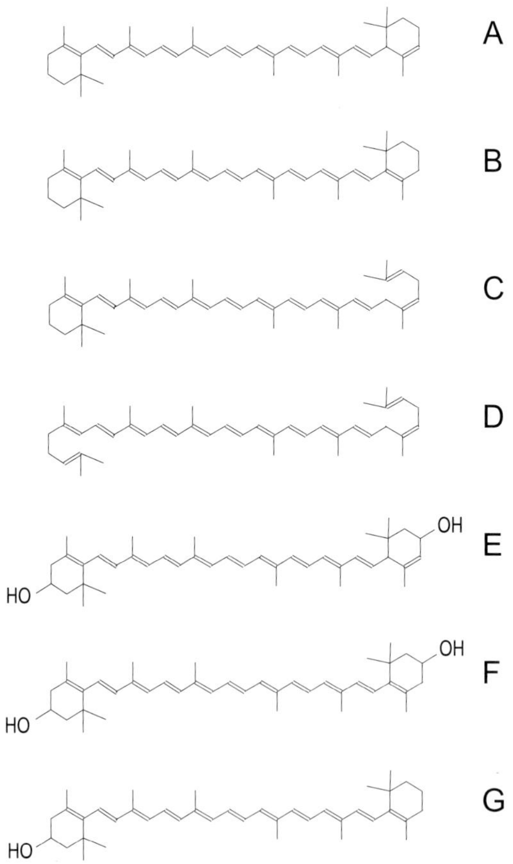

Fig. 12.1. Molecular structure of some important carotenoids. For identification see Table 12.1.

Table 12.1. Identification of the carotenoids in Fig. 12.1.

	Name	Molecular composition
(A)	α-Carotene	$C_{40}H_{56}$
(B)	β-Carotene	$C_{40}H_{56}$
(C)	γ-Carotene	$C_{40}H_{56}$
(D)	Lycopene	$C_{40}H_{56}$
(E)	Lutein (xanthophyll)	$C_{40}H_{56}O_2$
(F)	Zeaxanthin	$C_{40}H_{56}O_2$
(G)	Cryptoxanthin	$C_{40}H_{56}O$

of the original material had to be processed. For example, in 1912, Willstätter and Escher had to extract 6000 hen eggs to obtain a crude, crystalline pigment for further investigation.[4] It was Palmer's particular merit that in his book he pointed out the usefulness of chromatography, giving detailed description of the technique; however, until the work in Heidelberg, in 1930, evidently nobody followed Palmer's advice.

We should also mention that some of the interests in the carotenoids stemmed from nutrition studies (this is also how Palmer became involved) and from observations connecting the carotenoids to vitamin A activity. These investigations further intensified after Hans von Euler-Chelpin at Stockholm University, in 1928,[9] and Thomas Moore, at Dunn Nutrition Laboratory, in Cambridge, England, in 1930,[10] clearly demonstrated that vitamin A is metabolized from carotene. (See Chapter 9 for a more detailed description of Palmer's work.)

12.3. Edgar Lederer and the Rebirth of Chromatography

A native of Vienna, Edgar Lederer (1908–1988) studied chemistry at the University of Vienna, receiving his PhD in 1930. He then accepted a post-doctoral position at the Chemistry Institute of the Kaiser Wilhelm Institute for Medical Research, in Heidelberg. He was assigned to assist Alfred Winterstein who, with Kuhn, was engaged in carotenoid research. Their newest research showed similarities of the pigments present in egg yolk and the yellow "xanthophylls" of green leaves studied some 25 years earlier by Willstätter. The situation was further complicated by a most recent paper of Paul Karrer, at the University of Zurich (see next chapter), describing a new "xanthophyll" isolated from yellow corn which he named zeaxanthin.[11] The just-arrived Lederer was assigned to carry out laboratory investigations to clear up this question. The subsequent events are well documented in two recollections of Lederer.[12,13]

Since Lederer was not familiar with the field of carotenoids (his thesis work in Vienna dealt with indole alkaloids), he first had to read the existing literature. Meanwhile he carried out measurements of the absorption spectra, melting points, and optical rotations of various carotene preparations from carrots, lutein from egg yolk, xanthophylls from green leaves, and zeaxanthin from yellow corn; he found that the last two pigments had distinctly different properties, while lutein seemed to lie between the two. Discussing the results, Kuhn had a suggestion: perhaps the lutein of egg yolk might be a mixture of leaf xanthophyll and corn zeaxanthin. Naturally, this could be proven if he could separate the "lutein" of egg yolk into two components. But how?

During his literature study, Lederer read the carotenoid book of Palmer[8] and there he found information about the chromatographic adsorption technique of Tswett. But he also read the chlorophyll book of Willstätter and Stoll[7] expressing the authors' misgivings about the technique (see Chapter 4); and he also read the publications of Dhéré, Palmer, and Coward (see Chapters 8–10), reporting on its successful use. Somewhat confused he went to Kuhn, asking his advice. At that time Kuhn remembered the manuscript copy of the translation of Tswett's 1910 book[5] specially made for Willstätter some 20 years earlier, that his teacher gave him many years ago. Fortunately he could still find it and he gave the copy to Lederer for study. After reading it, Lederer decided to try to apply the chromatographic technique to solve his problem.

In his recollections[13] Lederer vividly describes the event on a day in December 1930, when he prepared a column packed with $CaCO_3$ according to Tswett's instructions, then poured the carbon disulfide solution of a mixture of 0.5 mg of pure lutein and 0.5 mg of zeaxanthin to this column, and washed it with the same solvent: soon a large orange band became visible on the column, with distinctly different hues in the upper and lower parts. He carefully scraped off the two layers, dissolved the pigments in methanol, and "lo-and-behold, the upper zone had the spectrum of lutein, and the lower the spectrum of zeaxanthin." This was then followed by the chromatography of the crude lutein extract of the egg yolk of 100 eggs (the white of which previously had been "transformed to a delicious cake") on a column of 7-cm diameter. After breaking the packing into zones, extracting the pigment from them and separately rechromatographing the extracts, pure substances could be obtained. As stated by Lederer, these experiments proved that egg yolk is indeed a mixture of lutein and zeaxanthin, and that Tswett's chromatographic method can be used for preparative separation. In this way he also confuted Willstätter who, just a few years earlier, specifically stated "we can consider the chromatographic adsorption analysis ... not suitable for work in larger scale, that means for preparative purposes".[14]

Meanwhile Kuhn became involved in a controversy with Paul Karrer concerning the chemical structure of carotene: some experiments carried out at that time by Lederer were in contradiction with some of Karrer's data (obtained using non-chromatographic methods). Further work has shown the validity of Kuhn's assumption about the existence of isomeric carotenes, and Lederer could then separate α- and β-carotene on a column packed with alumina.

Their chromatographic results were finally summarized in three fundamental papers. The first was a brief report dealing with the fractionation of the two carotene isomers; it was submitted to *Naturwissenschaften* on February 17, 1931, and it mentions the use of chromatography only in a single sentence.[15] This was then followed by two papers submitted on March 10 and 18, respectively, describing in detail the experiments about the separation of xanthophylls[16] and isomeric carotenes.[17] These three papers represent the rebirth of chromatography. Figure 12.2 depicts a page from Ref. 16, outlining the experiments described earlier.

In one of these three papers[16] Kuhn also proposed a change in the nomenclature for the carotenoids: to use the name *xanthophyll* as a

KUHN, WINTERSTEIN UND LEDERER, Zur Kenntnis der Xanthophylle. *(Hoppe-Seyler's Z. Physiol. Chem.*, 197 (1931) 158.)

12. Zerlegung des Dotterfarbstoffes in seine Komponenten.

(a) Fraktionierung eines künstlichen Gemisches von Lutein und Zeaxanthin. 0.5 mg Lutein und 0.5 mg Zeaxanthin wurden in 10 cc Schwefelkohlenstoff gelöst. Diese Lösung saugten wir langsam durch ein mit Calciumcarbonat (gefällt, E. Merck) gefülltes Rohr von 15 cm Länge und 10 mm lichter Weite, das unten durch einen Wattebausch abgeschlossen war. Durch Nachwaschen mit Schwefelkohlenstoff (insgesamt 20–30 cc) gelang es, die Adsorptionsschichten mehrere Zentimeter auseinander zu ziehen. Das erhaltene Chromatogramm ist in der Figur dargestellt. Die Schichten 1–4 wurden getrennt herausgenommen und mit Methanol eluiert. Nach dem Filtrieren und Verdampfen fanden wir in Schwefelkohlenstoff folgende Absorptionsbanden:

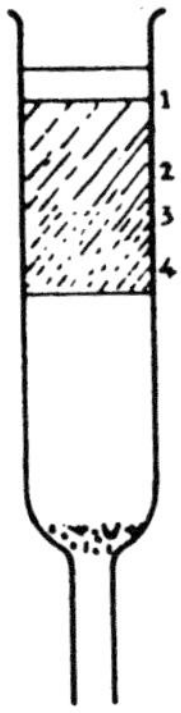

1. Schicht:	508	476	3. Schicht:	513	481
2. Schicht:	509	477	4. Schicht:	516	483
Lutein:	508	475	Zeaxanthin:	517	483

In den Schichten 1 und 4 waren je 10–15% des gesamten Farbstoffes vorhanden.

(b) Fraktionierung des Xanthophyllpräparates aus bulgarischen Hühnereiern (Spektrum: 509.5, 476 mμ). Adsorption und Elution wurden wie bei dem künstlichen Farbstoffgemisch ausgeführt.

1. Schicht:	507.5	476	3. Schicht:	510	477
2. Schicht:	510	477	4. Schicht:	512	478

(c) Fraktionierung eines Xanthophyllpräparates aus holländischen Hühnereiern (Spektrum: 509, 476 mμ).

1. Schicht:	507.5	477	3. Schicht:	510.5	478.5
2. Schicht:	509.5	477	4. Schicht:	514	480.5

In etwas grösserem Massstab wurde ein Präparat aus bulgarischen Eidottern von $[\alpha]_{Cd}^{20}$ = +100° (Chloroform) fraktioniert. Für 30 mg Farbstoff, die in 500 cc Schwefelkohlenstoff gelöst waren, verwendeten wir ein Adsorptionsrohr von 7 cm Durchmesser. Die Lutein- und Zeaxanthinzone wurden 5 cm auseinandergezogen, die mittleren Schichten eluiert, an frischem Calciumcarbonat adsorbiert. Die Mittelschichten des zweiten Rohres wurden wieder eluiert und der darin enthaltene Farbstoff ein drittes Mal durch ein Calciumcarbonat-Rohr geschickt.

Die vereinigten Schichten 1 und die vereinigten Schichten 4 enthielten etwa je 10% des angewandten Farbstoffes. Nach dem Eluieren mit Methanol und Überführen in Schwefelkohlenstoff fanden wir:

Schichten 1: 508 476 Schichten 4: 513.5 479 mμ

Die Luteinfraktion (1) zeigte nach dem Umkrystallisieren

$$(\alpha)_{Cd}^{18} = (+0.16° \cdot 100) : (0.022 \cdot 0.5) = +145° \text{ (Chloroform)},$$

was einem Luteingehalt von etwa 90% entspricht.

Die Zeaxanthinfraktion (4) war kaum optisch aktiv. Sie schmolz in noch unreinem Zustande bei 195–196° (korr).

Fig. 12.2. Page from the first publication of Kuhn, Winterstein and Lederer.[16]

generic name for all oxygenated C_{40} carotenoids and retain the name *lutein* only for the major constituent of leaf xanthophylls. Eventually this nomenclature was adopted by IUPAC.

12.4. Further Activities

Unfortunately, the happy and productive group of Kuhn in Heidelberg (Fig. 12.3) did not last for long: After Adolf Hitler came to power in Germany, in 1933, major changes were forced to occur. Soon after Hitler's ascent to power, Edgar Lederer had to flee because of his leftist political activities and Jewish origin: He was evidently forwarned because within 4 days the Gestapo came to arrest him. Since his wife was French, they went to France where he worked for two years in various research laboratories. In 1935 he moved to the Soviet Union as the research director of the Vitamin Institute in Leningrad, but two years later he returned to France and became associated with the Centre National de la Recherche Scientifique (CNRS). Lederer survived the war and German occupation in the countryside, escaping police and

Fig. 12.3. A happy day in Kuhn's laboratory in Heidelberg, sometime in 1931.[13] Identified persons are (1) R. Kuhn; (2) Mrs Kuhn; (3) A. Wassermann (later emigrated to England); (4) H. Brockmann (became professor in Göttingen); (5) Miss G. Stein (became Mrs Brockmann); (6) M. Hoffer (emigrated to the United States); (7) Th. Wagner-Jauregg (emigrated to the United States, but later returned to Switzerland); (8) H. Roth (became professor at Greifswald and Braunschweig); and (9) E. Lederer. (Courtesy: Prof. Edgar Lederer.)

Fig. 12.4. Richard Kuhn (right) and Edgar Lederer, in 1931 (author's collection).

Gestapo raids a number of times. He returned to Paris after the war, advancing within the CNRS, and starting to teach at the Sorbonne. From 1960 on, he has been the director of the Institut de Chimie des Substances Naturelles of the CNRS. In 1963 Lederer received the Wilhelm von Hoffmann Gold Medal of the German Chemical Society (G.d.Ch.) and on that occasion he again met Richard Kuhn, his boss 30 years earlier (Fig. 12.4), who, as president of the G.d.Ch., personally presented him the medal with the following words[18]:

> In Heidelberg, in my Institute, you separated α- and β-carotene, and resuscitated the method of chromatography.

In his whole life Lederer continued to be active in the leftist political movements and in 1967 he presented the Report on Chemical Warfare in Vietnam at the International War Crimes Tribunal against the United States, organized by Bertrand Russell and held in Stockholm.

Winterstein was able to remain in Heidelberg until 1934, but then he had to return to his native Switzerland where he joined Hoffmann-La Roche as their chief chemist. In the 1930s he often served as a roving ambassador of chromatography: AJP Martin remembers that while a student at Cambridge University, he heard a lecture given by Winterstein where he also demonstrated the separation of crude carotene into its components.[19] Winterstein had a distinguished career at Hoffmann-La Roche, the largest pharmaceutical house in the world; he died in 1960 when attending a congress in Tokyo.

Richard Kuhn remained in Heidelberg for his whole life, and in 1937 became director of the whole Kaiser Wilhelm Institute for Medical Research. He continued to contribute significantly to our knowledge of carotenoids, isolating a number of previously unknown compounds and establishing their structures. In 1938 he received the Nobel Prize in Chemistry "for his results in the field of carotenoids and vitamins," but by then Hitler did not permit Germans to accept the prize. At the 1962 International Symposium on Gas Chromatography held in Hamburg, he gave one of the keynote lectures on "Comments to the Development of Separation Methods".[20] Kuhn died from cancer in 1967.

References

1. A. Stocklasa, V. Berdik and A. Ernest, *Ber. Dtsch. Botan. Ges.* **27**, 10–20 (1909).
2. L. Zechmeister and L. Cholnoky, *Die Chromatographische Adsorptionsmethode* (Springer Verlag, Wien, 1937; 2nd enlarged edition: 1938).
3. R. Willstätter and W. Mieg, *Ann. Chem.* **355**, 1–28 (1907).
4. R. Willstätter and H. H. Escher, *Z. Physiol. Chem.* **76**, 214–225 (1912).
5. M. S. Tswett, *Khromofilly v Rastitel'nom i Zhivotnom Mire* (*Chromophylls in the Plant and Animal World*) (Karbasnikov Publishers, Warsaw, 1910).
6. M. Tswett, *Ber. Dtsch. Botan. Ges.* **29**, 630–636 (1911).
7. R. Willstätter and A. Stoll, *Untersuchungen über Chlorophyll: Methoden und Ergebnisse* (Springer Verlag, Berlin, 1913), pp. 234–235.
8. L. S. Palmer, *Carotinoids and Related Pigments. The Chromolipids* (Chemical Catalog Co., New York, 1922).

9. B. V. Euler, H. V. Euler and H. Hellström, *Biochem. Z.* **203**, 370–384 (1928).
10. T. Moore, *Biochem. J.* **24**, 692–702 (1930).
11. P. Karrer, J. Salomon and H. Wehrli, *Helv. Chim. Acta* **12**, 790–792 (1929).
12. E. Lederer, *J. Chromatogr.* **73**, 361–366 (1972).
13. E. Lederer, in *75 Years of Chromatography — A Historical Dialogue*, eds. L. S. Ettre and A. Zlatkis (Elsevier, Amsterdam, 1979), pp. 237–245.
14. R. Willstätter, *Untersuchungen über Enzyme*, Vol. 1 (Springer Verlag, Berlin, 1928), p. 295.
15. R. Kuhn and E. Lederer, *Naturwiss.* **19**, 306 (1931).
16. R. Kuhn, A. Winterstein and E. Lederer, *Z. Physiol. Chem.* **197**, 141–160 (1931).
17. R. Kuhn and E. Lederer, *Ber. Dtsch. Chem. Ges.* **64**, 1349–1357 (1931).
18. *Nachrichten aus Chemie und Technique* **12**, 286 (1964).
19. A. J. P. Martin, in *Gas Chromatography in Biology und Medicine* (*1969 CIBA Foundation Symposium*), ed. R. Porter (Churchill, London, 1969), p. 240.
20. R. Kuhn, in *Gas Chromatography 1962* (*Hamburg Symposium*) ed. M. Van Swaay (Butterworths, London, 1963), pp. xxiii–xxvi.

Chapter

13

The Rapid Spreading of the Technique*

We have seen that after Tswett's twin papers that were published in 1906 and described the chromatographic separation technique, practically nobody took up the method. This is contrary to the situation 25 years later, after the publications of the three papers by Kuhn's group in 1931. Tswett's papers were published eight years before the start of the First World War, and it is a coincidence that the papers of Kuhn's group were published eight years before the outbreak of the Second World War: thus it is interesting to compare these two eight-year periods. In the eight years following Tswett's publications, only three scientists successfully applied chromatography: Kränzlin (Chapter 7), Dhéré (Chapter 8), and Palmer (Chapter 9), for a total of nine publications. At the same time the second edition of the chromatography textbook of Zechmeister and Cholnoky,[1] which gave the

*Partly based on the articles by L. S. Ettre and Cs. Horváth, published in *Anal. Chem.* **47**(4), 422A–444A (1975), and by L. S. Ettre, published in *Anal. Chem.* **61**, 1315A–1322A (1959); **62**, 71A (1990).

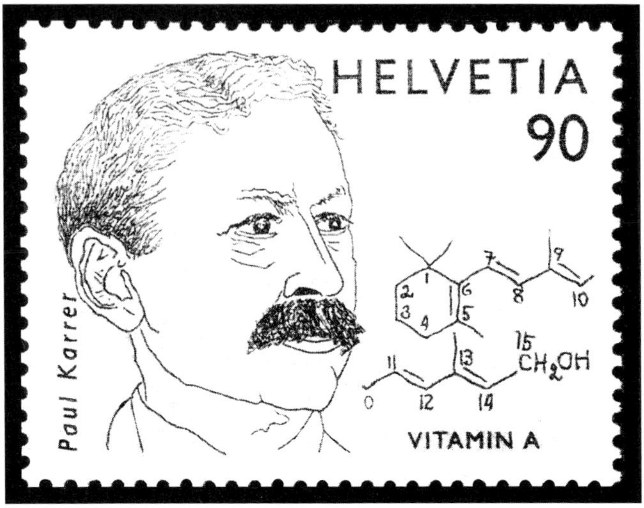

Fig. 13.1. Paul Karrer. A stamp issued in 1977 by Switzerland.

Lavoslav Ružicka (1887–1976) was born in Croatia, in the Austro-Hungarian Monarchy. He studied chemistry at the Technical University of Karlsruhe, Germany, and obtained his PhD under Hermann Staudinger, in 1910. When Staudinger moved to the ETH in Zurich, Ružicka went with him as his assistant. In the next decade he started to have some association with the Swiss perfume manufacturers and this connection turned his interest to terpenes, the constituents of the fragrant oils of vegetable origin. Between 1921 and 1930 Ružika had other positions but in 1930, he returned to the ETH as a professor of organic chemistry. This started the most brilliant period of his career when, in addition to fundamental investigations of terpenes, he also developed the first successful synthesis of the sex hormones androstrone and testosterone. Ružicka retired in 1957 but continued to be active in the laboratory for a number of years. He was succeeded at the ETH by his assistant Vladimir Prelog (1906–1998) who eventually also received the Chemistry Nobel Prize in 1975, for his achievements in the field of natural compounds and stereochemistry. Another former assistant of Ružicka was Tadeus

Reichstein (1897–1996) who in 1938 became professor at the University of Basel. His earlier work was related to the study of the flavoring substances in roasted coffee; in Basel he had been working on various biochemical subjects, most importantly on the synthesis of ascorbic acid (vitamin C) and the isolation of cortisone. For his achievements in this field he received in 1950 the Nobel Prize for Physiology and Medicine.

Both Karrer and Ružicka were *users* of chromatography. Karrer started to apply the technique almost immediately after Kuhn's first papers and their neck-to-neck competition is best illustrated by the number of their publications in which chromatography was used: the book of Zechmeister and Cholnoky lists 52 papers by Kuhn and 49 papers by Karrer. Ružicka started to report on the use of chromatography only by 1936, but this might be due to the well-known secrecy of the perfume companies.

Looking at it as a chronicler of chromatography, the main merit of the members of the Zurich schools was to prove the universal applicability of chromatography for the separation of complex mixtures and its superiority in obtaining pure substances. In the latter question Karrer said the most decisive words in 1939[5]:

> It would be a mistake to believe that a preparation purified by crystallization should be purer than one obtained from chromatographic analysis. To all recent investigations chromatographic purification widely surpassed that of crystallization.

It was also Karrer who, in a plenary lecture at the 1947 Congress of the International Union of Pure and Applied Chemistry (IUPAC) clearly established Tswett's invention as one of the key developments of the 20th century[6]:

> No other discovery has exerted as great an influence and widened the field of investigation of the organic chemist as much as Tswett's chromatographic adsorption analysis. Research in the field of vitamins, hormones, carotenoids and numerous other natural compounds would never have progressed so rapidly and achieved such a great results if it had not been for this new method.

13.2. Activities of Zechmeister

Besides Kuhn and Karrer, the third scientist pioneering in carotenoid research in the 1930s was Zechmeister, professor at the University of Pécs, in Hungary. In addition to pioneering work in this field, he was also the author of the first comprehensive chromatography textbook.

László Zechmeister (1889–1972) (Fig. 13.2) was born in Györ, a Hungarian city on the Danube, and studied at the Swiss Federal Technical University (ETH) in Zurich, graduating in 1911 as a chemical engineer. He continued his graduate studies there under Richard Willstätter. This period of Willstätter is usually identified with his chlorophyll research; however, he also had a few other pet projects, one of which was the investigation of the cellulose and lignin of trees, and this was the subject of Zechmeister's doctoral thesis.[7,8] As seen earlier, Willstätter moved in 1912 from Zurich to Berlin–Dahlem, as director of the Kaiser Wilhelm Institute for Chemistry, and he invited Zechmeister to join him as his assistant. However, he could stay there only for a short time. At the outbreak of the First World War, Zechmeister enlisted in the Hungarian Army He was soon wounded and became a prisoner of war in Russia, returning to Hungary only in 1919. By that

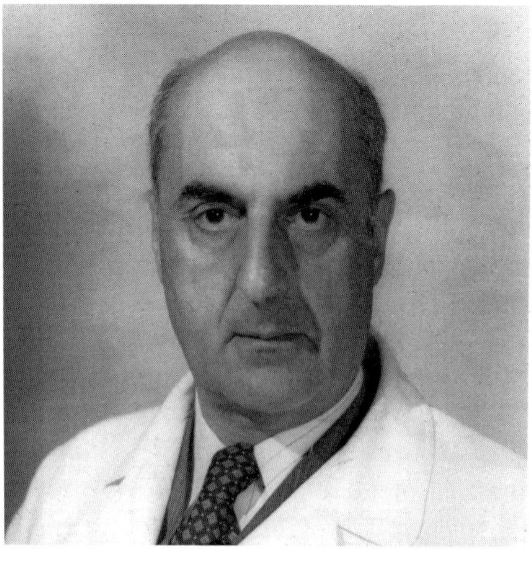

Fig. 13.2. László Zechmeister (authors' collection).

time there was a chaos in both Hungary and Germany and he could only find temporary positions; then, in 1920, he had the opportunity to move to Copenhagen, where he joined Professor Niels Bjerrum at the Royal Danish Veterinary Agricultural Academy. Finally, in 1923, he was recalled to Hungary, where he was offered a permanent position. In the peace treaty after the war, Hungary lost two-thirds of its former territory to the successor states and there were two Hungarian universities in these areas. Now, after the political situation consolidated, the Hungarian government re-established these schools within Hungary proper. One of these, formerly located in the city of Pozsony (now Bratislava, in Slovakia), was settled in Pécs, an ancient city in the southern part of Hungary where in medieval times a university had already existed. Zechmeister was appointed a full professor and head of the Chemical Institute within the Faculty of Medicine of the newly established school.

On paper this was a great honor to Zechmeister. He was only 33 years old and in Hungary such a young person never received a cathedra. However, the task he accepted was formidable: he had to start from scratch. The "transfer" from Pozsony to Bratislava only meant a continuum in name, everything remained in Bratislava. Buildings had to be erected, laboratories established and equipped, staff hired, and the country was very poor. We cannot compare Zechmeister's situation with Kuhn in Heidelberg, or Karrer in Zurich, who had generous funds, well-equipped laboratories, and a large staff. In this respect, it is worthwhile to compare Kuhn's large staff shown (partially) in Fig. 12.3, with Zechmeister's situation in Pécs, where he had only one associate professor — László Cholnoky (1899–1967), his close collaborator and eventually his successor — and a maximum of two to three graduate students.

Zechmeister used the first years in Pécs to write textbooks on analytical chemistry and organic chemistry; he could finally start laboratory investigations in the second half of the 1920s. Evidently he remained in contact with Willstätter (now in Munich) and his first research projects were done in association with people in Willstätter's lab. However, he soon also entered the field of carotenoids and his particular interest was on the investigation of the pigments of Hungarian

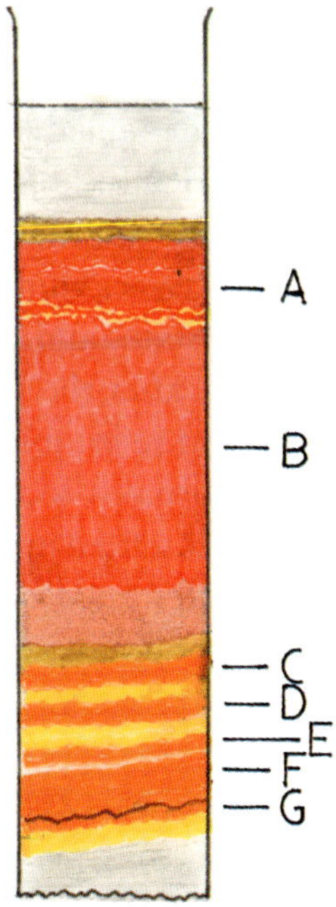

Fig. 13.3. Chromatogram of the pigments from the skin of Hungarian red paprika. Composite column: upper half, $CaCO_3$, lower half, $Ca(OH)_2$. Solvent and eluent: benzin (60°C–80°C BP fraction of gasoline). (A) capsorubin, (B) capsantin, (C) zeaxanthin, (D) cryptoxanthin, (E) γ-carotin, (F) β-carotin, (G) α-carotin. Figure prepared from information provided in Ref. 1.

red paprika: he published more than a dozen papers on this subject, the first in 1927, and the last in 1937. Figure 13.3 shows a typical chromatogram of such a sample.[1] Zechmeister was considered such an expert on carotenoids that he was asked to contribute a chapter on these compounds for the *Handbook of Plant Analysis* published in 1932[9] which soon he expanded into a 340-page monograph published two years later.[10]

It is difficult to establish when Zechmeister started to use chromatography. Although both he and Kuhn were Willstätter's students (Kuhn was 11 years younger than Zechmeister), it is very unlikely that they had any connection. Probably Zechmeister knew about Tswett and his work while still in Zurich with Willstätter, but as mentioned, he was not involved there in any chromatography work, and his detailed knowledge about chromatography came most likely from Palmer's carotenoid book published in 1922.[11] In his book chapter on carotenoids,[9] Zechmeister already had discussed in detail the chromatographic technique and its possible use for plant pigment investigations, and from the text it is clear that he already had some experience in its use. However, his first journal paper specifically mentioning the use of the technique was published only in 1934.[12]

It is interesting to speculate about the reasons for this hiatus: in my opinion the reason for it was a lack of instrumentation in Pécs. At that time the identification of pure compounds had to be corroborated by elemental analysis (without that, a paper was not a "paper"), however, Zechmeister's lab could not purchase a microbalance (then a very expensive item!) until the early 1930s, and this definitely would have been needed to perform such determinations on the very small chromatography fractions. (I learned this personally from Dr Cholnoky: he showed me proudly in the mid-1950s the balance acquired 20 years earlier.) However, from then on, Zechmeister continuously published a large number of papers using chromatography: the bibliography section of the second edition of his book[1] lists 33 titles published by him before the summer of 1938.

Zechmeister's most important contribution to chromatography was his monograph on the chromatographic adsorption method.[1] This was the right book, published at the right time, and it became an instant best seller: the original edition published in 1937 was followed within a year by a greatly enlarged, second edition. About one-third of the text dealt with fundamentals and methodology, while two-third discussed applications, and a complete bibliography of papers describing the use of chromatography was added. This book contained detailed instructions on the analysis of a large number of sample types and also standardized the chromatographic hardware (Fig. 13.4). Columns as shown in Zechmeister's book were in use in many laboratories until the end of the 1950s.

By the end of the 1930s Zechmeister was well established and respected both at home and abroad: he had received a number of major awards, and had been invited a number of times to lecture at various universities, both in the European countries and in the

Fig. 13.4. Chromatographic system of Zechmeister.[1]

United States. However, political events interrupted his life: the advent of Nazi dominance in Europe made his stay there intolerable for him and he accepted an invitation to join the faculty of the California Institute of Technology, in Pasadena. He sailed to the United States in the early 1940s with the last ship from an Italian port.

In Pasadena Zechmeister remained active in the study of complex natural substances and the use of (classical) chromatography. In 1948 he finished the continuation of his chromatography textbook and it was published in 1950.[13] This book received excellent reviews, calling him "a master in the application of chromatography" who can make "chromatography interesting and easy to use for every biochemist".[14]

In the last decade of his active life he concentrated on the chromatographic separation of stereoisomers: his last publication was a monograph on cis- and trans-isomeric carotenoids published in 1962.[15]

13.3. Beginnings of Inorganic Chromatography

Both Tswett and the pioneers who followed his work utilized chromatography for the separation of complex, naturally occurring organic substances. Then, in 1937, the technique was extended to the separation of inorganic ions by G.-M. Schwab at the University of Munich.

Georg-Maria Schwab (1899–1984) graduated in 1923 at the University of Berlin. After two years of postdoctoral work at the University's Institute of Physical Chemistry he joined the University of Würzburg and, finally, in 1928, he moved to the University of Munich becoming associated with the Chemical Institute headed by Professor Heinrich Wieland (1877–1957) who received the 1927 Chemistry Nobel Prize for his work on biologically important compounds, particularly steroids. At Wieland's institute Schwab's main field of activities involved reaction kinetics, catalysis, and solid-state reactivity.

In the early 1930s one of the graduate students in Wieland's institute was Gerhard Hesse (1908–1997) who later became professor at the University of Erlangen-Nürnberg and one of the best known German chromatographers (see Chapter 19). His thesis subject was the investigation of the poisons of certain toads. When reading the first publications of Kuhn's group in Heidelberg, Hesse naturally tried to use chromatography for the separation of the constituents of toad skin extracts, using aluminum oxide as the adsorbent. However, he found that chemical changes and even decomposition occurred on the column.[16] Hesse asked the help of Schwab (working in the same institute, in the next lab) and he found an explanation: technical Al_2O_3 always contained impurities serving as active centers. Based on their cooperation eventually Hesse improved the production methods of aluminum oxide, developing various chromatographic-grade adsorbents; at the same time Schwab developed the chromatographic

Fig. 13.6. H. Urey with one of his long (over 30 ft) columns packed with zeolite and used for the separation of lithium and potassium isotopes. (Source: M. Cohn, *Chem. Heritage* **23**(4), 8–11, 48 (2005/06); **24**(1), 2 (2006)). Courtesy[25] of Emilio Segré Visual Archives, American Institute of Physics.

In the second half of the 1930s, a new method started to gain in acceptance, the so-called *Durchflusschromatogramm*, or flow-through chromatogram (it was also callled a liquid chromatogram). In this technique the individual sample components did not remain on the column but were continuously washed out of the column with the eluents (or, as it is called today, the mobile phase).

Weil[26] credited the new technique to Koschara, with reference to his 1936 paper.[27] However, this reference is most likely incomplete. The cited paper entitled Adsorption Analysis in Aqueous Solutions' represents a summary of methodology employed by Koschara in his investigations of separation from aqueous solutions, and the flow-through technique is mentioned matter-of-fact, as one of the methods: it seems improbable that this technique was described there for the first time. When speaking about the technique of flow-through chromatography, Zechmeister and Cholnoky[28] also refer to Koschara's paper, and credit it with the detailed description of methodology but, however, not with the invention of the technique. Actually, Tswett, in his 1910 book,[29] had already suggested such an operation. When speaking about the xanthophylls, he stated that

> Owing to the very fast migration of xanthophyll α and β bands in benzene, these pigments can be washed out of the column and isolated separately as benzene solutions, by passing benzene through the column for a sufficiently long time.

In fact, the system described by Engler may also be considered as an early version of flow-through chromatography (see Chapter 3). Thus, it is most likely that Koschara's merit is a thorough study of the variables and optimum conditions. Starting by the end of the 1930s flow-through chromatography had been used with increasing frequency, for example also by Ružicka and Reichstein.

Flow-through chromatography had a number of advantages. It permitted the separation of colorless substances, by properly monitoring the effluent from the column. Also, various difficult-to-separate compounds could be analyzed in this way, by eluting the analytes from the column using different solvents in a series, by successively increasing the developing and elution powers.[28] We may consider this method of operation as the precursor of gradient-elution chromatography.

In flow-through chromatography, column eluent may be collected in consecutive small fractions. Next, the amount of the analyte(s) in each fraction can be individually determined by plotting these amounts against the serial number of the fractions or the accumulated volume of the eluent, and real "chromatograms" can be obtained. Examples will be shown in Chapter 17.

A further improvement in flow-through chromatography was the introduction of continuous monitoring of the column effluent by some means. Here Arne Tiselius (1902–1971), professor at the University of Uppsala, Sweden, and the winner of the 1948 Chemistry Nobel Prize, carried out some pioneering work by continuously measuring the column effluent's refractive index,[30] and developing automatic and self-recording detectors.[31,32] Eventually refractive index detectors became an integral part of modern liquid chromatography systems (see Chapters 28 and 29).

References

1. L. Zechmeister and L. Cholnoky, *Die chromatographische Adsorptionsmethode: Grundlagen, Methodik und Anwendungen* (Springer Verlag, Vienna, 1937; 2nd, enlarged edition: 1938).
2. G. M. Schwab and K. Jockers, *Angew. Chem.* **50**, 546–553 (1937).
3. H. Willstaedt, *L'Analyse chromatographique et ses applications* (Hermann & Cie, Paris, 1938).
4. J. N. Wilson, *J. Am. Chem. Soc.* **62**, 1583–1591 (1940).
5. P. Karrer, *Helv. Chim. Acta* **22**, 1149–1150 (1939).
6. E. Lederer and M. Lederer, preface, *Chromatography — A Review of Principles and Applications* (Elsevier, Amsterdam, 1955).
7. L. Zechmeister, Contributions to Our Knowledge About Cellulose and Lignin, PhD thesis, ETH, Zurich, 1913.
8. R. Willstätter and L. Zechmeister, *Ber. Dtsch. Chem. Ges.* **46**, 2401–2412 (1913).
9. L. Zechmeister, in *Handbuch der Pflanzenanalyse*, ed. G. Klein (Springer Verlag, Vienna, 1932), Vol. 3, pp. 1239–1350.
10. L. Zechmeister, *Carotinoide: Ein biochemischer Bericht über pflanzliche und tierische Polyenfarbstoffe* (Springer Verlag, Berlin, 1934).
11. L. S. Palmer, *Carotinoids and Related Pigments: the Chromolipids*, (Chemical Catalog Co., New York, 1922).
12. L. Zechmeister and L. Cholnoky, *Ann. Chem.* **509**, 269–287 (1934).
13. L. Zechmeister, *Progress in Chromatography 1938–1947* (J. Wiley & Sons, New York, 1950).
14. *Enzymologia* **1951**, 324.
15. L. Zechmeister, *Cis-trans Isomeric Carotenoids, Vitamin A and Aryl Polyenes* (Springer Verlag, Vienna, 1962).
16. G. H. Hesse, in *75 Years of Chromatography — A Historical Dialogue*, eds. L. S. Ettre and A. Zlatkis (Elsevier, Amsterdam, 1979), pp. 131–140.

17. G. M. Schwab, in *75 Years of Chromatography — A Historical Dialogue*, eds. L. S. Ettre and A. Zlatkis (Elsevier, Amsterdam, 1979), pp. 375–380.
18. G. M. Schwab and G. Dattler, *Angew. Chem.* **50**, 691–692 (1937).
19. G. M. Schwab and G. Dottler, *Angew. Chem.* **51**, 709–711 (1938).
20. G. M. Schwab and A. N. Ghosh, *Angew. Chem.* **52**, 666–668 (1939).
21. G. M. Schwab and A. N. Ghosh, *Angew. Chem.* **53**, 39 (1940).
22. T. I. Taylor and H. C. Urey, *J. Chem. Phys.* **6**, 429–438 (1938).
23. O. Samuelson, *Z. Anal. Chem.* **116**, 328 (1939).
24. O. Samuelson, *Svensk. Kem. Tdskr.* **51**, 195–206 (1939).
25. M. Cohn, *Chem. Heritage* **23**(4), 8–11, 18 (2005/2006); **24**(1), 2 (2006).
26. H. Weil, *Petroleum* (London) **14**(1), 5–12, 16 (1951).
27. W. Koschara, *Z. Physiol. Chem.* **239**, 89–96 (1936).
28. L. Zechmeister and L. Cholnoky, *Die chromatographische Adsorptionmethode: Grundlagen, Methodik und Anwendungen*, 2nd edn. (Springer Verlag, Vienna, 1937), pp. 77–78.
29. M. S. Tswett, *Khromofilly v Rastitel'nom i Zhivotnom Mire* (Chromophylls in the Plant and Animal World) (Karbashnikov Publishers, Warsaw, 1910).
30. A. Tiselius, *Ark. Kem. Mineral. Geol.* **14B**(22), 1–5 (1940).
31. A. Tiselius, *Ark. Kem. Mineral. Geol.* **14B**(32), 1–8 (1941).
32. A. Tiselius and S. Claesson, *Ark. Kem. Mineral. Geol.* **15B**(18), 1–6 (1942).

Part Five

The Evolution of the Chromatographic Techniques

Chapter

14

The Development of Partition Chromatography*

On June 7, 1941, the (British) Biochemical Society held its 214th meeting in London, at the National Institute for Medical Research. At this meeting two young chemists, A. J. P. Martin and R. L. N. Synge (Martin was 31 and Synge 26) presented a paper on the separation and determination of the monoamino monocarboxylic acids present in wool, using a new method.[1] This lecture and its subsequent detailed publication[2] represent the birth of partition chromatography: first as an improved way to carry out liquid chromatography, then in the simplified form of paper chromatography, and finally, 10 years later, extending it to the field of gas chromatography.[3] These achievements were finally crowned by the 1952 Chemistry Nobel Prize.

The invention of liquid–liquid partition chromatography and the development of gas–liquid partition chromatography are fascinating stories. Fortunately they are recorded fairly well in the Nobel lectures

*Based on the article by L. S. Ettre, published in *LCGC* (North America) **19**, 506–512 (2001).

of Martin[4] and Synge,[5] and also in the autobiographical treatments of the two scientists.[6,7] This chapter is based on these summaries.

14.1. The Start at Cambridge University

Both Archer John Porter Martin (1910–2002) and Richard Lawrence Millington Synge (1914–1994) studied at Cambridge University, in England: Martin graduated in 1932 and Synge in 1936. However, they belonged to different colleges and probably did not even know each other during their undergraduate time.

When still in high school Martin became fascinated by fractional distillation and even built in the shed of their house some long distillation columns from empty coffee cans, soldered together. He went to Cambridge University on a chemical engineering scholarship, but changed to biochemistry upon the influence of J. B. S. Haldane, Reader of biochemistry at Cambridge. Synge had already been interested as a teenager how living things functioned, and thus he also majored in biochemistry at Cambridge. After graduation both remained at the university as graduate students, but first had separate activities.

Martin joined the Dunn Nutritional Laboratory and was involved in the research on vitamin E. Their task was to separate carotenes, and Martin built a very complicated laboratory machine for countercurrent extraction; it consisted of 45 5-ft long tubes connected to one another and serving as extraction funnels. Details of the elaborate machine have not been published (it was only described in Martin's PhD thesis), but from descriptions we have some idea how complicated it was: 90 ball valves rattled loudly on their seats preventing the liquid from dropping back to the previous tube.

Synge had been active in studying glycoproteins and utilized derivative formation and liquid–liquid extraction in his work. Then in 1938 he was offered an unusually generous fellowship by the International Wool Secretariat. This organization was maintained by the wool producers of Australia, New Zealand, and South Africa, and among others they also funded research on various aspects of wool. The aim of Synge's scholarship was to study in detail the amino acid composition of wool and improve the methods of amino acid analysis. Synge started by measuring the partition coefficients of acetylated amino acids between two phases, chloroform and water, and the planned next step was to carry out separation by liquid–liquid extraction. At that time he was proposed to contact Martin, whose unorthodox large all-glass machine for countercurrent extraction standing in the entrance hall of the Dunn Nutritional Laboratory was widely known in Cambridge. With this contact, their five-year long, very successful cooperation started.

14.2. The Birth of Partition Chromatography

Martin's existing apparatus was not suitable for use with a chloroform–water system; therefore, he designed a completely different machine, now for this solvent pair and for continuous flow, liquid–liquid countercurrent extraction. Meanwhile, Martin and Synge moved from Cambridge to the laboratories of the Wool Industries Research Association, in Leeds, and they took the new machine with them. It was also a very complicated system: the two operators had to watch its operation in 4-h shifts, continually battling drowsiness due to leaking chloroform vapor. They could achieve some success with the separation of the monoamino monocarboxylic acids present in wool, and a preliminary report on their work was published;[8] however, it was obvious that the system is too complicated and unreliable for future systematic research.

In 1940, Martin had a radically different idea: to pack a glass tube with a mixture of wool and cotton, with the fibers parallel to the axis of the tube, and to have chloroform flow above, and water below the packing, in opposite direction. The idea was that the fibers would separate the two flows, and the amino acids would distribute differentially between the two solvent flows. However, the system did not work as hoped. Martin realized that the problem was related to creating equilibria in the two liquids moving continuously in opposite direction. Then, suddenly, he found a solution: it was not necessary to move both liquids, but only one, and keeping the other stationary in the tube. This was the birth of partition chromatography.

Martin and Synge decided that chloroform should serve as the moving phase, and water as the stationary phase. They impregnated silica gel with water, packed it in a glass tube (the column), added the acetyl amino acid mixture to the top, and poured chloroform down the column. In the first experiment they could separate acetylproline and acetylleucine, collecting the respective fractions (Fig. 14.1).

This sounds simple. However, it took them months of hard work to establish the optimum conditions and refine the system. In addition Martin also developed the theory of partition chromatography, applying the theoretical plate concept from distillation (he learned it while still in high school, when building the distillation columns).

Here we have an interesting question. As explained in this summary, Synge started to use liquid–liquid extraction, partitioning between two solvents; the cooperation with Martin changed this to countercurrent extraction, and then the final system fixed one of the solvents. The interesting question is how did Martin and Synge realize that the final process they selected is not anymore a modified countercurrent extraction system but a version of chromatography? The answer to this question lies in the fact that both Martin and Synge were aware of chromatography although they did not use it earlier. Martin specifically mentions that after he saw in 1933 a demonstration of chromatography by Winterstein of Kuhn's laboratory (see Chapter 12), he realized the existence of a relationship between chromatography and distillation,[6] and Synge mentions that when they changed from two countercurrent moving solvents to the final system, they realized that now, the system worked "as it were a chromatogram".[7]

As mentioned earlier Martin and Synge presented the first report on their results in June 1941. They were slow with finishing the final manuscript for publication. It was submitted only in November to the editor of *Biochemical Journal*; however, its publication was almost instantaneous and it was already included in the December 1941 issue of the journal. This paper, entitled A New Form of Chromatogram Employing Two Liquid Phases, consisted of two parts: the first presented the theory of partition chromatography, while the second reported on the separation of the monoamino acids present in proteins.[2]

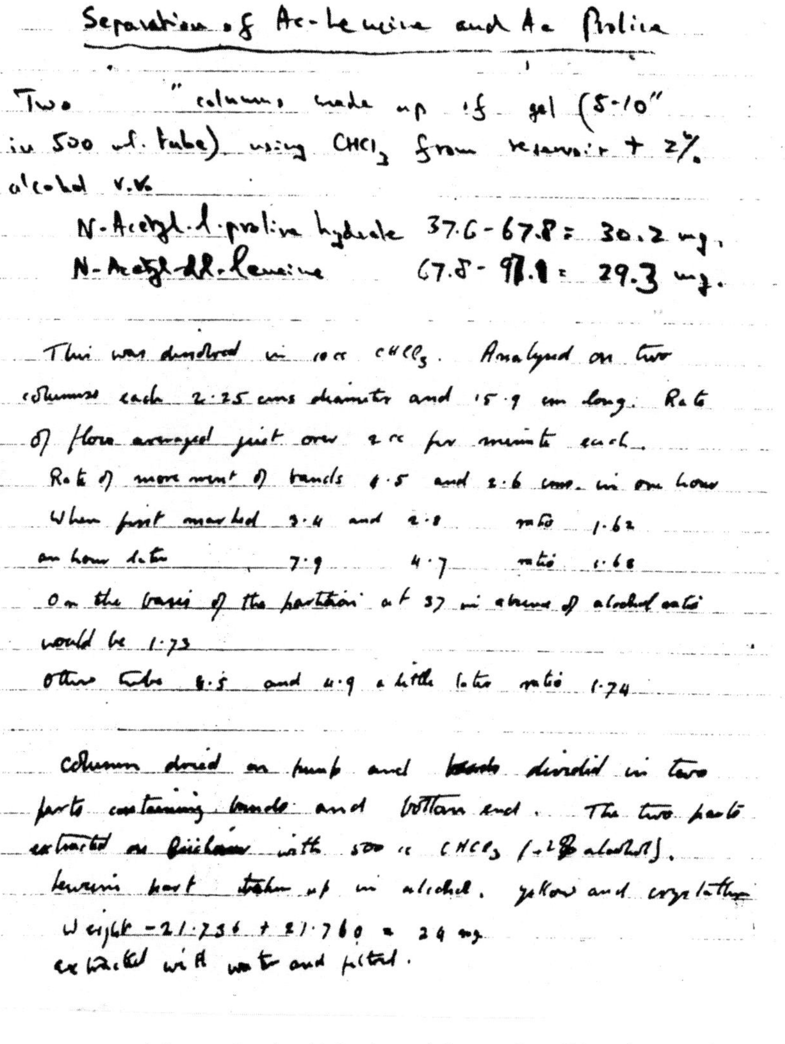

Separation of Ac-leucine and Ac Proline

Two " columns made up of gel (5-10" in 500 ml. tube) using $CHCl_3$ from reservoir + 2% alcohol v.v.

N-Acetyl-l-proline hydrate 37.6-67.8 = 30.2 mg.
N-Acetyl-dl-leucine 67.8-97.1 = 29.3 mg.

This was dissolved in 10 cc $CHCl_3$. Analysed on two columns each 2.25 cms diameter and 15.9 cm long. Rate of flow averaged just over 2 cc per minute each.
Rate of movement of bands 4.5 and 2.6 cms. in one hour
When first marked 3.4 and 2.1 ratio 1.62
an hour later 7.9 4.7 ratio 1.68
On the basis of the partition at 37 in absence of alcohol ratio would be 1.73
Other tube 8.5 and 4.9 a little later ratio 1.74

Column dried on pump and bands divided in two parts containing bands and bottom end. The two parts extracted on Soxhlet with 500 cc $CHCl_3$ (+2% alcohol).
leucine part taken up in alcohol, yellow and crystalline
Weight - 21.736 + 21.760 = 24 mg.
extracted with water and filtered.

Fig. 14.1. Page of the notebook of Martin and Synge, describing the experiment on the separation of acetylleucine and acetylproline (From Ref. 6).

It may be interesting to mention that in their paper, Martin and Synge spoke about liquid–liquid chromatography and did not use the term "partition chromatography." However, in subsequent years this expression became more and more used, and the citation of their 1952

Chemistry Nobel Prize specifically mentions that it was awarded for the "invention of partition chromatography" (Figs. 14.2 and 14.3). It is, however, interesting to note that the British Postal Service was evidently completely ignorant of this term, and in a stamp issued in

Fig. 14.2. A. J. P. Martin receiving the 1952 Nobel Prize in Chemistry from King Gustav VI Adolphus of Sweden. (Courtesy: Nobel Foundation, Stockholm.)

Fig. 14.3. R. L. M. Synge receiving the 1952 Nobel Prize in Chemistry from King Gustav VI Adolphus of Sweden. (Courtesy: Nobel Foundation, Stockholm.)

Fig. 14.4. British stamp honoring the 1952 Nobel Prize of Martin and Synge. On the occasion of the centenary of the Royal Institute of Chemistry, the British Postal Service issued four stamps on March 2, 1977, honoring British achievements in chemistry; this stamp was one of the four. Note that the text on the stamp incorrectly mentions "starch chromatography."

1977 honoring the invention of chromatography (Fig. 14.4) called the technique "starch chromatography."

Partition chromatography on a column using silica as the support for the stationary phase permitted Martin and Synge to determine the monoamino monocarboxylic acids. However, soon they found out that it cannot be used for the analysis of dicarboxylic acids: they were permanently adsorbed on the silica support. Thus, another support had to be found. This was accomplished by using filter paper as the support. This work, the development of paper chromatography, represents the second part of the story and will be discussed in the next chapter.

Meanwhile Synge left Leeds in 1943, and joined the Lister Institute of Preventive Medicine in London. In 1946–1947 he was a visiting scientist at Uppsala University in Sweden, in the laboratory of Arne

Tiselius. Then, in 1948, he joined the Rowett Research Institute, in Aberdeen, Scotland, and finally, in 1967 he became associated with the Food Research Institute in Norwich. Martin left Leeds in 1946 to become the head of the biochemistry division in the research department of Boots Pure Drug Co., in Nottingham, but in 1948, he joined the British Medical Council, first at the Lister Institute (Synge just left it), and then in 1950 he went over to the Council's National Institute for Medical Research, at Mill Hill, in London. It was there that the third stage of this story took place: the development of gas–liquid partition chromatography.

14.3. Gas–Liquid Partition Chromatography

When speaking about the two-phase system, one stationary and the other mobile, the famous 1941 paper of Martin and Synge[2] contained the following statement:

> The mobile phase need not be a liquid but may be a vapour. ... Very refined separations of volatile substances should therefore be possible in a column in which permanent gas is made to flow over a gel impregnated with a nonvolatile solvent.

This prediction clearly indicates the possibility of gas–liquid partition chromatography (GLPC). However, evidently nobody picked up this suggestion. This seems to be strange, but it can be explained easily. Let us not forget that in 1941, Second World War was raging and England was in her most difficult period. Also, British journals could be received only in a few countries, and the whole continental Europe was under German occupation: communication between scientists was almost non-existent. In fact even after the war, the 1940–1945 issues of many journals were missing from libraries. When communication was finally restored, paper chromatography was the most exciting new technique, and people did not (or could not) go back to the basic 1941 publication.

It was finally up to Martin to prove the validity of their original prediction. We have seen above that by 1950 he moved to the British National Institute for Medical Research and there, one of his colleagues mentioned to him that he would need a more refined method than paper chromatography for separating fatty acids. This query initiated Martin's interest to go back to their 1941 prediction.

At the Institute Martin was joined by Anthony Trafford James (born 1922), a young scientist who had been associated with Synge for a short time at the Lister Institute.[9] It was with him that Martin started to investigate the feasibility of GLPC. They took Celite (a diatomaceous earth material) as the support, coated its fine particles with a silicone oil to serve as the stationary phase, and packed this in a column; nitrogen was used as the mobile phase. First they tried to separate the lower fatty acids, but this gave considerable trouble due to dimerization on the column. They found out that this can be prevented by adding 10% of a long-chain nonvolatile acid (they used stearic acid) to the stationary phase. For detection of the individual fatty acids, the column's end was dipped into a test tube containing an indicator solution, and the amounts of the eluted compounds were determined by titration, first manually, but later Martin constructed a very elegant (albeit, somewhat complicated) automatic titrator for this purpose. In this way a chromatogram could be constructed (Fig. 14.5).

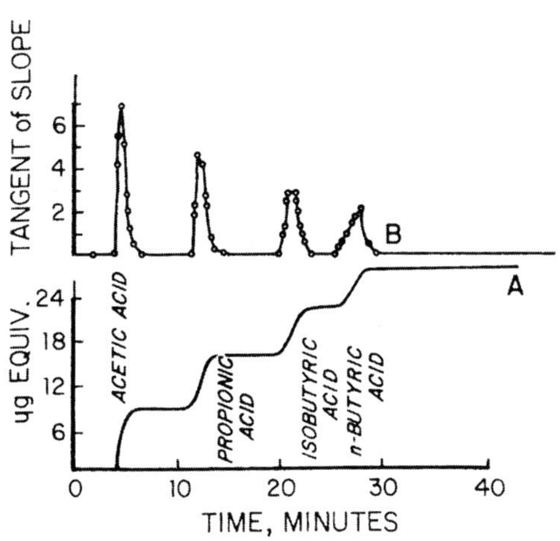

Fig. 14.5. Separation of acetic, propionic, isobutyric, and n-butyric acid by gas chromatography. Column length, 4 ft; liquid phase: DC-550 silicone oil containing 10% stearic acid, coated on Celite; mobile phase, nitrogen, at 33 mL/min; column temperature, 100°C. (A) experimental curve, (B) differential of the experimental curve (From Ref. 6).

Chapter

15

Paper Chromatography*

Paper chromatography occupies a very important place in the evolution of the chromatographic techniques: it was through it that chromatography became everybody's tool. Its development is generally credited to A. J. P. Martin and his group, representing the second stage of the development of partition chromatography (see previous chapter). The new technique was first reported on 25 March 1944, at the Annual Meeting of the (British) Biochemical Society[1] and then published in the Society's journal.[2] However, they were not the first who carried out some kind of separation on filter paper: we should refer here to the work of Runge and Goppelsroeder. Even Tswett utilized filter paper to initiate the process occurring in plants (see Chapter 4). In his Nobel Prize lecture[3] Martin mentioned that he was aware of the use of filter paper by dyestuff chemists to check the quality of the dyes, and he stated that this gave him the idea of using this media.

*Based on the articles by L. S. Ettre published in *Chromatographia* **54**, 409–414 (2001) and *LCGC (North America)* **19**, 506–512 (2001).

However, none of these investigations led to the actual development of a viable separation method: this was the merit of Martin's group.

In this chapter we shall mention the precursors of paper chromatography, particularly the so-called capillary analysis of Goppelsroeder and its modification by Liesegang which resembles the most paper chromatography, and then outline the steps which led Martin's group to the development of the technique which, in the second part of the 1940s, revolutionized the way biochemical investigations were carried out.

15.1. The Precursors

15.1.1. *F. F. Runge*

In the chromatography literature one can often find the German physician–chemist Friedlieb Ferdinand Runge (1794–1867) mentioned as the precursor, or even as the "inventor" of paper chromatography. His life story and his investigations are fascinating and we have devoted a previous chapter (Chapter 2) to their discussion. It is important to emphasize that Runge's aim was not to carry out separation: he wanted to create unique pictures on filter paper and demonstrate in this way the interaction of various compounds and the existence of a special natural force. In other words, his beautiful multicolored pictures mirrored his fantasy and not chromatography, and can be characterized as "the most original scientific playing."[4]

15.1.2. *Capillary Analysis*

This technique started in 1861 with the observation of Christian Friedrich Schoenbein (1799–1868), professor at the University of Basel, and the discoverer of ozone, that when dipping a filter paper strip into an aqueous solution, the solvent (water) and the dissolved substances will travel up in the paper at different speeds, the solvent being the fastest.[5] Schoenbein's report was immediately followed by Friedrich Goppelsroeder (1837–1919), his student, who described his own observation, claiming the possibility of recognizing individual dyes in their mixtures.[6]

From then on Goppelsroeder devoted practically his whole productive life exploring various aspects of this method which he named *Kapillaranalyse* (capillary analysis); he published scores of papers reporting on his observations when investigating the widest possible variety of natural substances. He also collected these papers in a few books, with some additional comments.[7–9] However, as noted by Grüne[10]

> ...it is usually very tiring to read (these publications), because they report on a large number of individual observations, without any clear line and without any observable advances.

Goppelsroeder used long, narrow filter paper strips, with their lower end immersed in the sample solution. In other words, instead of adding a finite amount of the sample solution onto the strip, the sample was continuously fed to it. Usually a dozen such strips were affixed to a rack, sometimes the system was placed in a vacuum chamber or under pressure. Periodically Goppelsroeder also extracted various colored zones from the paper strip with alcohol and then repeated the investigation with this solution. He always carefully recorded the height and color of the advanced zones, but without any identification of the individual zones and without any conclusions.

Schoenbein clearly attributed the movement of the sample components on the filter paper due to capillary action; however, Goppelsroeder thought that a number of physical phenomena are involved. His assumptions were aptly characterized by Synge who said that[11]

> Goppelsroeder hopelessly confused adsorption, surface tension, diffusion and other effects, and arrived at no satisfying explanations for his phenomena.

The fundamental shortcoming of Goppelsroeder's capillary analysis was his methodology: the sample was continuously fed and thus, there was no separation of the zones of the individual components. This problem was well characterized by Newesely[12]:

> If Goppelsroeder would have tried only once to add a finite sample to the paper and then washed it with the pure solvent, then he might be named the father of modern chromatography, and paper chromatography would have been invented 80 years earlier.

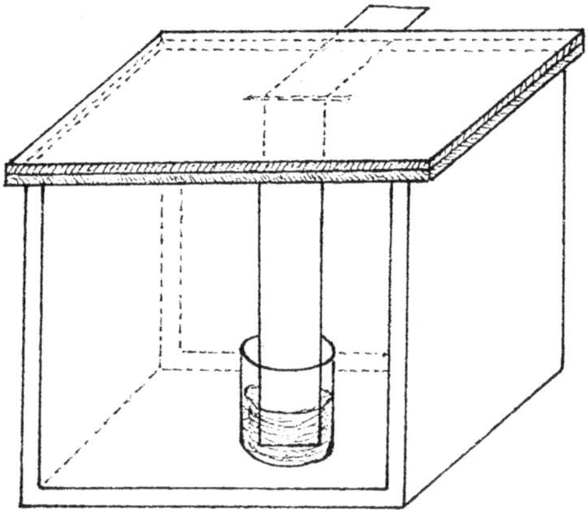

Fig. 15.1. Capillary analysis arrangement as shown by Platz.[13] The container is covered by two wooden plates: the lower one has a slit and the filter paper strip is pulled through it. The purpose of the top plate is to hold the paper in place.

After Goppelsroeder's death capillary analysis was further widened. As stated by Platz[13] it was actually included in 1922 in the Dutch and somewhat later, in the German Homeopatic Pharmacopeia. Platz dealt in detail with methodology which was practically unchanged since Goppelsroeder's original work: Fig. 15.1 shows the arrangement for multiple strips as described by Platz (which is very similar to Goppelsroeder's system). The methodology in the Pharmacoppeias remained unchanged, even well after the advent of paper chromatography.[14] In this respect it is interesting to refer to a photograph in the autobiographical text of Egon Stahl, the developer of thin-layer chromatography[15]: this photo shows him in 1948, as a college student, in his home laboratory and as the text explains, he earned money by testing drugs, tinctures, and extracts of medicinal plants for a pharmaceutical wholesale company, in Karlsruhe. On the wall behind him one can see a rack with six filter strips, just as shown in Fig. 15.1, and he personally told me that he had regularly used capillary analysis in his testing work.

The methodology of capillary analysis was finally changed by Raphael Eduard Liesegang (1869–1947). First he placed the filter strips into a closed chamber (Fig. 15.2) so that the atmosphere surrounding it was saturated with the solvent vapor.[16] A further radical

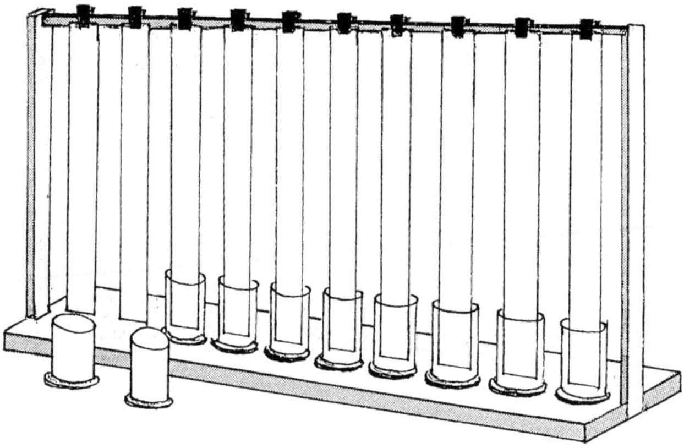

Fig. 15.2. The capillary analysis system described by Liesegang in 1927.[16] It is essentially the same as originally described by Goppelsroeder. The whole system was usually in a closed chamber.

change was made by Liesegang in 1943, when he spotted the sample on the paper strip and then developed it, by dipping the end of the strip into pure solvent.[17] This, of course, now completely differed from Goppelsroeder's "capillary analysis" technique, and was practically identical to paper chromatography. In the same year Liesegang also introduced another variant he called *Kreuzkapillaranalyse* (cross capillary analysis), with development in two different directions, practically identical to two-dimensional paper chromatography described one year later by Martin's group (see below).

There is no question that the 1943-modification of capillary analysis as developed by Liesegang was very close — in fact, almost identical — to paper chromatography as developed by Martin's group. However, neither of the two groups was aware of the other's activities. Let us not forget that at that time there was no communication between Germany and England, and even after the end of the war, activities in Germany during the war had little impact on future scientific development. Also Liesegang was an independent, colloid chemist (in the literature he is often mentioned as an "amateur scientist") without any academic affiliation; his work was not continued after his death and thus, represented a dead-end street in the evolution of chromatography. Paper chromatography revolutionizing biochemical analysis was clearly the result of Martin's group first published in 1944.

15.2. The Invention of Paper Chromatography

As discussed in the previous chapter, the partition chromatographic columns of Martin and Synge permitted the separation of the monocarboxylic monoamino acids. However, they could not be used for the analysis of dicarboxylic acids: the silica serving as the support for the water stationary phase permanently adsorbed them. Thus, other support material had to be found. Martin's first thought was to use filter paper. He, with Synge, placed a drop of the solution of two amino acids in the center of a piece of paper impregnated with water (the stationary phase). Butanol (the mobile phase) moved up the paper by capillary action, eventually reaching its edge, moving the two amino acids at different speeds. This brief experiment indicated that the technique would work.

Amino acids are colorless: thus some way had to be found to reveal them in the filter paper piece. A. H. Gordon, a new addition to their team, had the task to search for a suitable color reaction: he found in *Beilstein's Handbuch der Organischen Chemie* a description of the reaction of the amino acids with ninhydrin, and this was adopted for their purpose. As the next step they developed a convenient set-up in which filter paper strips were placed in a closed container in which air was saturated with water vapor and the tops of the strips were dipped into troughs containing the mobile phase; the samples were spotted at the top of the strips. (Today we call such a system as descending development.)

Martin's team tried many different solvents as the mobile phase, however, no single solvent was able to resolve a mixture of all common amino acids. Then they successfully tried what we now call *two-dimensional chromatography*. After developing the chromatogram on the paper strip in one dimension, with a certain solvent, they turned the paper 90° and used a different solvent to further separate the spots formed in the first development (Fig. 15.3). Today, as electrophoresis enjoys its renaissance, it may be interesting to mention that in the very

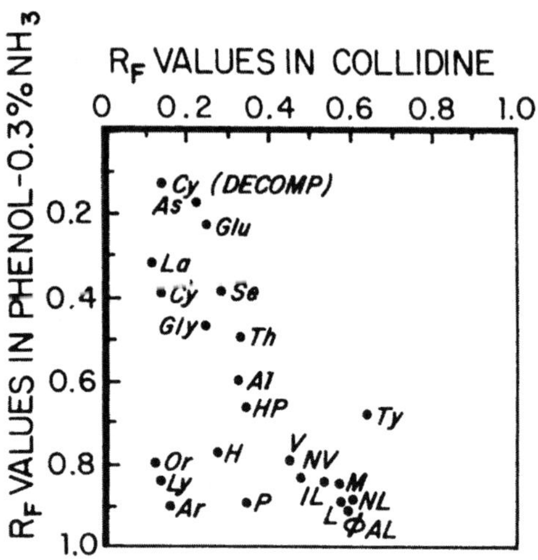

Fig. 15.3. Diagram showing the positions of amino acid spots on a two-dimensional paper chromatogram, prepared with phenol–ammonia (0.3%) and collidine solvents.[19] For the identification of the spots, see Table 15.1.

Table 15.1. Identification of the spots in the paper chromatogram shown in Fig. 15.3.

Al	Alanine	M	Methionine
Ar	Arginine	NL	Norleucine
As	Aspartic acid	NV	Norvaline
Cy	Cystine	Or	Ornithine
Glu	Glutamic acid	ØAl	Phenylalanine
Gly	Glycine	P	Proline
H	Histidine	Se	Serine
HP	Hydroxyproline	Th	Threonine
IL	Isoleucine	Tr	Tryptophan
La	Lanthionine	Ty	Tyrosine
L	Leucine	V	Valine
Ly	Lysine		

first experiments of two-dimensional chromatography, the first development was done by electrophoresis. However, they did not pursue this technique.

Synge participated in the initial work[20] but he left Leeds in 1943 to join the Lister Institute of Preventive Medicine in London, thus ending his participation in the final development of the technique.

Three scientists authored the main report on paper chromatography: Martin, Gordon, and R. Consden, a young chemist who meanwhile joined the team. As mentioned earlier this report was first presented on 25 March 1944, at the Annual Meeting of the Biochemical Society, and then published in the society's journal.[2] This classic article represents the start of paper chromatography.

While at the Lister Institute, Synge remained in contact with his former colleagues. By then his interest turned to the investigation of the amino acid composition of the antibiotics tyrocidin[21] and gramicidin S.[22–24] The latter work was particularly important: Synge was the first who was able to elucidate the amino acid sequence in a polypeptide, and for this he used mainly paper chromatography. This work represented the basis of the more elaborate investigations of F. Sanger, determining the entire peptide sequence of insulin, for which he received the 1958 Nobel Prize in Chemistry. An excellent summary of these investigations was given in Synge's Nobel Prize Lecture.[25]

Although liquid–liquid partition chromatography carried out in a column had at that time only relatively few followers, the use of paper chromatography advanced very rapidly. This was mainly due to the remarkable simplicity of the method. At that time, filter papers of standardized quality were commercially available and the necessary setup was within the reach of every laboratory.

Naturally, it took some time for Martin's group until paper chromatography could became everybody's tool. The situation was amply characterized by Consden, in the preface he wrote 10 years later to the English edition of F. Cramer's textbook on paper chromatography[26]:

> Like other established methods, paper chromatography was not brought into the world without considerable birth pangs, and much could be written about these early adventurous days.

Using paper chromatography, separation required only a relatively short time and surpassed any techniques known at that time. A good characterization of the impact of paper chromatography was given by W. J. Whelan of the Department of Biochemistry and Molecular Biology at the University of Miami who, from 1945 to 1948, was a graduate student at the University of Birmingham, with Professor

N. Haworth, the winner of the 1937 Nobel Prize in Chemistry, where he extensively used the technique[27]:

> The technological advance the technique represented was astonishing. Amino acids, which were formerly separated by laborious techniques of organic chemistry and where large quantities of protein hydrolysates were needed, could now be separated in microgram amounts and visualized ...(Paper chromatography) would allow one within the space of a week to carry out first a test of homogeneity and then a structural analysis of an oligosaccharide, which until then could very well have occupied the three years of a Ph.D. dissertation using Haworth's technique of exhaustive methylation, hydrolysis, and identification of the methylated monosaccharides.

Today, paper chromatography is almost completely superseded by thin-layer chromatography (TLC). However, the two are closely related, and TLC is a logical extension of paper chromatography, providing the possibility for the use of different stationary phases, with increased sample size, while still maintaining the simplicity of the technique. Its development is discussed in the next chapter.

References

1. R. Consden, A. H. Gordon and A. J. P. Martin, *Biochem. J. Proc.* **38**, ix (1944).
2. R. Consden, A. H. Gordon and A. J. P. Martin, *Biochem. J.* **38**, 224–232 (1944).
3. A. J. P. Martin, in *Nobel Prize Lectures — Chemistry 1942–1962* (Elsevier, Amsterdam, 1964), pp. 355–375.
4. H. Bechhold, *Z. Phys. Chem.* **52**, 185–199 (1905).
5. C. F. Schoenbein, *Verh. Naturforsch. Ges. Basel* **3**, 249–255 (1861).
6. F. Goppelsroeder, *Verh. Naturforsch. Ges. Basel* **3**, 268–275 (1861).
7. F. Goppelsroeder, *Capillaranalyse beruhend auf Capillaritäts- und Adsorptionserscheinungen* (Capillary Analysis, Based on Capillarity and Adsorption Phenomena) (Emil Birkhäuser Verlag, Basel, 1901).
8. F. Goppelsroeder, *Anregung zum Studium der auf Capillaritäts- und Adsorptionserscheinungen beruhenden Capillaranalyse* (Stimulus to the Study of Capillary Analysis Based on Capillarity and Adsorption Phenomena) (Helbing & Lichtenhalm Verlag, Basel, 1906).

9. F. Goppelsroeder, *Capillaranalyse beruhend auf Capillaritäts- und Adsorptionserscheinungen* (Capillary Analysis, Based on Capillarity and Adsorption Phenomena) (Steinkopff Verlag, Dresden, 1910).
10. A. Grüne, *Österr. Chem. Z.* **60**, 301–311 (1959).
11. R. L. M. Synge, in *British Biochemistry Past and Present* (Biochem. Soc. Symposium No. 30), ed. T. W. Goodwin (Academic Press, London, 1970), pp. 175–182.
12. M. Newesely, *Chromatographia* **30**, 595–596 (1990).
13. H. Platz, *Über Kapillaranalyse, und ihre Anwendung im pharmazeutischen Laboratorium* (On Capillary Analysis and its Application in the Pharmaceutical Laboratory) (W. Schwabe, Leipzig, 1922).
14. W. Schwabe, *Homeopatisches Arzneibuch* (Book of Homeopatic Medications) (Private Publication, Berlin, 1950).
15. E. Stahl, in *75 Years of Chromatography — A Historical Dialogue*, eds. L. S. Ettre and A. Zlatkis (Elsevier, Amsterdam, 1979), pp. 425–435.
16. R. E. Liesegang and H. Schmidt, *Kolloidchemische Technologie* (Technology of Colloid Chemistry) (Steinkopff Verlag, Berlin, 1927).
17. R. E. Liesegang, *Z. Anal. Chem.* **126**, 172–177, 334–336 (1943).
18. R. E. Liesegang, *Naturwiss.* **31**, 348 (1943).
19. A. J. P. Martin, in *75 Years of Chromatography — A Historical Dialogue*, eds. L. S. Ettre and A. Zlatkis (Elsevier, Amsterdam, 1979), pp. 285–296.
20. A. H. Gordon, A. J. P. Martin and R. L. M. Synge, *Biochem. J. Proc.* **37**, xiii–xiv (1943).
21. A. H. Gordon, A. J. P. Martin and R. L. M. Synge, *Biochem. J.* **37**, 313–318 (1943).
22. A. H. Gordon, A. J. P. Martin and R. L. M. Synge, *Biochem. J.* **37**, 86–92 (1943).
23. R. L. M. Synge, *Biochem. J.* **39**, 363–367 (1945).
24. R. Consden, A. H. Martin, A. J. P. Martin and R. L. M. Synge, *Biochem. J.* **41**, 596–602 (1947).
25. R. L. M. Synge, in *Nobel Lectures — Chemistry 1942–1962* (Elsevier, Amsterdam, 1964), pp. 372–387.
26. F. Cramer, *Paper Chromatography* (Macmillian, London, 1954).
27. W. J. Whelan, *FASEB J.* **9**, 287–288 (1995).

Chapter

16

The Evolution of Thin-Layer Chromatography*

We have seen in the previous chapter that chromatography can also be carried out on a planar surface, and in fact paper chromatography introduced in 1944 represented the first widespread application of partition chromatography. Then, toward the end of the 1950s, thin-layer chromatography (TLC) practically replaced paper chromatography and is still one of the most popular routine chromatographic techniques.

16.1. The Beginnings

The technique of thin-layer chromatography was first used in 1937–1938 at the Institute of Experimental Pharmacy of the State University of Kharkov, Ukraine, by Nikolai A. Izmailov (1907–1961), the young head of this Institute, and Maria S. Shraiber (1904–1992), his graduate student. As described in her recollections[1] they were searching for

*Based on the article by L. S. Ettre and H. Kalász published in *LCGC* (North America) **19**, 712–721 (2001).

appropriate methods for the rapid analysis of galenic pharmaceutical preparations (plant extracts). They were aware of classical column chromatography, but such an analysis would have required too much time. They were also aware of Goppelsroeder's capillary analysis, but remembered Tswett's criticism in his book[2] that the basic shortcoming of Goppelsroeder's technique is not using finite sample volumes. Thus, they thought that using an open, flat surface but with a thin adsorbent layer would accelerate the separation process and at the same time will have the needed sample capacity. Accordingly they coated microscope slides with a suspension of various adsorbents (calcium, magnesium, and aluminum oxide), deposited one drop of the sample solution on this layer, and added one drop of the solvent one would use in a column to develop separation. The test was successful: the separated sample components appeared as concentric rings that fluoresced in various colors under a UV lamp.

In the paper summarizing their results,[3] Izmailov and Shraiber demonstrated that the sequence of the concentric multicolored rings on the plate was identical to the sequence of the colored rings one would have obtained on a regular chromatographic column containing the same adsorbent; however, the time needed for analysis was much shorter. They called the new variant "spot chromatography" and the result on the microscope slides "ultrachromatograms." In their paper Izmailov and Shraiber showed a number of idealized drawings of these ultrachromatograms (Fig. 16.1) and tabulated the colors of the rings for a few plant extracts used as medications. It was also their intention to compile a manual presenting color drawings of the ultrachromatograms of a large number of galenic preparations; however, the war interfered with these plans.

The paper of Izmailov and Shraiber was published in a Russian pharmaceutical journal, practically unknown outside the USSR. However, its abstract was included in a Russian review journal and through it, in *Chemical Abstracts*.[4] This was read by M. O. L. Crowe of the New York State Department of Health who then adapted the technique for his use. Crowe prepared the adsorbent layer in a Petri dish, added a drop of the sample solution to the center and then the

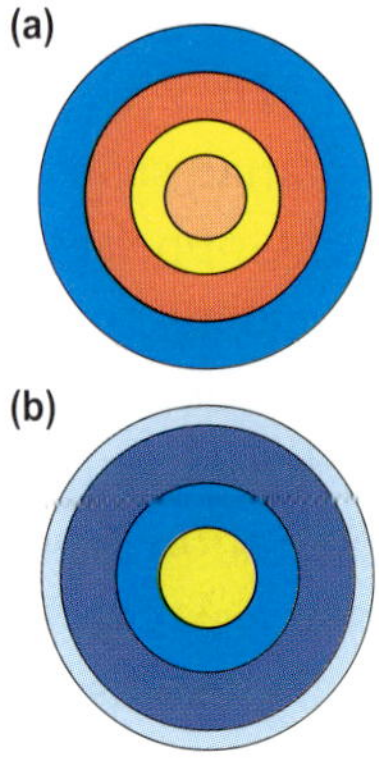

Fig. 16.1. Idealized "ultrachromatograms" of extracts of belladonna (*Atropa belladonna*) and digitalis (*Digitalis purpurea*), according to Izmailov and Shraiber.[2]

developing solvent dropwise until sufficient separation was obtained. According to his brief note published toward the end of 1941, he had used this technique in the two prior years to scout for the best developing solvent to be used in column chromatography.[5]

In 1947 T. I. Williams, in his textbook on chromatography, described a further improvement of the method of Izmailov and Shraiber.[6] He now prepared the adsorbent-coated glass plates in the form of a sandwich: the adsorbent layer was covered by a second glass plate with a small hole through which the sample (and solvent) drops could be applied.

In the development of TLC the next step was the work of Meinhard and Hall at the University of Wisconsin. They now used a binder (corn starch) to hold the coating on the glass plate and added a small amount of Celite powder to the adsorbent particles to improve the consistency of the layer. Meinhard and Hall called the technique "surface chromatography" and used it for the separation of inorganic ions.[7]

As mentioned, in these early investigations development of the sample spot was carried out by adding one or a couple of drops of a solvent. This type of chromatography strongly resembled the "spot-test analysis" technique (Tüpfelanalyse) of Fritz Feigl, an Austrian scientist, developed in the 1920s and 1930s.

16.2. TLC Matures

Modern TLC was started 50 years ago by Justus G. Kirchner (1911–1987), working at the US Department of Agriculture Fruit and Vegetable Laboratory, in Southern California. He investigated the flavor components of the juices of citrus fruits. According to his personal recollections[8] very large volumes of fresh juices had to be processed (3000 gallons of orange and 2760 gallons of grapefruit juice!) because the amount of the flavor material was exceedingly small. The next problem was to find an analytical method for the investigation of the composition of the juice concentrates. Column chromatography would have been adequate, except that the compounds to be separated did not have a distinct color or were colorless, thus their identification would have been quite difficult. Paper chromatography (by then well established) seemed to be suitable, particularly because of the convenience of visualizing the separated spot by spraying the paper with selective reagents. However, Kirchner soon found that paper was too mild an adsorbent for the separation of terpenoid compounds present in the juices. Paper impregnated with silica gel had some promises, but its sample capacity was too small. Then one day, Kirchner remembered the abstract of the paper of Meinhardt and Hall[7] he had read in *Chemical Abstracts*, and decided to follow it. He coated a layer of silicic acid (using starch as the binder) on strips of glass, but instead of adding just a drop of the developing solvent (as it was done by the earlier investigators), developed the plates in the so-called ascending mode used in paper chromatography. In this technique the spotted plates are placed in a closed chamber, dipping their lower side into the solvent (the mobile phase) which would then ascend through capillary action, carrying with it the sample components which are separated during their passage on the plate. The experiments proved to be successful and their publication can be considered as the start of modern TLC.[9] Soon after the possibility of quantitative analysis was also demonstrated, using absorbance measurement of the separated spots.[10]

Kirchner introduced the term "chromatostrips" for the adsorbent-coated glass plates. It should be noted that his group used not only narrow glass strips but also square coated plates, permitting multiple samples to run simultaneously on a single plate. They also demonstrated the possibility of two-dimensional chromatography on such plates, a technique already well-known at that time in paper chromatography. In addition, for identification purposes they also used reactions carried out on the plate.

Kirchner also developed another variation of TLC: he coated silicic acid (bound with gypsum) on a glass rod. In this way a silica column was created without the containing envelope (i.e. the tube); we may also consider these "chromatobars" (Kirchner's name for them) as a thin adsorbent layer wrapped around a glass rod. Development of the chromatograms was carried out similar to the chromatostrips, in the ascending manner, dipping the end of the rod into the solvent, and the separated zones could also be identified in a similar manner as in paper chromatography or TLC, by spraying with various reagents.[11] It is interesting to note that Kirchner's "chromatobars" were recently reinvented and are now even produced under the name "chromatorod." The only difference is that while Kirchner used standard glass rods and prepared a fairly thick adsorbent layer around it, present-day "chromatorods" utilize a thin (0.9 mm) quartz wire as their core and a thin (75 μm) coating around it.[12,13]

After Kirchner's publications, a limited number of laboratories started to use his technique. Among these investigators we may mention Reitsema[14] who utilized such "chromatoplates" (his term) for the analysis of a wide variety of essential oils. In spite of this, however, it took over a decade until TLC became a generally accepted, major variant of chromatography.

16.3. The Activities of Egon Stahl

TLC became a universally accepted analytical technique, a full-fledged variant of chromatography, mainly due to the activities of Egon Stahl (1924–1986), associated first with the University of Mainz and, from 1958 on, with the University of Saarbrücken, in Germany.[15] In fact, he was the first who consistently used the term *Dünnschicht-Chromatographie* (thin-layer chromatography) to characterize the technique and his choice of this name was almost immediately universally accepted.

Stahl was involved in the investigation of various essential oils and tried to use adsorbent-coated glass plates, following the recommendations of Kirchner, and he could obtain good results. However, he soon found that neither the method nor the adsorbent to be used had been optimized; also that the commercially available adsorbents had to be modified and treated in various ways before they could be used for coating the plates. Therefore, Stahl started systematic investigations of the operational parameters and the preparation of the proper adsorbents. He first reported his preliminary findings in 1956, in a German pharmaceutical journal,[16] however, this publication was largely ignored by the scientific public. Meanwhile, Stahl continued his efforts to standardize the method, construct simple equipment that permitted the application of a uniform thin layer, and have standard adsorbents commercially available which can be directly coated on the plates without further preparation. His efforts were finally fulfilled by the spring of 1958: the necessary basic instrumentation was introduced by DESAGA, while E. Merck introduced "Silica Gel G According to Stahl for TLC"; both were first shown at the International ACHEMA Exhibition of chemical equipments, in Frankfurt am Main. Simultaneously Stahl also published an informative article dealing with the use of this system and showing a wide range of applications.[17] Now the situation changed: the standardized method aroused a wide interest and within a few years, TLC became a widely used laboratory method. Stahl significantly contributed to the meteoric rise of its application, by further improving the technique and expanding its application fields. His activities culminated in 1962 with the publication of a very useful and highly popular handbook of TLC[18] that was translated into a number of languages.

16.4. High Performance TLC

Even though TLC soon enjoyed a wide application, it was essentially considered as a qualitative technique for the analysis of relatively simple mixtures. Further advances were directed in three ways: instrumentation permitting more precise spotting of the sample onto the plates, the quantitative evaluation of the separated spots, and improvements in the technique itself, resulting in higher separation power and faster analysis. Analogous to the name change of liquid chromatography to "high-performance liquid chromatography" (HPLC) characterizing the significant change in performance capabilities, this improved TLC was also given the name high-performance TLC (HPTLC) by R. E. Kaiser, who was instrumental in its development.[19]

The main difference between conventional and HPTLC was in the particle size and in the range of the adsorbent. While the original "silica gel for TLC according to Stahl" had a fairly broad particle size range (10–60 μm), with an average about 20 μm, the material for HPTLC had a narrower range and an average particle size of only about 5 μm. The plates were also smaller, 10×10 cm as compared to the conventional 20×20 cm plates, and the sample volume was reduced by an order of magnitude. The method of sample application was also improved with the design of mechanical applicators ("dosimeters") permitting a reduction in the diameter of the starting spots. As a result of these improvements, the time needed for an analysis was significantly reduced, with a simultaneous increase in the separation efficiency.

The use of very fine particles, however, resulted in some additional problems, one of them being that the movement of the mobile phase on the plate significantly slowed down after a relatively short distance. On the other hand, as emphasized by Guiochon *et al.*,[20–24] a fast and constant flow velocity of the mobile phase is needed to obtain an optimum efficiency. To overcome this problem, Kaiser started to apply pressure to the TLC plate. This then led to the development of the so-called forced-flow TLC.

16.5. Forced-Flow TLC

Developing the TLC plates in the conventional developing chambers has a shortcoming, due to the fact that in addition to the stationary phase (on the plate) and the liquid mobile phase (ascending by capillary

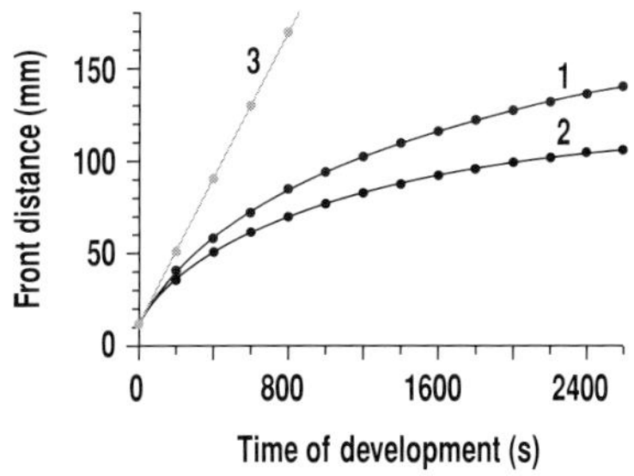

Fig. 16.2. Front distance vs time of development. Conventional TLC using (1) saturated and (2) unsaturated development chamber; (3) FFTLC.

action), a third phase is also present: the vapor of the mobile phase. During development molecules of the mobile phase will condense from the vapor phase onto the plate, both above the ascending mobile phase front (where the plate is dry) and below the front (where it is wet). At the same time molecules of the mobile phase will evaporate from the wet part of the plate. Due to this constant condensation–evaporation process the speed of the movement of the mobile phase front will depend on the degree of saturation of the vapor phase (Fig. 16.2). Also, with the increase of the front distance, the upward movement of the mobile phase on the plate will slow down; this is a direct consequence of the increasing weight of the developing solvent on the plate. In the case of mixed mobile phases, there will also be a solvent composition gradient on the plate, due to differences in the ascending speed of the mobile phase components and in their vapor pressures.

In order to overcome the problem with the changing velocity of the mobile phase in the plate and eliminate the presence of the vapor phase in the TLC system, Tyihák, Minchovics, and Kalász developed the so-called over-pressured TLC (OPTLC) or forced-flow TLC (FFTLC).[25,26]

In the FFTLC system the samples are spotted on the dry plate which then is placed inside a pressurized development chamber. There the stationary phase layer is tightly covered and sealed on its sides by an elastic membrane (a plastic sheet), pressurized by an inert gas or water filling up the "cushion" above the layer (Fig. 16.3). The mobile phase is delivered with the help of a pump through a slit in the membrane, at

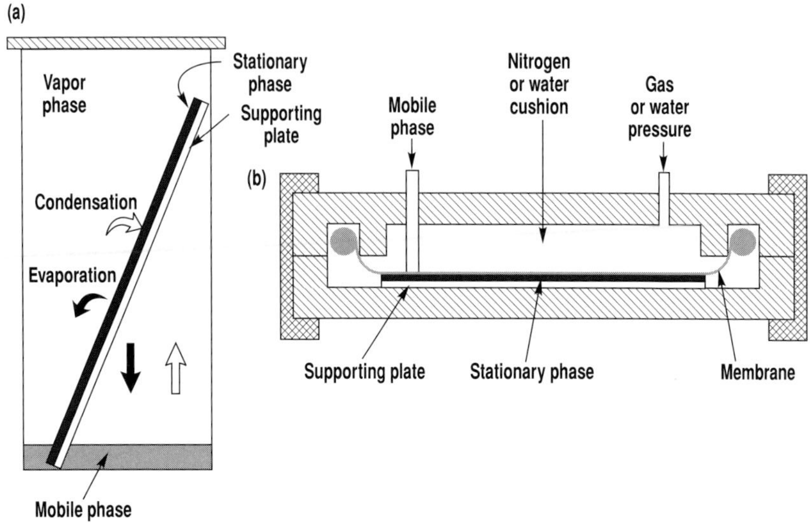

Fig. 16.3. Functional schematics of (left) ascending development chamber for conventional TLC and of (right) FFTLC chamber.

a constant velocity, directly to the stationary phase layer. Depending on the construction and location of the solvent inlet, various configurations can be handled, such as having a linear or circular front, or having the entry point either at the lower edge or at the middle of the plate (Fig. 16.4).

It is interesting to compare the mobile phase flow profile in liquid column chromatography with its profile in conventional and FFTLC. In column LC the stationary phase particles are wet, because the mobile phase is continuously flowing through the column, and the flow is pressure-driven; here, the profile of the microflow amongst the particles is convex. In TLC the stationary phase is dry before the advancing front of the developing solvent. In conventional TLC the solvent flow is propagated in the dry plate by capillary action and thus, has a meniscus representing a concave profile. In FFTLC the flow is again pressure-driven as in column chromatography, but now, moving on a dry plate; here the two effects — the convex front and the meniscus — compensate one another, resulting in a straight flow profile (Fig. 16.5).

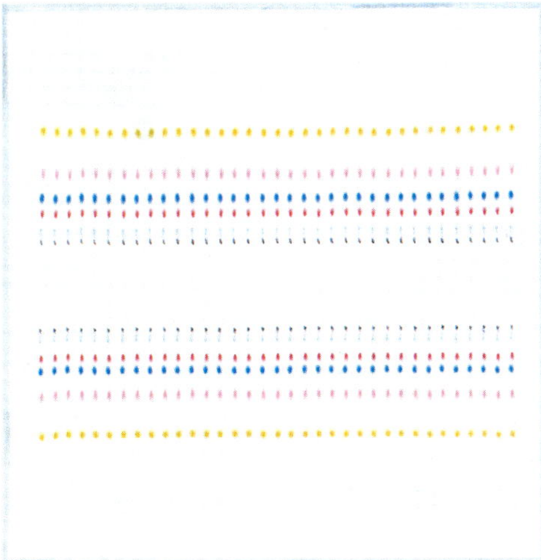

Fig. 16.4. TLC of a dye mixture using FFTLC. Plate: 20 × 20 cm silica; particle diameter: 5 μm; Mobile phase: toluene; Sample: toluene solution of CAMAG Test Substance II + Ceres Violet; Sample loading: manual, with a 1-μL syringe, in the center of the plate; 2 × 35 spots, loading time: 15 min; cushion pressure (water): 50 atm; Running time: 3 min. (Courtesy: Dr E. Mincsovics.)

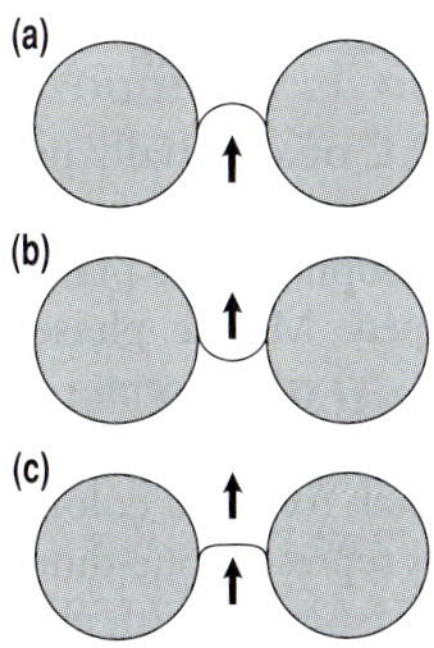

Fig. 16.5. The shape of the microflow profiles between the stationary phase particles. (A) HPLC; (B) conventional TLC; (C) FFTLC.

16.6. Newer Developments

In the last decades a number of further improvements have been introduced in TLC; these are related to the selection of the stationary phase and the way the plates are developed.

7. J. E. Meinhardt and N. F. Hall, *Anal. Chem.* **21**, 185–188 (1949).
8. J. G. Kirchner, in *75 Years of Chromatography — A Historical Dialogue*, eds. L. S. Ettre and A. Zlatkis (Elsevier, Amsterdam, 1979), pp. 201–208.
9. J. G. Kirchner, J. M. Miller and G. J. Keller, *Anal. Chem.* **23**, 420–425 (1951).
10. J. G. Kirchner, J. M. Miller and R. G. Rice, *J. Agr. Food. Chem.* **2**, 1031–1033 (1954).
11. J. M. Miller and J. G. Kirchner, *Anal. Chem.* **23**, 428–430 (1951).
12. R. G. Ackman, C. A. McLeod and A. K. Banerjee, *J. Planar Chromatogr.* **3**, 450–462 (1990).
13. E. D. Hudson, R. J. Helleur and Ch. C. Parrish, *J. Chromatogr. Sci.* **39**, 146–152 (2001).
14. R. H. Reitsema, *Anal. Chem.* **26**, 960–963 (1954).
15. E. Stahl, in *75 Years of Chromatography — A Historical Dialogue*, eds. L. S. Ettre and A. Zlatkis (Elsevier, Amsterdam, 1979), pp. 425–435.
16. E. Stahl, *Pharmazie* **11**, 633–637 (1956).
17. E. Stahl, *Chemiker Ztg.* **82**, 323–329 (1958).
18. E. Stahl, ed., *Dünnschicht-Chromatographie, ein Laboratoriumshandbuch* (Thin-layer Chromatography, a Laboratory Handbook) (Springer, Berlin, original German edition: 1962; second German edition: 1967; English edition: 1969).
19. A. Zlatkis and R. E. Kaiser eds., *HPTLC: High-Performance Thin-Layer Chromatography* (Elsevier, Amsterdam, 1977).
20. G. Guiochon, A. Siouffi, H. Engelhardt and I. Halász, *J. Chromatogr. Sci.* **16**, 152–157 (1978).
21. G. Guiochon and A. Siouffi, *J. Chromatogr. Sci.* **16**, 470–481 (1978).
22. G. Guiochon and A. Siouffi, *J. Chromatogr. Sci.* **16**, 598–609 (1978).
23. G. Guiochon, F. Bressole and A. Siouffi, *J. Chromatogr. Sci.* **17**, 368–386 (1979).
24. G. Guiochon, G. Körösi and A. Siouffi, *J. Chromatogr. Sci.* **18**, 324–329 (1980).
25. E. Tyihák, E. Mincsovics and H. Kalász, *J. Chromatogr.* **174**, 75–81 (1979).
26. H. Kalász, *Chromatographia* **18**, 628–632 (1984).

Part Six

Ion-Exchange Chromatography

Chapter

17

Preparative Ion-Exchange Chromatography and the Manhattan Project*

On 22–24 September 1949, the (British) Faraday Society held a symposium entitled General Discussion on Chromatographic Analysis. It was a true review of the state of chromatography and probably the most important symposium on the technique in the first half of the 20th century. Everybody who counted in this field was there and presented a summary of their results (see Chapter 30). Two of these papers are of interest here: they were presented by F. H. Spedding of Iowa State College (the present Iowa State University), Ames, IA,[1] and by E. R. Tompkins of Clinton National Laboratories (the present Oak Ridge National Laboratory), Oak Ridge, TN.[2] In these papers they summarized the elaborate investigations carried out in their laboratories on the separation of rare earths.

*Based on the article by L. S. Ettre published in *LCGC* (*North America*) **17**, 1104–1109 (1999). The help of Dr. Jack E. Powell, professor emeritus at Iowa State University, who was a member of the original Ames team, is greatly acknowledged.

The activities of these two groups were carried out in conjunction with the Manhattan Project, the development of the atomic bomb, and were part of the so-called Metallurgical Project involving the chemistry of the production of uranium and plutonium. Their activities complemented each other: while those in Oak Ridge mainly studied the fission products and their impact, the Ames group was involved in the production of pure uranium metal and rare earths. For us the latter is particularly interesting.

In the first four decades of its evolution, chromatography had already been used for "preparative" purposes: however, this only meant the collection of small amounts — a maximum of a few grams the most — of a separated pure substance, for further investigations by other techniques.[3] The activities of the Ames group were the first demonstration of the use of chromatography on a process scale, for production of larger quantities of pure compounds. In fact, in the second stage of their activities, in the 1950s, these led to very large scale, truly industrial production. Their work was rightly praised by E. N. Lightfoot, in a review of the invention and development of process (liquid) chromatography, as a major advance in this evolution.[4]

It should be noted that because of the association with the Manhattan Project, the work carried out at Oak Ridge and Ames was highly classified and its publication forbidden, even for a few years following the war: results were only summarized in restricted-circulation internal reports. The first time they were permitted to publicly report on their activities was at the fall 1947 National Meeting of the American Chemical Society where, on 17th September, a special Symposium on Ion-Exchange Separations was organized for this purpose. Just prior to this meeting a summary of the activities of the two groups was published in *Chemical & Engineering News*.[5] Following this Symposium detailed scholarly papers reporting on their results were published in the November 1947 issue of the *Journal of the American Chemical Society*. Due to their historical importance the titles of these papers are listed in Table 17.1. Finally, one year later the two summary reports on the whole project were presented at the Faraday Society Conference.[1,2] We should also list here the names of the scientists in the two groups who were involved in these activities (Table 17.2).

Table 17.1. Papers published in the November 1947 (Vol. 49) issue of the *Journal of the American Chemical Society* on the separation of rare earths by ion-exchange chromatography.[a]

Ion exchange as a separation method

E. R. Tompkins, J. X. Khym and W. E. Cohn (ORNL): The separation of fission-produced radioisotopes, including individual rare earths, by complexing elution from Amberlite resin.

D. Harris and E. R. Tompkins (ORNL): Separations of several rare earths of the cerium group (La, Ce, Pr, and Nd).

E. R. Tompkins and S. W. Mayer (ORNL): Equilibrium studies of the reactions of rare earth complexes with synthetic ion-exchange resins.

The exchange adsorption of ions from aqueous solutions by organic zeolites

G. E. Boyd, J. Schubert and A. W. Adamson (ORNL): Ion-exchange equilibria.

G. E. Boyd, A. W. Adamson and L. S. Myers, Jr. (ORNL): Kinetics.

G. E. Boyd, L. S. Myers, Jr. and A. W. Adamson (ORNL): Performance of deep adsorbent beds under nonequilibrium conditions.

B. H. Ketelle and G. E. Boyd (ORNL): The separations of yttrium-group rare earths.

The Separation of rare earths by ion exchange

F. H. Spedding, A. F. Voigt, E. M. Gladrow and N. R. Sleight (ISC): Cerium and yttrium.

F. H. Spedding, A. F. Voigt, E. M. Gladrow, N. R. Sleight, J. E. Powell, J. M. Wright, T. A. Butler and P. Figard (ISC): Neodymium and praseodymium.

F. H. Spedding, E. I. Fulmer, T. A. Butler, E. M. Gladrow, M. Gobush, P. E. Porter, J. E. Powell and J. M. Wright (ISC): Pilot-plant scale separations.

Miscellaneous

J. A. Marinsky, L. E. Glendenin and C. D. Coryell (ORNL): The chemical identification of radioisotopes of neodymium and element 61.

J. A. Ayres (ISC): Purification of zirconium by ion-exchange columns.

S. W. Mayer and E. R. Tompkins (ORNL): A theoretical analysis of the column separation process.

[a]ORNL: Oak Ridge National Laboratory; ISC: Iowa State College.

This discussion is based mainly on the 1947 papers and particularly on the two reports, and on information I personally obtained from Professor Jack E. Powell, professor emeritus of Iowa State University.

17.1. Background

Rare earths represent a group of some 15 elements that start with atomic number 57 and extend through atomic number 71, to which we also add yttrium, element 39 (Table 17.3). These elements have

Table 17.2. Members of the two teams working on rare earth separation by ion-exchange (1942–1948).

Institute of Atomic Research, Iowa State College,[a] Ames, Iowa	Clinton National Laboratory, Oak Ridge, Tennessee[b]
D. H. Ahmann	A. W. Adamson
J. A. Ayres	R. H. Beaton
V. Bulgrin	G. E. Boyd
T. A. Butler	A. R. Brosi
A. H. Daane	W. E. Cohn
V. A. Fassel	C. D. Coryell
P. Figard	L. E. Glendenin
E. I. Fulmer	D. H. Harris
E. M. Gladrow	B. H. Ketelle
M. Gobush	J. X. Khym
C. F. Miller	J. A. Marinsky
P. E. Porter	S. W. Mayer
J. E. Powell	L. S. Myers, Jr.
N. R. Sleight	G. W. Parker
F. H. Spedding	E. R. Russell
A. D. Tevebaugh	J. Schubert
R. Q. Thompson	J. A. Swartout
A. F. Voigt	E. R. Tompkins
E. J. Wheelwright	
H. A. Wilhelm	
J. M. Wright	
I. S. Yaffe	

[a]Today: Iowa State University.
[b]From the fall of 1943 on; previously at the Metallurgical Laboratory of the University of Chicago.

almost identical chemical and physical properties and up to the 1940s, were the least understood elements the separation of which was an almost insurmountable task. The situation was best characterized by Spedding in the following way[1]:

> In the past the best means of separating these elements have been the well-known but laborious method of fractional crystallization, fractional decomposition, etc. ...These processes were tedious and required a great deal of drive and patience. ...The same operations had to be repeated many times and in some of the rarer and more difficult -to-separate rare earths, it was required up to 20,000 operations to accomplish purifications. Therefore, except to the few stout

Table 17.3. The rare earth metals.[a]

Atomic no.	Name	Symbol	Atomic mass
39	Yttrium	Y	88.91
57	Lanthanum	La	138.91
58	Cerium	Ce	140.12
59	Praseodymium	Pr	140.91
60	Neodymium	Nd	144.24
61	Promethium	Pm	144.91
62	Samarium	Sm	150.36
63	Europium	Eu	151.97
64	Gadolinium	Gd	157.25
65	Terbium	Tb	158.93
66	Dysprosium	Dy	162.50
67	Holmium	Ho	164.93
68	Erbium	Er	167.26
69	Thulium	Tm	168.93
70	Ytterbium	Yb	173.04
71	Lutetium	Lu	174.97

[a]Various names are used to group these elements. Elements 57–71 are the *lanthanides*; within this series elements 57–62 belong to the *cerium group*, elements 63–66 to the *terbium group*, and elements 67–71 to the *yttrium group*. The reason for the name of the last group is that these elements are always found in nature together with yttrium. Sometimes distinction is made between the *lighter rare earths*, meaning the cerium-group elements, and the *heavier rare earths*, referring to those of the yttrium group. When speaking generally about the rear earth compounds, usually the symbol R is used in structure formulae, such as R_2O_3.

souls who were willing to devote a life to this sort of study, the rare earths were not generally available.

A few minerals contain a higher amount of the rare earths such as gadolinite, monazite, bastnaesite, and xenotime.[6] From these some rare earth concentrates were produced such as "didymium" which was essentially a mixture of samarium, neodymium, and praseodymium oxides, with a small percentage of other lanthanide elements, and "neodymium carbonate" and "crude yttrium oxalate" of the (now defunct) Lindsay Light and Chemical Co. which was located in West Chicago, IL. The former was a mixture of Nd, Sm, Pr, and Gd compounds while the latter was a wider mixture of rare earth compounds.

It should be mentioned that promethium (element 61) does not have a stable isotope and thus, it does not appear in the earth's crust and in minerals. Its existence had been indicated in 1926, but it was first positively identified in Oak Ridge among the fission products of ^{235}U by Marinsky *et al.*,[8] through separation by ion-exchange chromatography. Later, kilogram amounts of Pm_2O_3 have been separated from reactor wastes at Hanford.[6]

The sudden interest in the rare earths was related to the Manhattan Project, the crash program during Second World War aiming at the development of the atomic bomb. It was found that when uranium atoms undergo fission, intensely radioactive isotopes of various elements — among them the rare earths — are formed as fragments: as characterized by Spedding[1] "they are among the ashes of the atomic reaction." In order to be successful in the production of plutonium, processes had to be developed to dispose the spent fuel elements and to separate the plutonium produced from uranium and the various fission products.

At that time very little information was available on the rare earth elements although this was necessary for better understanding of the fission process. For the study of their nuclear properties and their chemical behavior the availability of pure rare earths was needed. This was also desirable so that mixtures of the non-radioactive elements closely approximating the characteristics of the radioactive isotopes could be used for the investigations. However, as mentioned, pure rare earth metals and salts were generally not available.

Meanwhile, the first polymeric ion-exchange resins were developed in England in 1935 by B. A. Adams and E. L. Holmes at the National Chemical Laboratories, in Teddington, soon followed by Rohm & Haas Research Laboratories, in Bridesburg, PA, introducing Amberlite IR-1 sulfonated phenol–formaldehyde-type cation-exchange resin.[9] This was followed by the improved Amberlite IR-100 and then by Dowex 50 of Dow Chemical Co., Midland, MI, a sulfonated styrene–divinylbenzene copolymer.[10]

Starting in 1942 G. E. Boyd and his associates, working at the so-called Metallurgical Laboratory of the University of Chicago, observed that these synthetic organic ion-exchange resins would adsorb some of the products of the fission of uranium and that these ions could be differentially eluted from the adsorbent under controlled conditions. By the summer of 1943 a program was initiated by Dr. Waldo E. Cohn to obtain the individual fission products in as pure a form as possible, mainly to be used for biological studies. In the autumn of 1943 both groups moved to Clinton National Laboratories, in Oak Ridge, TN (the present-day Oak Ridge National Laboratories, ORNL). In the next year their work continued there and succeeded in selectively eluting the rare earths and other fission products in groups, in tracer and micro quantities, and — as mentioned — element 61 (promethium) was also positively identified among the fission products.

It is interesting here to compare the pioneering work of the two American groups with the activities in Germany related to atomic research. As noted by Kumin, one of the American pioneers in the development of ion-exchange resins[9]

> Oddly enough, Otto Hahn, Lise Meitner and Fritz Strassmann ... completely ignored the use of ion exchange and used very difficult techniques to separate the nuclides that eventually led to the discovery of nuclear fission, even though ion-exchange resins were manufactured and used for many applications in Germany during this period.

Members of the ORNL team also carried out theoretical studies on the ion-exchange process: e.g., S. W. Mayer and E. R. Tompkins were the first to apply the plate theory which was introduced in 1941 by Martin and Synge for partition chromatography, to describe the efficiency of the ion-exchange separation process.

17.2. The Rare Earth Project at Ames

The development of methods for the separation and isolation of pure rare earths and eventually their production in sizeable quantities was carried out at the Institute of Atomic Research at Iowa State College (ISC) — the present-day Iowa State University — in Ames, IA. This Institute was formed as part of the Manhattan Project and from 1942 on, it was involved in the preparation of pure metallic uranium, based on the "Ames process" developed under the direction of Harley A. Wilhelm and Frank H. Spedding. In this process uranium tetrafluoride (the so-called "green salt") — produced elsewhere — was reacted with magnesium metal at high temperatures.[11] Within a few years a large amount of pure uranium metal was produced at Ames.

Spedding had also served as the director of analytical chemistry for the uranium project at the University of Chicago and he regularly attended the inter-group meetings of the Metallurgical Project. It was at the 19 December 1944 meeting in Chicago that W. E. Cohn of ORNL reported on their preliminary results on the ion-exchange separation of small quantities of rare earth groups. Hearing this report Spedding immediately raised the question whether this separation could not be further refined to separate the individual elements, and used for handling macro-quantities. Returning to Ames he initiated such investigations and within a few months they succeeded in separating gram quantities of Y and Ce, and of Pr and Nd. Subsequently, the technique was further refined and large scale separation of the individual rare earths was carried out.

A number of chemists — mostly recent Ph.D.'s and graduate students — were already affiliated with the Institute at Ames when the rare earth project started. They were now organized into two groups, one dealing with the separation of the cerium group of elements and yttrium, while the other was busy studying the separation of the elements belonging to the yttrium group. Soon others also joined the program and its success was really a team effort. The names of the scientists active in this program are listed in Table 17.2.

17.2.1. *Methodology*

For the earlier laboratory investigations glass columns of about 2-cm diameter, in lengths of 1–6 ft were used; these were then scaled up in the pilot plant operation to 4-in. diameter tubes. Most of the early work was carried out using Amberlite IR-100 ion-exchange resin; in later work it was replaced by the more efficient Dowex-50. The resin packed into the columns was washed with 5% hydrochloric acid solution to put it in the acid cycle; next, a slightly acidic solution of the rare earth chloride mixture was added to the top of the column where they were adsorbed as a band. The chloride ions were washed out of the column with distilled water and the column was developed with help of an ammonium citrate solution at a controlled pH.

Considerable work was done to optimize the concentration of the citrate solution and its pH. In general higher citrate concentrations resulted in better separation on a tracer scale; however, for large-scale operation the cost of citric acid became prohibitive. In this respect we should not forget that — as will be shown below — many hundred liters of eluant were used in a single run. With regard to the pH, lower values resulted in improved separation, however, the price to be paid was significantly increased elution time. Thus, for the first pilot plant work the Ames group finally settled at a citrate concentration of 0.1% and a pH of about 6.0.

It may be interesting to mention that switching to lower citrate concentrations (dictated by the very limited solubility of the hydrated neutral light rare earth–citrate species) and higher pH values created an unexpected problem: a mold started to grow in the column, attacking the citrate solution and preventing separation. This difficulty was finally overcome by adding 0.1–0.2% phenol to the essentially diammonium hydrate citrate solution.

The elution process was very slow: in almost all examples shown in the publications the linear velocity was only 0.5 cm/min. It is intuitive to calculate the flow rate from this, using the general relationship of

$$F = u \cdot \varepsilon \cdot r_c^2 \pi,$$

where F is the flow rate, u is the linear velocity, r_c is the column radius, and ε is the interparticle porosity; assuming that $\varepsilon = 0.40$ (a value established as an average in column chromatography), the calculated volumetric flow rate is 0.76 mL/min for a 2.2-cm i.d. column and 16.82 mL/min for a 4-in. i.d. column. Since — as seen in the chromatograms shown below — many hundreds of liters of eluant were needed for a complete run, the time for one cycle was considerable: with a flow rate of 16.82 mL/min 1000 L eluant represents

43 days! Thus, it is understandable that, as noted by Spedding, the productivity of the separation was characterized as grams of pure rare earth salts obtained per man–month of labor.

The effluent from the column was collected in fractions and the rare earths precipitated with oxalic acid. The purity of the fractions was established by emission spectroscopy. The achieved separation was characterized by plotting concentration (always expressed in R_2O_3) vs. the volume of the eluate.

To prepare the pure rare earth elements, the oxalates were transformed to the corresponding halides which then were reduced at high temperatures with calcium metal in a special crucible or furnace. This was a very complex procedure and was different for the individual rare earths or rare earth groups. It may be interesting to note that at the time when Spedding submitted his paper to the Faraday Society Conference, this process was still classified; it could be described in details only three years later.[12,13] A good summary of the techniques involved was later published by Spedding.[6]

When exposed to air the rare earth metals react readily with oxygen; however, the oxidation takes only place on the surface, thus, the metal bar will be encrusted with an oxide layer. To prevent corrosion, the bars should be kept in mineral oil.

17.2.2. *Separation of the Individual Rare Earths*

Within a short time after the start of the investigations at Ames it was found that yttrium and cerium could be separated easily and reasonably well. Figure 17.1 illustrates the result of a typical early investigation on a 190-cm long column, with 2.2-cm diameter, using a high citrate concentration (5%), low pH (2.77), and high velocity, (5 cm/min).[14] In this sample cerium was present in a 25-fold excess and in one single run 60% of cerium and 80% of yttrium could be obtained in spectroscopic purity. Material eluting in the overlapping region was then recycled. If the sample consisted of a 1:1 mixture of the two rare earths, over 90% of each could be obtained in pure form. As mentioned, in subsequent work the citrate concentration was reduced and the pH increased.

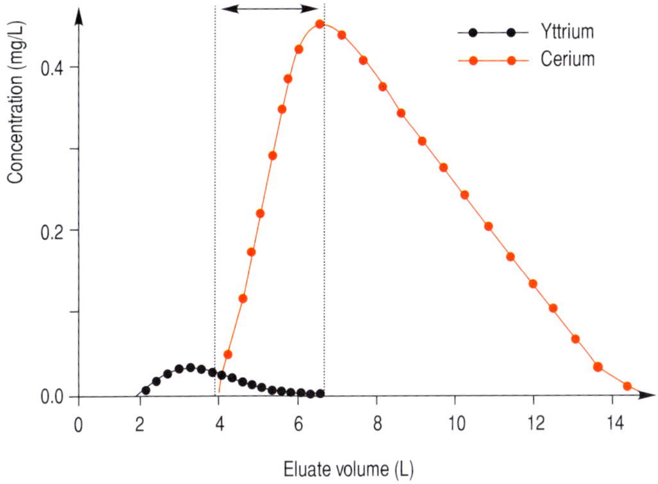

Fig. 17.1. Separation of small amount of yttrium in the presence of a large amount of cerium. Column: 190 cm × 1.6 cm containing Amberlite IR-1 resin (<20 mesh). Eluant: 5% citrate solution at pH 2.77 and a velocity of 5 cm/min. Composition of the original sample: 10 mg Y + 250 mg Ce. The broken vertical lines indicate the overlap region. After Ref. 14.

Investigations of the possibility of separating the components (Pr, Nd, and Sm) of "didymium" (a commercially available concentrate) also yielded positive results. Figure 17.2 shows a typical separation obtained on a 120-cm × 2.2-cm column, now using an eluent with low (0.5%) citrate concentration, higher pH (5.30), and slow velocity (0.5 cm/min). As indicated an appreciable fraction of these rare earths could be obtained in a pure state with a single pass through the column. However, the situation was more complicated with the heavier rare earths where the overlap was found to be significant and very little separation could be obtained in a single run. Therefore, fractions had to be recycled many times until individual pure rare earths could be obtained.

After the initial investigations and the optimization of the conditions the separation process was scaled up to pilot plant size, using a battery of 4-in. diameter columns. Figure 17.3 shows the photo of the pilot plant, with 10-ft long columns. A total of 12 columns were

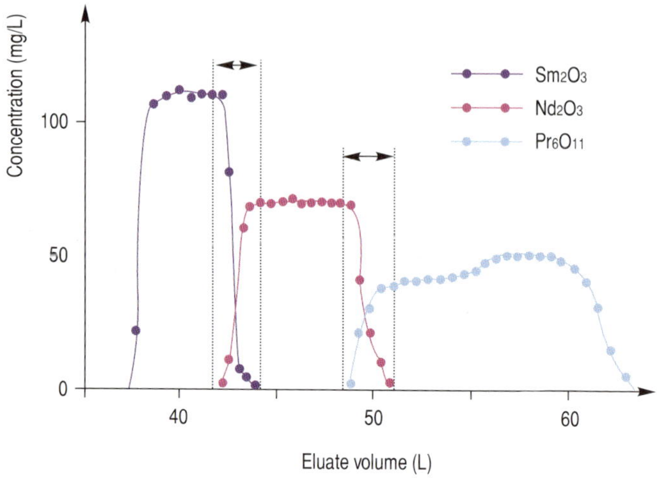

Fig. 17.2. Separation of the main components of didymium concentrate. Column: 120 cm × 2.2 cm, containing Amberlite IR-100 resin (30/40 mesh). Eluant: 0.1% citrate solution at pH 5.30 and a velocity of 0.5 cm/min. The broken vertical lines indicate the overlap region. After Ref. 1.

installed here, and 50–100 g of mixed rare earth chlorides were introduced into each column. As already mentioned a very low velocity was used (0.5 cm/min); thus, one separation cycle needed a considerable time, many days. The fractions were collected in 45-L bottles that were changed every 12 h. In two examples given by Spedding 24 and 36 fractions are shown which represent 288 and 432 h or 12 and 18 days, respectively, and to this one has to add the time needed until the elution of the first fraction. In addition we should not forget that — as mentioned earlier — fractions had to be recycled a number of times until pure rare earths of the yttrium group metals could be obtained. This was the reason for having a number of columns operated parallel: after all — as pointed out by Spedding[1] — "it takes (for the operator) very little more energy to watch 20 columns than it does to watch one."

Figure 17.4(a) is an example for the elution of a sample consisting of 100 g of concentrate obtained from gadolinite. Here, 10 rare earths were identified and their distribution in this chromatogram is typical: the yttrium group elements elute first and those of the cerium group

Fig. 17.3. Pilot plant installed at Iowa State College for the separation of rare earths by ion-exchange chromatography. The columns are 10 ft × 4 in. The large tank in the left background was used to prepare the citrate solution used as the eluant.[1] (Courtesy of Professor J. E. Powell.)

last, while yttrium elutes between the two fractions, with dysprosium overlapping with the yttrium band. Figure 17.4(b) shows the elution of the heavy-earth fraction from the previous run, after re-introducing it into the column. As seen the separation is somewhat improved, but — as noted by Spedding[1] — up to 10 consecutive passes were needed until these rare earths could be obtained in 99.5% purity.

By 1947 the separation of yttrium and cerium, and of the cerium-group elements was well established. As indicated in their publication,[15] besides yttrium and cerium, sizeable amounts of other

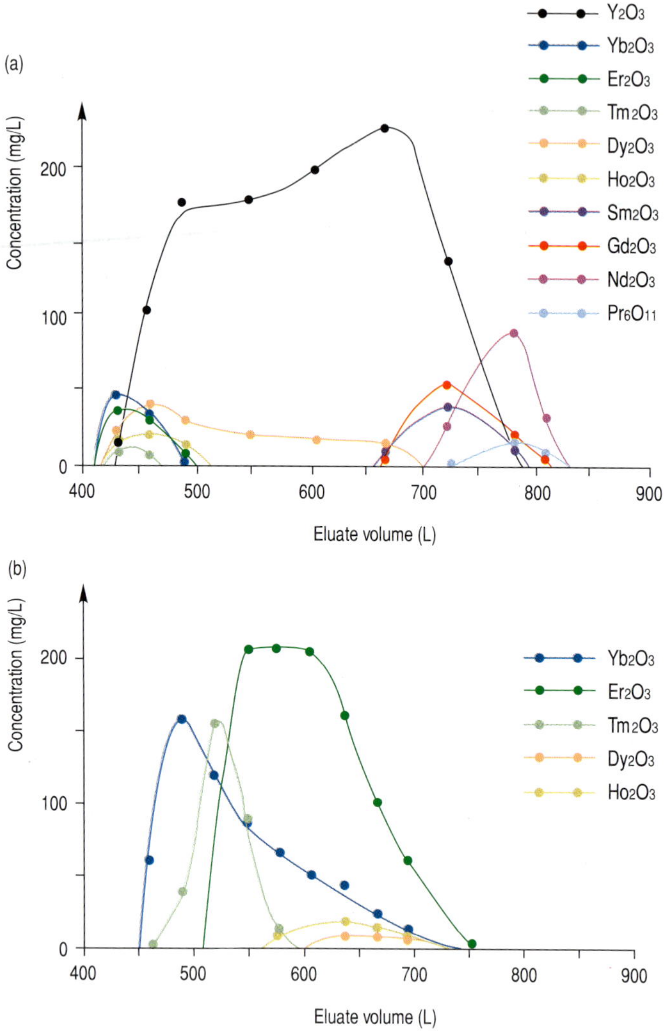

Fig. 17.4. (a) Elution of 100 g rare earth concentrate obtained from gadolinite ore; (b) Elution of 67 g of the heavy rare earth fraction from the previous run, after re-introducing it into the column. Column: 4 ft × 3/4 in. containing Amberlite IR-100 resin (30/40 mesh). Eluant: 0.1% citrate solution at pH 6.00 and a velocity of 0.5 cm/min. After Ref. 1.

Table 17.4. Pure cerium-group elements prepared at the Institute of Atomic Research of Iowa State College, prior to 1949.[15a]

Element	Amount	Purity	Impurities
Praseodymium	35 g	99% Pr_6O_{11}	1% Nd_2O_3
	160 g	90% Pr_6O_{11}	10% Ce, La, and Ca oxides
Neodymium	800 g	99.9% Nd_2O_3	<0.1% Pr_6O_{11}
	770 g	98% Nd_2O_3	2% Pr_6O_{11}
Samarium	160 g	>99.9% Sm_2O_3	<0.01% Eu_2O_3, 0.05% CaO
	600 g	99% Sm_2O_3	0.5% Eu_2O_3, 0.5% CaO

[a]Elements: Ca = calcium, Ce = cerium, Eu = europium, La = lanthanum, Nd = neodymium, O = oxygen, Pr = praseodymium, Sm = samarium.

very pure metals were prepared; Table 17.4 lists the produced amounts. It should be noted that the calcium impurity is due to the final process in the preparation of the metal in which the halides are reduced with calcium.

In the printed text of Spedding's paper[1] — submitted a little over two months prior to the Conference — it is mentioned that he was planning to exhibit metal bars of pure La, Ce, Pr, and Y, as well as some salts of the rarer rare earths. That this indeed took place was confirmed in a retrospective article[16] by R. L. M. Synge who was present at the Conference. I can imagine the surprise of the participants who most likely have never seen before such exotic materials!

Figure 17.4(a) and (b) demonstrate that the separation of the heavy rare earths was much more difficult than of the cerium-group elements, and by 1947–1948 only partial success was achieved. Still, as indicated by Spedding[1] they succeeded in isolating 300 g of pure ytterbium, 15 g spectrographically pure lutetium, and almost an equal amount of "very rich" thulium.

17.2.3. *Displacement Ion-Exchange Chromatography*

The group at Ames continued its activities after the war. With respect to the lighter rare earths the separation process using the ammonium citrate elution process was further refined by increasing the column diameter in the pilot plant to 6 in. and further increasing the pH of the 0.1% ammonium citrate eluent to pH 8.0.[17–21] This improved system had been used to produce ultrapure Y, Sm, Nd, and Pr in multi-kilogram quantities. It also permitted a better separation of the yttrium-group rare earths, however, it was still not enough for being able to produce them in sizeable quantities. The real breakthrough to solve this problem occurred around 1953 with the introduction of displacement ion-exchange chromatography. For this, some modifications were carried out.

First, a complexing agent stronger and more selective than citrate was sought to move the sorbed rare earths along the column system. For this purpose N,N,N',N'-ethylenediamine tetraacetic acid (EDTA) was found to be preferable but it could not be utilized on hydrogen-cycle cation-exchange beds (where H^+ ion served to retain rare earths in the case of high pH and dilute citrate elutions to foster displacement-type elution). It was at that time suggested by Powell that either trivalent iron[22] or divalent copper[23,24] ions should be used in lieu of H^+ to retain the rare earth cations on the cation-exchange beds, while using NH_4^+–HEDTA (hydroxy-EDTA) eluant to displace them. Of the two metal ions copper proved to be superior for this purpose. In these systems a continuous removal–redeposition process takes place and band-spreading is minimized. Therefore, very long columns can be used without any increase in the band's length.

The combined elution–displacement technique quickly supplanted the original dilute citrate elution process and also permitted the scaling up of the system. The new plant-size unit at Ames consisted of 12 10-ft long, 30-in. diameter columns used in series; the height of the resin bed in each column was 9 ft. This system permitted the separation of the cerium and terbium group elements plus ytterbium[25,26] which could then be obtained in very pure form from the respective fractions.

Eventually, the process was modified by replacing the cupric ions as retainer with zinc ions: the Zn–EDTA complex is less stable and can be split by simply acidifying the effluent. Another modification was that the yttrium-group (Lu, Yb, Tm, Er, and Ho) fraction obtained in the first step of the operation (which also contained the small inherent Dy–Y overlap region), was further developed through a series of 6-in. diameter columns (total length over 80 ft), using 0.018 M NH_4^+–HEDTA at pH 8.4.

According to a report from Iowa State University,[27] 3000 pounds of high-quality yttrium (99.9% Y_2O_3) plus many kilograms of the other rare earths (all at least 99.9% R_2O_3) were produced in the period of 1956–1958 in this plant from the mineral xenotime.

17.3. Postscript

In the late 1950s a number of industrial companies also began to produce pure rare earths — mainly yttrium — by the ion-exchange chromatographic method developed at Ames. The head and tail cuts from this industrial process were used by the Ames plant to prepare some of the rarer rare earths in high-purity (>99.9 %).

In the second part of this century the need for rare earths continuously increased: according to Spedding[6] in 1979 the US consumption of rare earths (expressed as R_2O_3) was approximately 15,000 metric tons and was steadily rising. A more recent review[28] estimated the world's consumption of lanthanides in 1990 as approximately 35,000 metric tons which, by 2000, was expected to grow to roughly 56,000 tons. It is important to note that this production was made possible by the activities of the Ames group in the 1940s.

In Ames the large-scale plant slowly ceased operation but smaller-scale production of special-purity rare-earth oxides (>99.99%) continued well into the 1980s. At that time the displacement technique was further modified and used at 92°C at which temperature the solubility of EDTA is much improved.

We have discussed the activities of the Ames group in more detail, because it illustrates well the capabilities of chromatography, that it can be used not only for minute quantities and for the determination of trace amounts but also for large-scale production. In the literature one can also find mentions of other production facilities, based on chromatographic principles. Thus, the Ames experience again demonstrates how wrong Willstätter was when he stated that "the chromatographic method ... appears unsuitable for preparative work".[29]

After the war the experience gained in Oak Ridge in the investigation of fission products was soon turned toward the solution of problems in biochemistry. Here, Waldo E. Cohn's pioneering work in 1949–1950 on the separation of nucleic acid constituents by ion-exchange chromatography is particularly noteworthy. These investigations led to new techniques which, within a few decades, changed the way biochemical investigations are carried out. Thus, what started as a project important to the war effort opened new fields in both the industry and science, and significantly contributed to the advancement of chromatography.

References

1. F. H. Spedding, *Disc. Faraday Soc.* **7**, 214–231 (1949).
2. E. R. Tompkins, *Disc. Faraday Soc.* **7**, 232–237 (1949).
3. E. Geeraert and M. Verzele, *Chromatographia* **11**, 640–644 (1978).
4. E. N. Lightfoot, *Amer. Lab.* **31**(12), 13–23 (June 1999).
5. W. C. Johnson, L. L. Quill and F. Daniels, *Chem. Eng. News* **25**, 2494 (1947).
6. F. H. Spedding, Rare-Earth Elements, in *Kirk-Othmer Encyclopedia of Chemical Technology*, 3rd edn. (Wiley, New York, 1983), Vol. 23, pp. 502–547.
7. J. A. Marinsky, L. E. Glendenin and C. D. Coryell, *J. Am. Chem. Soc.* **69**, 2781–2785 (1947).
8. J. A. Marinsky and L. E. Glendenin, *Chem. Eng. News* **26**, 2346–2348 (1948).
9. R. Kunin, *Chem. Heritage* **17**(2), 8–9, 36–38 (Summer 1999).
10. W. C. Bauman and J. Eichhorn, *J. Am. Chem. Soc.* **69**, 2830–2836 (1947).
11. F. Weigel, Uranium and Uranium Compounds, in *Kirk-Othmer Encyclopedia of Chemical Technology*, 3rd edn. (Wiley, New York, 1982), Vol. 19, pp. 833–854.
12. F. H. Spedding and A.H. Daane, *J. Am. Chem. Soc.* **74**, 2783–2785 (1952).
13. F. H. Spedding, H. A. Wilhelm, W. H. Keller, D. H. Ahmann, A. H. Daane, C. C. Hach and R. P. Ericson, *Ind. Eng. Chem.* **44**, 553–556 (1952).
14. F. H. Spedding, A. F. Voigt, E. M. Gladrow and N. R. Sleight, *J. Am. Chem. Soc.* **69**, 2777–2781 (1947).
15. F. H. Spedding, E. I. Fulmer, T. A. Butler, E. M. Gobush, P. E. Porter, J. E. Powell and J. M. Wright, *J. Am. Chem. Soc.* **69**, 2812–2818 (1947).

16. R. L. M. Synge, A Retrospect on Liquid Chromatography, in *British Biochemistry Past and Present*, ed. T. W. Goodwin (Academic Press, London, 1970), pp. 175–182.
17. F. H. Spedding, E. I. Fulmer, T. A. Butler and J. E. Powell, *J. Am. Chem. Soc.* **72**, 2349–2354 (1950).
18. F. H. Spedding, E. I. Fulmer, J. E. Powell and T. A. Butler, *J. Am. Chem. Soc.* **72**, 2354–2361 (1950).
19. F. H. Spedding, E. I. Fulmer, J. E. Powell, T. A. Butler and I. S. Yaffe, *J. Am. Chem. Soc.* **73**, 4840–4847 (1951).
20. F. H. Spedding and J. E. Powell, *J. Am. Chem. Soc.* **76**, 2545–2550 (1954).
21. F. H. Spedding and J. E. Powell, *J. Am. Chem. Soc.* **76**, 2550–2557 (1954).
22. F. H. Spedding, J. E. Powell and E. J. Wheelwright, *J. Am. Chem. Soc.* **76**, 612–613 (1954).
23. F. H. Spedding, J. E. Powell and E. J. Wheelwright, *J. Am. Chem. Soc.* **76**, 2557–2560 (1954).
24. F. H. Spedding, J. E. Powell and E. J. Wheelwright, *J. Am. Chem. Soc.* **78**, 34–37 (1956).
25. F. H. Spedding and J. E. Powell, *Trans. Metall. Soc. Amer. Inst. of Mining & Metall. Eng.* **215**, 457–463 (1959).
26. J. E. Powell, Separation Chemistry, in *Handbook on the Physics and Chemistry of Rare Earths*, eds. K. A. Gschneider Jr. and L. Eyring (North-Holland Publishing Co., Amsterdam, 1979), Vol. I, pp. 81–109.
27. Iowa State University, *US Atomic Commission R&D Report: Chemistry General* (UC-4), TID 4500 (1 October 1958); pp. 175–181.
28. J.-L. Sabot and P. Maestro, Lanthanides, in *Kirk-Othmer Encyclopedia of Chemical Technology*, 4th edn. (Wiley, New York, 1995), Vol. 14, pp. 1091–1115.
29. R. Willstätter and H. Stoll, *Untersuchungen über Chlorophyll: Methoden und Ergebnisse* (Springer Verlag, Berlin, 1913). English translation: *Investigations on Chlorophyll: Methods and Results* (Science Press, Lancaster, PA, 1928). The quotation is from p.142 of the English edition.

Chapter

18

The Development of the Amino Acid Analyzer*

Most chromatographers do not realize that the first liquid chromatography instrument was not developed in the 1960s: rather, it was the automated amino acid analyzer developed in 1958 at the Rockefeller Institute for Medical Research.[1,2] The amino acid analyzer opened an entirely new field for research, by permitting the elucidation of the composition of proteins; we may even say that the rapid expansion of biochemistry would have been impossible without it.

The study of amino acids and proteins formed by hundreds of amino acids connected by peptide bonds has a long history. The first major explorer of this field was Emil Fischer (1852–1919), the great German chemist, recipient of the 1902 Chemistry Nobel Prize, who showed in the early 1900s how amino acids are bound to each other forming polypeptides, the building blocks of proteins;[3] he also used the fractional distillation of the amino acid esters prepared from protein

*Based on the article by L. S. Ettre and C. W. Gehrke published in *LCGC* (*North America*) **24**, 390–400 (2006).

hydrolyzates to obtain an understanding of their original composition. A number of other methods have also been introduced in the first decades of the 20th century for the determination of the individual amino acids, among others selective precipitation as insoluble salts, colorimetric analysis and microbiological assays, but these were complicated and time-consuming operations.[4]

Partition chromatography first described in 1941 by A. J. P. Martin and R. L. M. Synge for column chromatography[5] and then its paper chromatography version introduced in 1944 by Martin's group[6] represented a major breakthrough, permitting the separation and simultaneous determination of amino acids in a straightforward manner (See Chapters 14 and 15).

Using partition chromatography on columns packed with starch and also paper chromatography Synge started in 1944 to establish the sequence of amino acids in peptides, leading to the structure elucidation of gramicidin S, a cyclic decapeptide.[7] In the next 10 years Frederick Sanger at Cambridge University, in England, further advancing Synge's methodology, finally succeeded in the establishment of the sequence of the 51 amino acids forming the molecule of insulin;[8] his achievements were recognized with the 1958 Nobel Prize in Chemistry.

18.1. Amino Acid Research at the Rockefeller Institute

The research leading to the development of the amino acid analyzer had been carried out at The Rockefeller Institute for Medical Research, in New York City. This Institute was founded in 1901 by John D. Rockefeller (1839–1937) and built on a farmland along the East River. The first laboratory opened in 1904 and its hospital for the study of human diseases was established in 1910. Very soon the Institute developed into one of the principal research organizations in biomedical sciences. The Institute started to grant Ph.D. degrees in 1954, taking the status of a graduate university, and finally, in 1965, it changed its name to Rockefeller University. Since our story is related to the period prior to 1960, we shall use the name of Rockefeller Institute in our narrative.

When in March 1933 Hitler and his Nazi party came to power in Germany, a number of Jewish scientists soon had to leave the country, many of whom immigrated to the United States. A prominent scientist involved in this intellectual migration was Max Bergmann.

Max Bergmann (1884–1944) was a student and associate of Emil Fischer and had been involved in his amino acid research. He had a distinguished career in Germany, was a founder and director of the Kaiser Wilhelm Institute for Leather Research, in Dresden, which in the 1920s he developed into a world-renowned, leading center for protein chemistry research. He left Germany in 1933 to join the Rockefeller Institute where he soon established a laboratory, and he became a central figure in protein research in the US. Bergmann surrounded himself with the most talented young, postdoctoral scientists, launching them to become important members of the international group of protein chemists. Two who are the key figures in our story are William H. Stein and Stanford Moore (Fig. 18.1).

William H. Stein (1911–1980) studied at Columbia University, graduating in 1937 with a thesis on the analysis of the amino acids of the protein elastin; he then went directly to Bergmann as a postdoctoral associate. Stanford Moore (1913–1982) studied at Vanderbilt University and the University of Wisconsin, graduating with a thesis on the characterization of carbohydrates as benzimidazole derivatives. It is interesting to note that during his undergraduate years, Moore also took engineering courses (in fact, his first thought was to major in engineering) and that as a graduate student in the laboratory of Professor Karl P. Link, at Wisconsin, he received an extensive training in microchemistry, including Pregl's microanalytical methods. After graduation, in 1939, he joined Bergmann on the recommendation of Professor Link.

There were two mainlines of the investigation in Bergmann's laboratory: the field of proteolytic enzymes, and the structural chemistry of proteins. Both Moore and Stein have been involved in the second field and their task was to further improve gravimetric methods of amino acid fractionation through the formation of sparingly soluble salts.[9–11] America's entry into the war interrupted their work: Bergmann's group started to work for the Office of Scientific Research

Fig. 18.1. S. Moore (left) and W.H. Stein, with the prototype of their amino acid analyzer.[29]

& Development (OSRD), investigating the physiological effects of mustard gas and related compounds, while Moore left the laboratory and served with the OSRD in Washington and at the headquarters of the US Armed Forces in the Pacific Area. At the end of the war Moore returned to the Rockefeller Institute. Bergmann died suddenly in 1944 and his former group was dissolved, but the director of the Institute allocated part of Bergmann's laboratories to Moore and Stein to develop their own research program. This is how their close cooperation began; it lasted for 40 years, eventually leading to the quantitative analysis of amino acids and its automation, and culminating in the elucidation of the 124 amino acid sequence of ribonuclease, honored by the 1972 Chemistry Nobel Prize. As noted by Moore[12]

> We approached problems with somewhat different perspectives and then focused our thoughts on the common aim. If I did not think

> of something, he was likely to, and vice versa, and this process of frequent interchange of ideas accelerated our progress in research.

Their research activities can be followed from their autobiographical treatment[13] and their Nobel Prize lectures.[14]

When Moore and Stein returned to peacetime research, the wartime issues of the British *Biochemical Journal* had just became available in the United States and there they read about the work of Synge describing the separation of free amino acids by partition chromatography on columns containing starch as the stationary phase.[15,16] They immediately adapted the technique to their investigations, further improving it. Now, column effluent was collected in small fractions and, by establishing the amount of amino acids in each fraction, chromatograms showing the separated peaks could be constructed. Quantitation of each fraction was carried out by adapting the color reaction of the amino acids with ninhydrin. At the beginning the small fractions were collected manually, but this was much too tedious due to the large number of fractions. Therefore, they developed a very sophisticated, fully automated fraction collector.

In November 1946 The New York Academy of Sciences held a two-day Conference on Chromatography and there, Moore and Stein presented a report on the early results of their work (see Chapter 29). The publication of the proceedings of the meeting was delayed by more than one year and thus, in the printed text of their presentation they could also include a detailed description of their newly developed fraction collector.[17] This was followed by four publications reporting on their further results and the determination of the amino acid composition of β-lactoglobulin and bovine serum albumin, using partition chromatography on starch columns.[18–21]

The starch columns worked well, however, they were very slow. Therefore, Moore and Stein were looking for other possibilities. Meanwhile the detailed reports on the use of ion-exchange chromatography (IEC) became known (see Chapter 17) and these were followed by the pioneering work of Waldo E. Cohn at Oak Ridge National Laboratories on the separation of nucleic acid constituents by IEC.[22,23] Reports from England also involved IEC for the separation of amino acids in protein hydrolyzates, although in the displacement and not in the elution mode.[24] These reports initiated intensive activities by Moore and Stein who were joined by C. H. W. Hirs (born 1923), just graduating in 1949 from Columbia University. Dr. Hirs remained for 10 years at the Rockefeller Institute and participated in much of the amino acid research. He was the first young postdoctoral associate in Moore and Stein's group; in the next 20 years 20 additional young scientists have spent a few years with them, cooperating in their research and learning the intricacies of protein investigations.

Hirs provided a very vivid narrative of the intensive research carried out by the group of Moore and Stein.[25] Their first report on the use of ion-exchange resins for amino acid separation was published in 1951,[26] followed by half a dozen more papers in the first part of the 1950s. Studying these one can follow the careful, painstaking, and systematic investigations, approaching every aspect of IEC and the possibility not only of quantitative amino acid analysis, but also of separation on semi-preparative scale.

The 1972 Chemistry Nobel Prize was awarded to Moore and Stein (together with C. B. Anfinsen) for their "contribution to the understanding of the connection between chemical structure and catalytic activity of the active center of the enzyme ribonuclease." These investigations started in 1954 and aimed at the determination of the complete amino acid sequence of the molecule. In the course of these studies a very large number of protein hydrolyzate analyses had to be performed and the manual procedure, one complete analysis taking a few days, was just not satisfactory: analysis time had to be reduced and some kind of automation was needed. Particularly Moore always had been interested in engineering work — let us not forget that in his college time, he also took engineering courses! — and thus, they decided to investigate the possibility to develop an automated amino acid analyzer. In this work they had a new associate: D. H. Spackman (born 1924), who received his Ph.D. at the University of Utah in 1954, and moved from there to the Institute.

The development proceeded along two lines: further improvement of the ion-exchange separation process and the construction of an instrument. With respect to separation, they selected a two-column system: a long (150 cm) column for the separation of the acidic and neutral amino acids, and a separate, short (15 cm) column for the basic amino acids. Since in the standard run the basic amino acids remained on the first column, the column had to be regenerated after each use. To assure the possibility of continuous operation, two long columns were included in the system: while a sample was analyzed on one, the other was regenerated. In this way, when one analysis cycle was finished on one column, the second was ready for the next sample. Meanwhile improved sulfonated polystyrene resins became available; however, they further fractionated the commercial product to obtain a narrower particle size cut. This method based on hydraulic flotation was developed at that time by P. B. Hamilton at the A.I. DuPont Institute of the Nemours Foundation, in Wilmington, DE.[27] This improved packing permitted the use of higher flow rates without any loss of resolution.

The analyzer contained the three columns in a thermostat. Reciprocal pumps maintained a constant flow of the various buffer solutions used as eluent and for column regeneration. Other major components were photometric detectors used at 570 and 440 nm, connected to a multi-pen potentiometric recorder, and a vessel between the columns and the detectors in which column effluent was continuously mixed with a ninhydrin solution; in this way the amino acids reacted with ninhydrin, forming colored compounds. The system also included other devices such as e.g., deaerators, manifold, valves and gauges. Figure 18.2 shows a simplified schematic of the system. The sample was added manually by pipette to the top of the column, but from then on operation was unattended. One full analysis of a protein hydrolyzate took one day and of a more complex physiological fluid sample about two days.

A preliminary description of the system was given at the April 1956 meeting of the Federation of American Societies for Experimental Biology (FASEB), in Atlantic City, NJ;[28] this breadboard system was then used in a number of investigations at the Institute. Finally on 28 February 1958, the Rockefeller Institute team submitted two papers to *Analytical Chemistry*, describing in details the preparation

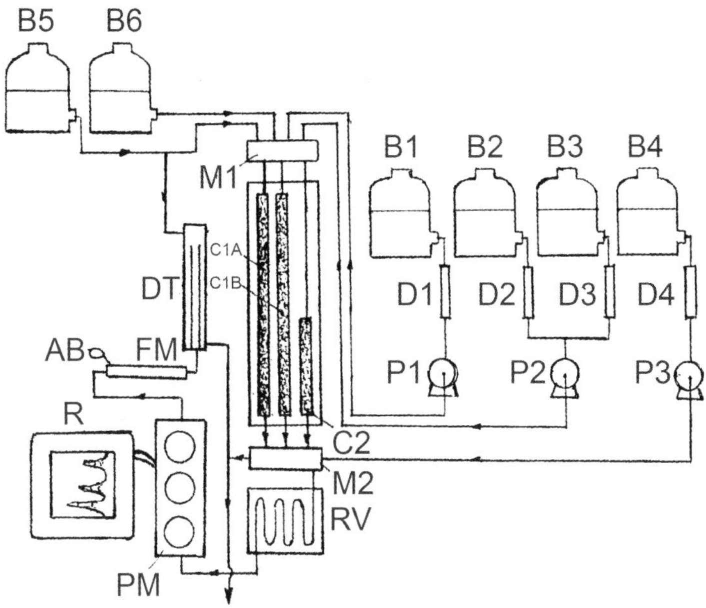

Fig. 18.2. Simplified schematic of the amino acid analyzer of Spackman, Moore and Stein. C1A and C1B are 150-cm long columns, C2 is the short (15 cm) column; B1 (pH 5.28), B2 (pH 3.25), and B3 (pH 4.25) are bottles containing the sodium citrate buffer solutions used as the eluent; B4 is the bottle containing the ninhydrin solution; B5 is a pH 3.25 buffer solution and B6 is a 0.2 N NaOH solution, both used for column regeneration and reequilibration; D1–D4 are deaerators; P1–P3 are pumps; M1 and M2 are multi-valve manifolds; RV is a reaction vessel containing a long, coiled tube in a boiling-water bath; PM contains three photometer units at different wavelengths; R is a three-pen potentiometer recorder; FM is a flow meter; AB is a rubber atomizer bulb to introduce an air bubble into the flow meter; DT is the drain tube. Various valves and pressure gauges and regulators are omitted from the schematic.

of the column packing, the operation of the column, and providing a very detailed description of the apparatus and its components, even giving an engineering drawing of the photometer consisting of three units. These two papers entitled Chromatography of Amino Acids on Sulfonated Polystyrene Resins[1] and Automatic Recording Apparatus for Use in the Chromatography of Amino Acids[2] were published in the July issue of the journal: they represent the start of automated amino acid analysis.

18.2. Production of the Amino Acid Analyzer

Although the two papers provided a detailed description of the instrument constructed in the workshops of the Institute, probably only a very few large laboratories would have been able to build it. This was also realized by the Rockefeller Institute team, and they turned to the Spinco Division of Beckman,[a] with which they already had a good relationship, to transfer the design to production.

Specialized Instruments Company ("Spinco") was founded in 1946 by M. C. Hanafin and E. G. Pickels, to commercialize the analytical ultracentrifuge developed at Rockefeller Institute under the direction of Dr. Pickels, constructed originally to aid the isolation of pure polio virus.[29] On 30 December 1954, Beckman Instruments acquired the company which, from then on, operated as the Spinco Division of Beckman.

Spinco built the first prototype of the amino acid analyzer — the so-called model MS (for Moore and Stein) — in the spring of 1958 (before the two papers were actually be published), but originally, it had many of the usual problems associated with the transfer of a complicated design from research to production. At that time, Darrel Spackman left Rockefeller Institute and joined Spinco. He remained with the company for over three years and had been involved in the improvements of the instrument; in 1962 he moved to the University of Washington and from then on, had academic affiliations. With Dr. Spackman's help the instrument was soon "debugged" and from then on, the production of the Model 120 Amino Acid Analyzer (its final designation) proceeded smoothly (Fig. 18.3) Spinco also maintained a continuing contact with Dr. Moore who regularly visited them.[29]

In the subsequent years the technique and the ion-exchange resins were further improved, permitting the use of a single column[30] and significant reduction of the analysis time. Spinco introduced the Model 120B in 1963 and the model 120C in 1966. Figure 18.4 shows a typical chromatogram from this period: a complex physiological sample could

[a]Information on Spinco and their activities related to the amino acid analyzer were obtained from Ms. Pat Ashton of the Heritage Exhibit of Beckman-Coulter, Fullerton, California. We appreciate her help.

Fig. 18.3. The Beckman–Spinco model 120M automated amino acid analyzer, introduced in 1963. The main differences from the original 1958 version were improvements in the ion-exchange resins and in the technology; differences in the appearance of the instruments were minimal. (Courtesy of Ms. Pat Ashton, Heritage Exhibit of Beckman-Coulter Co.)

now be analyzed in 11 h instead of the two days required originally in 1958; for a simpler protein hydrolyzate, the analysis time could be reduced to 4 h or even shorter. (In Figs. 18.4 and 18.5 the analytical conditions are not detailed.)

Spinco introduced additional models of the amino acid analyzer, further reducing analysis time. Soon other instrument companies also entered the field, most notably Hitachi (Japan) whose amino acid analyzer was mainly based on the investigations of H. Hatano (1924–1998), professor at Kyoto University.

The present stand of amino acid analysis by ion-exchange chromatography is best illustrated by Fig. 18.5 showing the chromatogram of a complex physiological fluid sample, with a total analysis time of 2 h. This chromatogram was obtained at the Chemical Laboratories

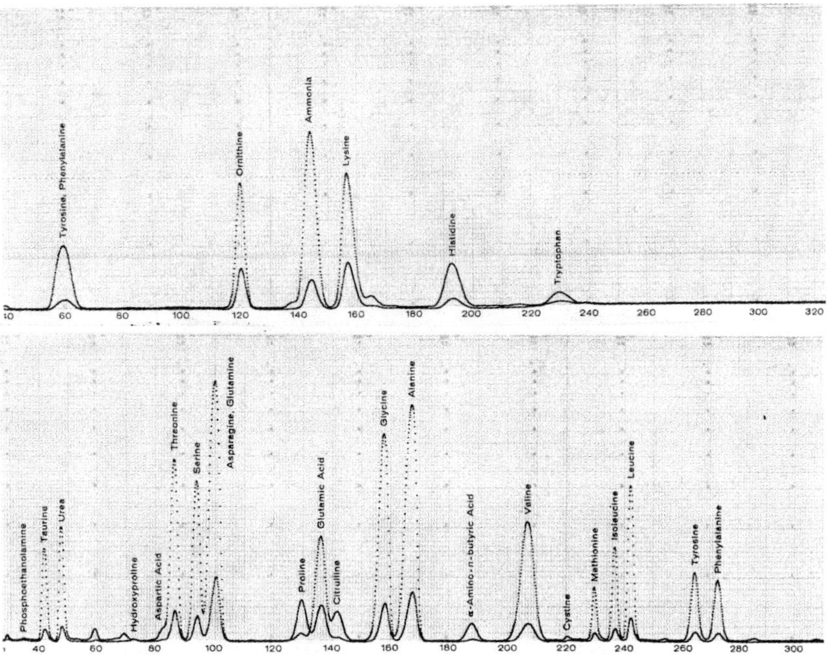

Fig. 18.4. Improved chromatogram of a physiological fluid, obtained with the Beckman–Spinco model 120C (1966). The (lower) main chromatogram was obtained at 570 nm, while the second (upper) chromatogram at 440 nm. Total analysis time was 11 h. At that time a simpler protein hydrolyzate could be analyzed in 4 h. (Courtesy of Ms. Pat Ashton, Heritage Exhibit of Beckman-Coulter Co.)

of Missouri State Experiment Station at the University of Missouri (Columbia). These laboratories handle about 1300 samples each month; depending on the sample complexity each run takes 10 min to about $2^1/_2$ h. These data illustrate how advanced automated amino acid analysis has become.

18.3. Other Methods

Today ion-exchange chromatography is not the only method used for the determination of amino acids: it is complemented by gas (GC) and liquid (LC) chromatography, and capillary electrophoresis (CE).

The use of GC for amino acid analysis became possible by the development of methods to prepare stable volatile derivatives, most

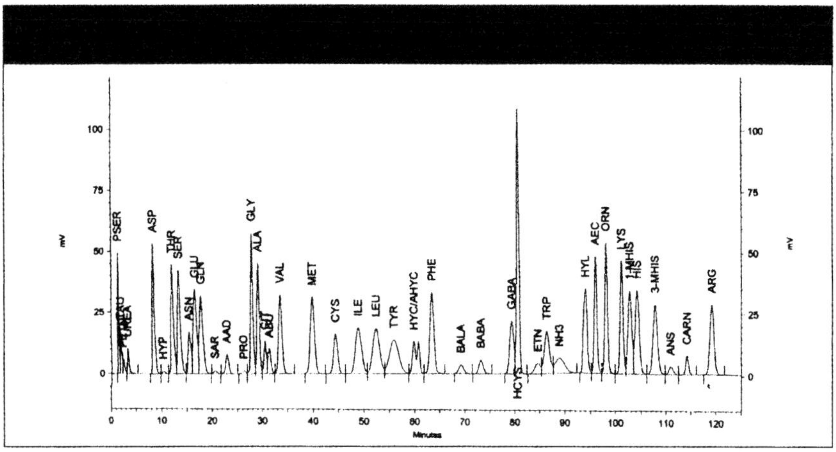

Fig. 18.5. Chromatogram of a complex physiological fluid, obtained on a Hitachi L8800 Amino Acid Analyzer. This figure shows the recording at 570 nm; the parallel chromatogram recorded at 440 nm is not shown. The analysis time was 2 h. For peak identification see Table 18.1. (Courtesy of Dr. Thomas P. Mawhinney, Missouri State Experiment Station, Columbia, Missouri.)

Table 18.1. Identification of the amino acids in Fig. 18.5.

Symbol	Amino acid	Symbol	Amino acid
PSER	Phosphoserine	LEU	Leucine
TAU	Taurine	TYR	Tyrosine
PETN	Phosphoethanolamine	HYC	Cystathionine
UREA	Urea	AHYC	Allocystathionine
ASP	Aspartic acid	PHE	Phenylalanine
HYP	Hydroxyproline	BALA	β-Alanine
THR	Threonine	BABA	β-Amino isobutyric acid
SER	Serine	GABA	ε-Amino butyric acid
ASN	Asparagine	HCYS	Homocystine
GLU	Glutamic acid	ETN	Ethanolamine
GLN	Glutamine	TRP	Tryptophan
SAR	Sarcosine	NH_3	Ammonia
AAD	α-Amino adipic acid	HYL	Hydroxylysine
PRO	Proline	AEC	S-2-Amino ethyl *L*-cysteine
GLY	Glycine	ORN	Ornithine
ALA	Alanine	LYS	Lysine
CIT	Citrulline	1-MHIS	1-Methylhistidine
ABU	α-Amino butyric acid	HIS	Histidine
VAL	Valine	3-MHIS	3-Methylhistidine
MET	Methionine	ANS	Anserine
CYS	Cystine	CARN	Carnosine
ILE	Isoleucine	ARG	Arginine

notably the *N*-trifluoroacetyl *n*-butyl esters.[31,32] This work culminated in the investigation of lunar samples from Apollo 11 through 17 missions, for the possible presence of amino acids.[33,34]

In LC the key step in sample treatment involved the preparation of less polar derivatives permitting the use of reversed-phase chromatography and enhancing the possibilities of UV, fluorescence, and mass spectrometric detection. More recently, methods for the LC analysis of underivatized amino acids, with mass spectrometric detection, have also been developed. Today, automated LC systems for amino acid analysis are available.

The use of CE for amino acid analysis is relatively new and is still growing. CE is very tolerant to biological fluids and less or even no sample treatment is necessary. UV absorbance, fluorescence, and electrochemical detection as well as combination with mass spectrometry are used.

With their pioneering work in the 1950s Moore and Stein opened up an entirely new field for biochemists and it can rightly be considered as one of the great milestones in the century old evolution of chromatography.

References

1. S. Moore, D. H. Spackman and W. H. Stein, *Anal. Chem.* **30**, 1185–1190 (1958).
2. D. H. Spackman, W. H. Stein and S. Moore, *Anal. Chem.* **30**, 1190–1206 (1958).
3. B. Helferich, Emil Fischer, in *Great Chemists*, ed. E. Farber (Interscience, New York, 1961), pp. 981–995.
4. J. P. Greenstein and M. Winitz, eds., *Chemistry of the Amino Acids* (Wiley, New York, 1961), Vol. 2, pp. 1299–1365.
5. A. J. P. Martin and R. L. M. Synge, *Biochem. J.* **35**, 1358–1368 (1941).
6. R. Consden, A. H. Gordon and A. J. P. Martin, *Biochem. J.* **38**, 224–232 (1944).
7. R. L. M. Synge, in *Nobel Lectures Including Presentation Speeches and Laureates' Biographies — Chemistry 1942–1962* (Elsevier, Amsterdam, 1964), pp. 374–387.

8. F. Sanger, in *Nobel Lectures Including Presentation Speeches and Laureates' Biographies — Chemistry 1942–1962* (Elsevier, Amsterdam, 1964), pp. 544–556.
9. M. Bergmann and W. H. Stein, *J. Biol. Chem.* **128**, 217–232 (1939).
10. S. Moore, W. H. Stein and M. Bergmann, *Chem. Rev.* **30**, 423–432 (1942).
11. S. Moore and W. H. Stein, *J. Biol. Chem.* **150**, 113–130 (1943).
12. S. Moore, in *Biographical Memoirs of the National Academy of Sciences* (National Academy Press, Washington, DC, 1986), Vol. 56, pp. 415–439.
13. S. Moore and W. H. Stein, in *75 Years of Chromatography — a Historical Dialogue*, eds. L. S. Ettre and A. Zlatkis (Elsevier, Amsterdam, 1979), pp. 297–308.
14. S. Moore and W. H. Stein, in *Nobel Lectures Including Presentation Speeches and Laureates' Biographies — Chemistry 1971–1980*, eds. T. Frängsmyr and S. Forsén (World Scientific Publishing Co., Singapore, and River Edge, NJ, 1993), pp. 73–95.
15. R. L. M. Synge, *Biochem. J.* **38**, 285–294 (1944).
16. R. L. M. Synge, *Biochem. J.* **39**, 363–365 (1945).
17. S. Moore and W. H. Stein, *Ann. N.Y. Acad. Sci.* **49**, 265–278 (1948).
18. W. H. Stein and S. Moore, *J. Biol. Chem.* **176**, 337–365 (1948).
19. S. Moore and W. H. Stein, *J. Biol. Chem.* **176**, 367–388 (1948).
20. S. Moore and W. H. Stein, *J. Biol. Chem.* **178**, 53–77 (1949).
21. W. H. Stein and S. Moore, *J. Biol. Chem.* **178**, 79–91 (1949).
22. W. E. Cohn, *J. Biol. Chem.* **186**, 77–84 (1950).
23. W. E. Cohn, *J. Amer. Chem. Soc.* **72**, 1471–1478 (1950).
24. S. M. Partridge, *Biochem. J.* **44**, 521–527 (1949).
25. C. H. W. Hirs, Polypeptides and Proteins; Analysis, in *The Beckman Symposium on Biomedical Instrumentation*, ed. C. L. Moberg (Rockefeller University, New York, 1986), pp. 67–73.
26. S. Moore and W. H. Stein, *J. Biol. Chem.* **192**, 663–681 (1951).
27. P. B. Hamilton, *Anal. Chem.* **30**, 914–919 (1958).
28. D. H. Spackman, W. H. Stein and S. Moore, *Federation Proc.* **15**, 358 (1956).
29. M. J. Gordon, The Rockefeller University — Beckman Instruments Relationship, in *The Beckman Symposium on Biomedical Instrumentation*, ed. C. L. Moberg (Rockefeller University, New York, 1986), pp. 31–35.
30. P. B. Hamilton, *Anal. Chem.* **45**, 2055–2064 (1963).
31. W. M. Lamkin and C. W. Gehrke, *Anal. Chem.* **37**, 383–389 (1965).

32. C. W. Gehrke and D. L. Stalling, *Separ. Sci.* **2**, 101–138 (1967).
33. C. W. Gehrke, R. W. Zumwalt, K. C. Kuo, C. Ponnamperuma and A. Shimoyama, *Origin of Life* **6**, 541–550 (1975).
34. C. W. Gehrke, Chromatography in Space Sciences, in *Chromatography: a Century of Discovery 1900–2000*, eds. C. W. Gehrke, R. L. Wixom and E. Bayer (Elsevier, Amsterdam, 2001), pp. 83–97.

Part Seven

Gas Chromatography

Chapter

19

Early Development of Gas Adsorption Chromatography*

Books or papers on gas chromatography usually imply that the technique was invented by James and Martin, citing their fundamental paper published in 1952.[1] This is, however, only true if we consider GC based on partitioning between the two phases: gas *adsorption* chromatography, where separation is based on selective adsorption–desorption processes, already existed at least one decade before gas partition chromatography, although its field was fairly limited. Some of the early work was actually not called "chromatography": at that time this term was generally restricted to separation in small glass tubes according to Tswett's original methodology, with elution serving as the basis for the passage of the molecules. Only after the work of Arne Tiselius in the first part of the 1940s became it clear that

*Based on two articles by L. S. Ettre published in *Chromatographia* 55, 497–504 and 625–631 (2002), and on discussions with Professors S. Claesson (Uppsala University), G. Hesse (University of Erlangen/Nünberg), and Erika Cremer (University of Innsbruck).

"chromatography" actually encompasses three techniques: besides elution, also frontal analysis, and displacement.

Among the pioneers utilizing gas adsorption chromatography for the first time, we shall mention Turner and the instrument he helped to develop, as well as Claesson in Sweden, and Hesse in Germany. In addition, we shall discuss the activities of Erika Cremer at the University of Innsbruck, Austria, who for the first time developed a true "gas chromatograph" that had essentially the same components as our present-day instruments, and the pioneering work of C.S.G. Phillips of Oxford University.

19.1. Analysis of Natural Gas

Besides methane, natural gas also contains small amounts of other low-boiling hydrocarbons, and knowledge of their concentration is of interest to the petroleum chemists. For their determination the so-called *charcoal test* had been developed in the early 1920s in USA.[2] In this method the gas sample was passed through a tube containing charcoal; the adsorbed amounts were gradually displaced by glycerol and measured volumetrically, after collection. Later, N. C. Turner further advanced the technique and together with the Burrell Corporation of Pittsburgh, PA, developed an instrument, the *Turner–Burrell Adsorption Fractionator*, introduced in 1943.[3,4] (The Burrell Technical Supply Corporation — later called simply the Burrell Corporation — was founded after the First World War by Guy Burrell, aiming to supply gas analysis equipment for the industry.) This instrument was a floor-standing monster, including a 6-ft long vertical column filled with charcoal, a sample handling system, a thermal-conductivity detector, and a potentiometric recorder. The column's diameter was gradually reduced along its length, from 3/4 in. to 3/16 in., and it was surrounded by a small heater which could be moved from the bottom to the top. A mercury reservoir was attached to the heater, moving with it along the column (Fig. 19.1). The operation of the system was fairly complicated and we only summarize it very briefly. After introducing a 2–5 L gas sample into the column, the temperature of the heater was raised to 750 F (399°C) and it was very slowly moved, together with

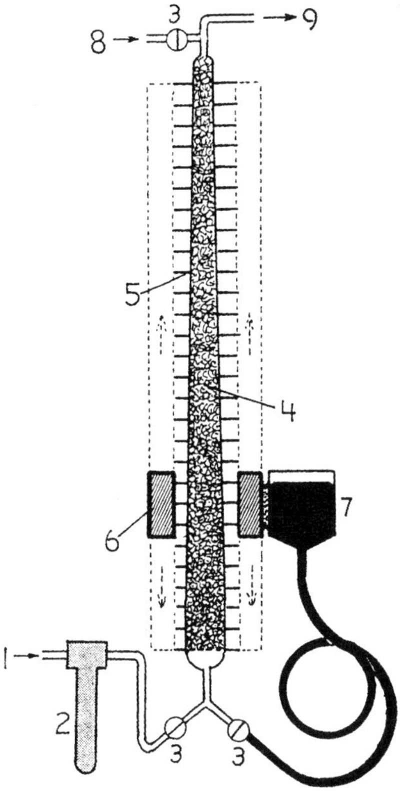

Fig. 19.1. The column assembly of the Turner–Burrell adsorption fractionator.[3,4] 1 = Sample inlet; 2 = flow meter; 3 = two-way valves; 4 = separation column (6-ft long; i.d., at bottom: 3/4 in., at top: 3/16 in.), filled with charcoal; 5 = circumferential fins for better heat exchange; 6 = movable coaxial heater; 7 = mercury reservoir, moving with the heater; 8 = hydrogen inlet (to fill the column during the cooling period); 9 = thermal-conductivity detector.

the mercury reservoir, upward: it took 8 h until it reached the top of the column. With the heater, the mercury level in the column also slowly raised, and the part under the heater evaporated (mercury's boiling point is 356.7°C), its vapor gradually displacing the adsorbed hydrocarbons. Since above the heater the column was unheated, the mercury vapor gradually condensed, flowing downward, while the mercury below the heater area remained liquid. The upward moving massive mercury volume slowly pushed the displaced gases out of the

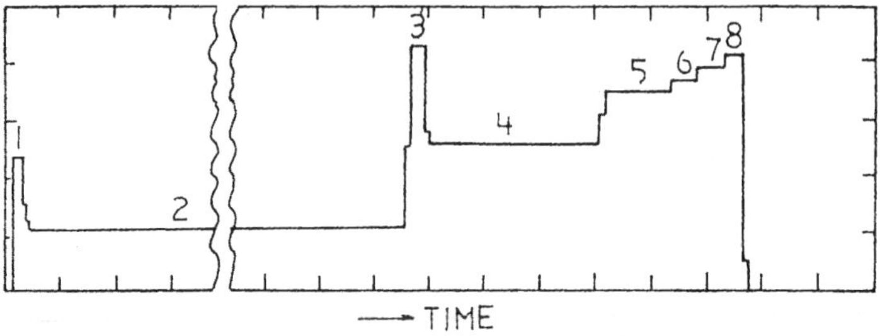

Fig. 19.2. Typical recording with the Turner–Burrell adsorption fractionator.[3,4] The individual plateaus correspond to 1 = air; 2 = methane; 3 = carbon dioxide; 4 = ethane; 5 = propane; 6 = isobutane; 7 = *n*-butane; 8 = pentanes. The ordinate (*y*-axis) represents relative thermal conductivities. In this particular case 5 L dry natural gas served as the sample, with the following composition (in mole%): methane: 86.39%, ethane: 9.00%, propane: 3.26%, *n*- and isobutane: 1.24%, pentanes: 0.1%. Total analysis time was about 8 h.

column and the column effluent was conducted to the thermal conductivity detector, recording a step-wise "chromatogram" (Fig. 19.2).

In the next decade the instrument was somewhat simplified by reducing the sample size to about 300–400 mL and replacing mercury with tetrachloroethylene (boiling point: 121°C), also permitting the heater's temperature to be reduced. Elimination of mercury vapor was certainly one of the reasons for changing the system, although I would not imagine that today they would permit the use of C_2Cl_4 vapors in a lab! The modified instrument was called the *Fracton* and was introduced in 1953; however, it was very short lived: it was replaced within a couple of years by gas chromatography.

19.2. Claesson's System

About the time of the development of the Burrell instrument another GC system based on displacement chromatography was developed at Uppsala University, Sweden, in the laboratory of Arne Tiselius, by Stig Claesson (1917–1988), his graduate student. Claesson's work involved detailed studies on adsorption, both from gas and liquid samples, and this system was constructed as a tool to facilitate his studies.

Claesson's "instrument" was also very elaborate, representing a typical graduate-student designed system (Fig. 19.3) with complicated

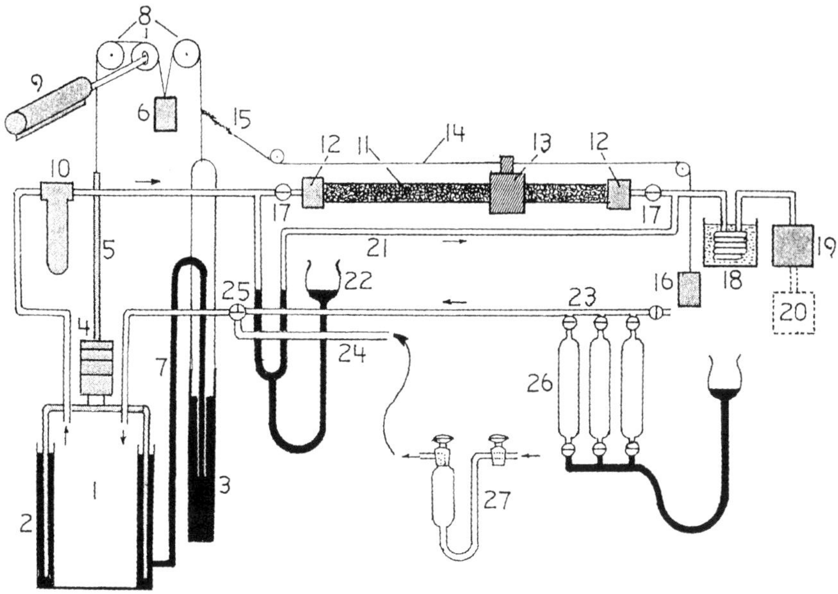

Fig. 19.3. Claesson's system for displacement-type analysis.[5] 1 = Gas holder (40 × 20 cm i.d.); 2 = mercury-filled trough; 3 = mercury-filled tube; 4 = weights placed on the top of the gas holder; 5 = rod attached to the gas holder; 6 = counterweight; 7 = siphon connecting tube No. 3 with the trough; 8 = wheels; 9 = drum covered with photographic paper, to register galvanometer deflection; 10 = flow meter; 11 = separation column (quartz, 400 mm × 5–18 mm i.d.) filled with activated charcoal; 12 = water-cooled metal holders of the column; 13 = movable small oven; 14 = cord connecting the movable heater to the spring; 15 = spring; 16 = weight; 17 = two-way stopcocks; 18 = cooling spiral in a 20°C thermostat; 19 = thermal-conductivity detector; 20 = gas density meter (optional); 21 = glass tube for column by-pass; 22 = vessel containing mercury; 23, 24 = glass tubes to transfer the sample (gas or vapor) to the gas holder; 25 = three-way stopcock; 26 = inlet system for gas samples, with removable gas burettes; 27 = system to evaporate a liquid sample or the displacer (connected to tube No. 24).

hand-made components. This is particularly true about the mobile phase supply (a large tank the top of which was "swimming" in a mercury trough), the flow regulation system which was synchronized with the movement of the oven surrounding the column, and the recording system which consisted of a thermal-conductivity detector, a galvanometer, and rotating photographic paper. Separation was carried out in 40-cm long, 5–18 mm i.d. columns filled with activated charcoal and surrounded by a small concentric heater which could be moved along the column to speed up displacement.

Claesson's thesis entitled Studies on Adsorption and Adsorption Analysis with Special Reference to Homologous Series was published in 1946 in English as a separate issue of *Arkiv för Kemi, Mineralogi och Geologi*, the journal of the Swedish Academy of Science[5]: it also contains a photo of the set-up, which occupied a whole room. According to my best knowledge, no further work was done with it after he finished his thesis work.

When I visited Dr. Claesson in Uppsala, around 1977–1979 (he remained associated with the university after graduation, and eventually become distinguished professor of physical chemistry), he told me an interesting story about the aftermath of his system. In 1946 or 1947 Dr. Arnold Beckman, the founder of Beckman Instruments, visited Professor Tiselius, to see whether they had any interesting developments his company could utilize, and Tiselius introduced Claesson who then explained his system in detail. Dr. Beckman was very interested and asked for copies of Claesson's reports so that he could take them back to California and discuss them with his development engineers. Claesson was already scheduled to fly to California within a couple of months on a post-doctoral fellowship, and so they agreed that after his arrival he should visit Beckman Instruments to discuss eventual cooperation. Indeed, soon after his arrival Claesson called Dr. Beckman who sent a car to pick him up. Claesson anticipated some nice consultantship, but was disappointed when Dr. Beckman very politely told him that after studying his reports, his engineers concluded that there is no practical future in gas chromatography and Claesson's system was just too complicated (which it was indeed...). Thus, the commercialization of GC had to wait one more decade...

19.3. Gerhard Hesse

Gerhard Hesse (1908–1997) was principally an organic chemist, who was mostly interested in complex substances present in plants and had been using (classical) liquid chromatography since his graduate student time. In 1938 he became professor at the University of

Marburg/Lahn, in Germany, moving in 1944 to the University of Freiburg/Breisgau, and then in 1952 to the University of Erlangen/Nürnberg where he was active for decades until his retirement.

While still in Marburg, Hesse had already written an important textbook on (classical) liquid chromatography[6] and after settling in Erlangen, he established intensive training courses in liquid chromatography for practicing chemists. Thus, he served as the teacher of dozens of German scientists who later became prominent in the various branches of chromatography.

Around 1940 Hesse was unsuccessful in the separation of the C_5–C_6 saturated and unsaturated fatty acid esters of α-lactucerol, a steroid. He tried to saponify the compounds and separate the obtained fatty acids by distillation (the routine tool of an organic chemist), but this was only partially successful. Based on his experience in liquid chromatography, he assumed that a similar technique but using a gas as the mobile phase instead of a liquid should be helpful in the analysis of volatile compounds (this was before the famous prediction of Martin and Synge in their 1941 paper!). This idea was in contradiction to the then prevailing opinion among German physico-chemists that, due to rapid longitudinal diffusion, one could not obtain permanently separated zones in a moving gas stream. It is said that this assumption originated from Arnold Eucken (1884–1950), professor at the University of Göttingen since 1929 and the doyen of German physicochemists. Ewald Wicke (1914–2000), one of Eucken's students and a long-time collaborator, specifically expressed this opinion in his lecture Fundamentals of Separation Techniques Based on Sorption held on 5 April 1940, stating that "the use of the chromatographic method with a gas as the means of elution seems without prospects owing to the mixing in the direction of flow." This lecture was held at the prestigious German State Office for Economic Development *(Reichsamt für Wirtschaftsausbau)*, and this emphasized even more its authenticity. (My attention to this lecture was drawn by Professor Cremer.)

Fortunately, Hesse was an organic chemist who always tried an experiment without first theorizing, and thus he set up a simple experiment to see whether he can separate bromine and iodine vapor on a glass column filled with starch, using nitrogen as the carrier gas and visually observing the color of the slowly separated zones. As he expected, the brown (bromine) and blue (iodine) zones became separated and remained so. This showed that he was correct and the assumptions of others were without foundation.

When investigating the systems constructed by the early researchers, they all reflect the main field of the individuals: thus, the system built by Hesse is typical for an organic chemist, with distillation flasks and condensers galore (Fig. 19.4). The center tube of a laboratory condenser served as the chromatographic column, and its jacket as the thermostat; the column was packed with silica gel. The sample was placed in a distillation flask and its vapor carried through the column by a continuously flowing inert gas (the mobile phase), usually carbon dioxide, and the separated fractions were collected in a cooled trap. As reported in Hesse's two papers published in 1941–1942,[7,8]

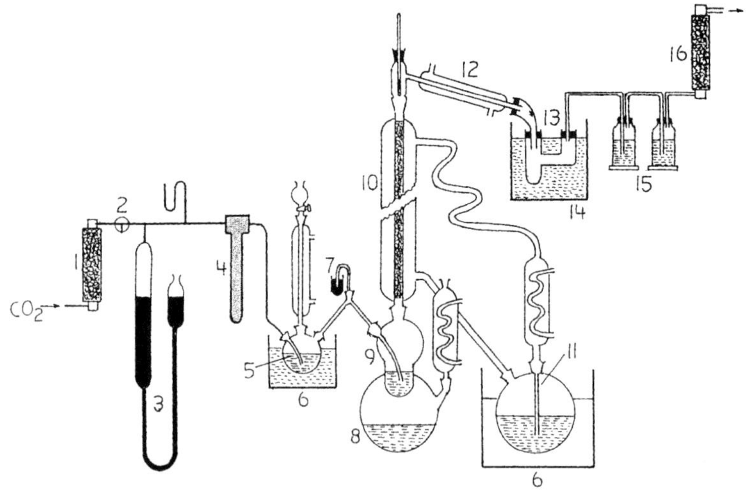

Fig. 19.4. The system of Hesse.[8] 1 = Adsorber to clean the carrier gas; 2 = three-way stopcock; 3 = mercury pressure regulator; 4 = flow meter; 5 = system to add vapor to the carrier gas (optional); 6 = constant temperature bath; 7 = mercury pressure valve; 8 = constant temperature vapor bath; 9 = flask containing the sample; 10 = separation column with vapor jacket; 11 = system to produce vapor to heat the separation column; 12 = condenser; 13 = fraction collector trap; 14 = Dewar flask; 15 = scrubbers; 16 = adsorbing the carrier gas and added vapor not to pollute the laboratory atmosphere (if needed).

he successfully separated volatile isomeric saturated and unsaturated fatty acids, various azeotropic mixtures, and substances having the same or very close boiling points. A particular merit of Hesse was to clearly recognize that what he did was a variant of "the analysis according to Tswett (chromatographic analysis)."

Looking with our present-day knowledge it is surprising that Hesse did not include a detector in his system. However, let us not forget that his aim was to obtain the individual sample components in pure form, and for this fraction collection was the logical solution.

I should mention an amusing story told me by Professor Hesse with respect to one of his papers.[8] Soon after submitting the manuscript to the editors of *Naturwissenschaften*, he was called to the army and served on the Russian front. However, the printer found him, and he received the proofs of the paper on the front through the field post!

19.4. The First Real Gas Chromatograph of Cremer

The ACHEMA is the world's largest chemical exposition held every third year in Frankfurt am Main, Germany. There, hundreds of companies show their process equipment and introduce new industrial techniques, with many tens of thousands of visitors from the whole world. The organizers of the 1952 show decided to try something new: they provided free booths to selected university laboratories where they could show their newly developed systems which might represent the basis of future industrial apparatus and processes. One of the participating laboratories was the Physico-Chemical Institute of the University of Innsbruck, lead by Professor Erika Cremer, and they exhibited a real gas chromatograph that consisted of a carrier gas source, a sample inlet system, a thermostatted separation column packed with silica gel, and a thermostatted thermal-conductivity detector.[9,10] It was undoubtedly the first gas chromatograph shown publicly: however, at that time nobody was interested in it.

Erika Cremer (1900–1996) was a physico-chemist by training who joined Innsbruck University in 1940. There she eventually became involved in research in the hydrogenation of acetylene. One of the problems encountered was the determination of the amounts of the two gases in a mixture. Their adsorption heats differ only very little and she realized that classical adsorption techniques (similar to those used at that time in the German industry for the separation of noble gases) could not help. While involved in these considerations, Cremer read the recently published book of Hesse on (liquid) chromatography[6] and this gave her the idea to investigate whether a chromatographic process similar to liquid chromatography, but using an inert gas as the eluent (the mobile phase), would not solve such problems and help in the determination of the adsorption heats of volatile substances with only very little difference. Being a theoretical chemist, she first summarized her thoughts in a short paper, with the appropriate equations, and in November 1944, submitted it to the editorial offices of the journal *Naturwissenschaften* (then located in Vienna). Her paper was accepted and in February 1945, she received the proofs which she then duly returned to the printer (she kept a copy until her death 50 years later). Shortly afterward the printing office in Vienna was destroyed in an air raid and her paper remained unpublished until 1976, when it was published as a historical document.[11]

In December 1944 buildings of the University of Innsbruck were heavily damaged in an air raid, and from then on regular activities became practically impossible. By the end of the war, life returned only very slowly to normal, but complications due to the different occupation zones (Innsbruck was now in the French zone of Austria) were severe for some time, and life returned to more-or-less normal circumstances only very slowly. The university could open in the fall of 1945, although still in temporary quarters and with today's standards, under primitive conditions; the students returned only slowly. Then a young man named Fritz Prior reported to Cremer with the aim of obtaining a Ph.D. degree, and she selected her 1944-idea as his thesis subject, with the aim of separating volatile compounds with small difference in their adsorption heats by gas chromatography. It took time to gather the items needed to assemble the necessary equipment — at that time it was not easy to find even the proper glass tubes — but what was most important, they found in the ruins of the institute parts of a thermal conductivity detector she started to build in 1944, and this became the most valuable part of Prior's system. Finally,

Fig. 19.5. Erika Cremer and Fritz Prior in 1977 (author's collection).

the experiments showed the validity of Cremer's assumption and the separation of simple binary gas mixtures could be achieved.

Prior finished his work in May 1947 (Fig. 19.5). Only its summary could be printed as a small booklet (the "little red book"), and the full text of his thesis entitled Determination of Adsorption Heats of Gases and Vapors, Using the Chromatographic Method in the Gas Phase was prepared only in typed copies deposited at the University's and the Institute's libraries (I have a copy of it). Figure 19.6 shows the system of Prior — essentially the same was exhibited at the 1952 ACHEMA — and Fig. 19.7 is one of the chromatograms of the Innsbruck group, showing the separation of air, ethylene, and acetylene. It actually had to be constructed from galvanometer readouts (represented by many dots), because no recorder was available at that time.

It should be mentioned that although the primary aim of Prior's thesis was to prove Cremer's assumption about determining adsorption heats and separation by GC, he went further: he also discussed the possibility of using GC for the quantitative determination of small

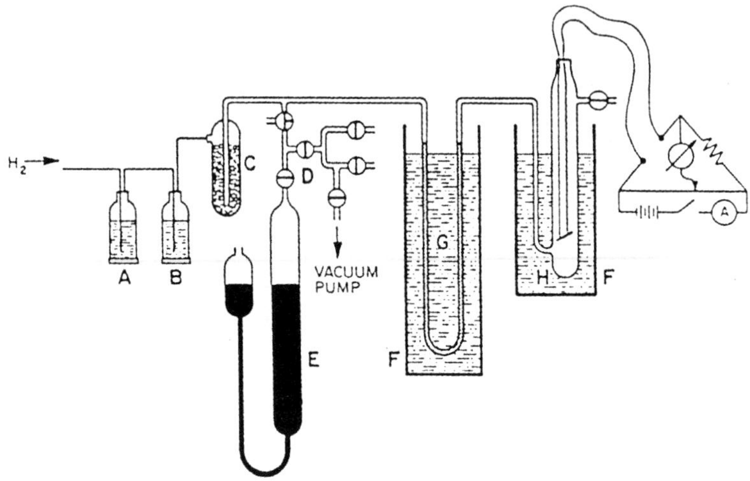

Fig. 19.6. The GC system constructed as part of Prior's thesis work.[14] A = Permanganate scrubber; B = concentrated sulfuric acid scrubber; C = adsorbent for carrier gas purification; D = sample inlet system; E = burette containing mercury with niveau glass, for sample introduction; F = thermostat; G = separation column filled with silica gel; H = thermal-conductivity detector.

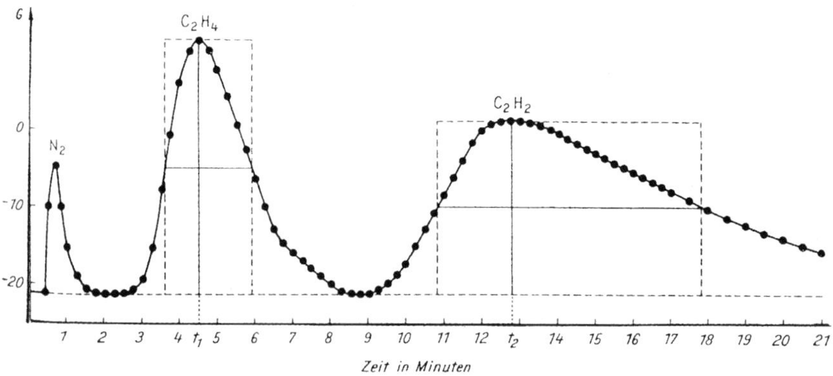

Fig. 19.7. Chromatogram from the work of E. Cremer's group.[13] Peaks: nitrogen (air), ethylene, acetylene. Ordinate: galvanometer deflection. Column packing: silica gel; carrier gas: hydrogen. The values of the retention times, peak heights and peak widths at half height are also indicated; the product of peak height and width at half height is proportional to the amount present.

amounts of volatile substances, and has shown calibration plots plotting peak height vs. sample amount.

Prior's work clearly showed that they found a new and promising method for physico-chemical measurements and qualitative and quantitative analysis, but it also left some unanswered questions. Meanwhile in 1947 another graduate student, Roland Müller, joined Professor Cremer and he then continued the investigations, now concentrating on the analytical aspects. In 1950 Müller finished his thesis entitled Application of the Chromatographic Method for the Separation and Determination of Very Small Amounts, reflecting this change in the emphasis.

In the first years after the war, publication of scientific papers and even organization of scientific meetings was practically impossible in Austria (and Germany). One of the first meetings in Austria was held by the newly organized Austrian Chemical Society in May 1949 in Linz, and there Cremer presented a paper summarizing Prior's results.[12] She also was able to participate at the meeting of the *Bunsengesellschaft* (the German Physico-Chemical Society) held in May 1950, in Marburg/Lahn, Germany, where again she presented a brief paper.[13] This was then followed by a more detailed presentation at the First Microchemical Congress held in July 1950, in Graz, Austria, summarizing the results of both Prior and Müller. By then they could submit manuscripts to scientific journals and their detailed reports were finally published in 1951 in the *Zeitschrift für Elektrochemie*, the journal of the *Bunsengesellschaft*,[14,15] and in the collection of the papers presented at the Graz meeting.[16]

It belongs to our story that Cremer's presentations received a negative response or no response at all. In retrospective we may explain this by the fact that she spoke to the wrong people and at the wrong meetings. In Austria, analytical chemistry by tradition meant mainly classical microchemistry of the Pregl-Emich type and the microchemists were certainly uninterested in the analysis of gases; also, the physicochemists assembled in the *Bunsengesellschaft* had other interests. We should also mention that the *Zeitschrift für Elektrochemie* published in German was little known outside Germany and Austria, particularly by analytical or petroleum chemists. Professor Cremer jokingly liked to mention one additional reason why nobody was interested in her presentation in Graz: she presented her paper in the afternoon of 5 July 1950, a day when the outside temperature reached the highest value ever recorded in Graz: 37.2°C (99F). The university lecture room, with its closed windows, resembled more an oven than a lecture room, and obviously not too many participants were willing to endure the conditions.

Professor Cremer's achievements were recognized only later, after the introduction of gas–liquid partition chromatography, when gas chromatography suddenly became the most exciting new method, but even then only slowly. This is best illustrated by the fact that the basic textbook on gas chromatography written by A. I. M. Keulemans and published in 1957[17] completely ignored her work (and in fact, even gas adsorption chromatography). This deficiency was finally corrected in the second edition with a new chapter aiming "to do fuller justices to the work of Professor Erika Cremer of Innsbruck, Austria, and her school," and she was then asked to write a special supplement to the German edition of the book.[18] Cremer remained active in gas chromatography until practically the end of her long life, and in 1990, on the occasion of her 90th birthday, a special symposium held in Innsbruck honored her achievements.

19.5. C. S. G. Phillips

The last scientist we should mention among those pioneering in gas adsorption chromatography is C. S. G. Phillips of Oxford University, England. He represents a direct transition to the work of A. T. James and A. J. P. Martin and gas–liquid partition chromatography (GLPC).

Courtenay S. G. Phillips (born 1924) started, also in 1946, investigations on the separation of hydrocarbons by adsorption–desorption on a charcoal column, the temperature of which was controlled by a vapor jacket, using a thermal-conductivity detector to monitor the column effluent. His original plan was to use elution chromatography, analogous to liquid chromatography; however, meanwhile he read Claesson's work using displacement chromatography and, after personally consulting him, he modified his original design to

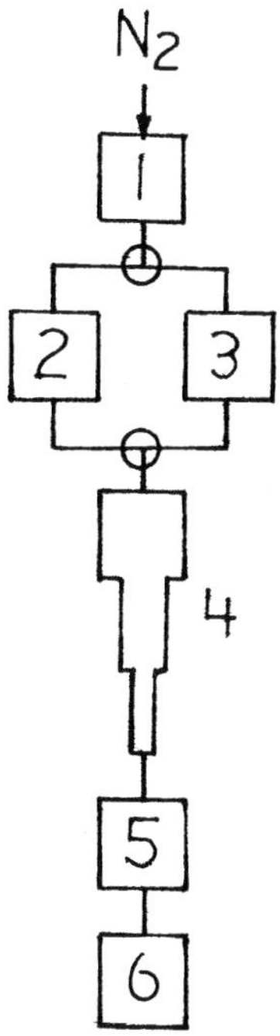

Fig. 19.8. Block diagram of the system used by C. S. G. Phillips. 1 = Flow controller; 2 = sample inlet; 3 = saturator; 4 = segmented column; 5 = thermal-conductivity detector; 6 = fraction collector.

accommodate this technique. Figure 19.8 shows the block diagram of Phillips' system: the saturator (a jacketed tube containing the liquid displacer through which the carrier gas bubbled through) was needed to add the displacer vapors to the carrier gas. Figure 19.9 shows his column arrangement with the unique segmented column.[19] The use of this was due to his observation that the smaller the column diameter the sharper the displacement fronts. In his system the last, small-diameter segment took care of this, while the earlier larger-diameter segments still permitted the introduction of reasonable sample amounts.

Phillips presented the first report on his work at the famous 1949 Conference of the (British) Faraday Society[20] (see Chapter 29). In it, he demonstrated both qualitatively and quantitatively the separation and determination of lower hydrocarbons, and thus we may consider

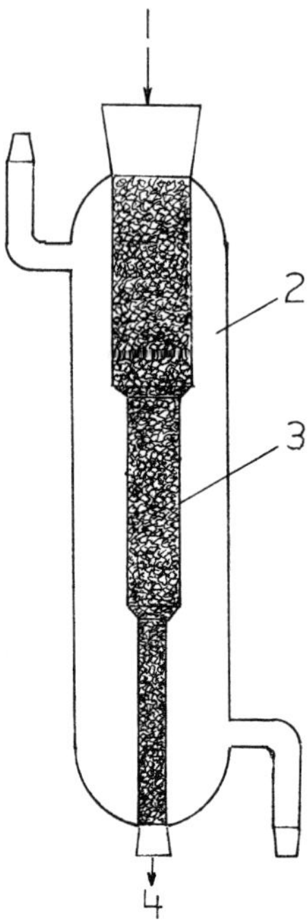

Fig. 19.9. The column assembly used by C. S. G. Phillips.[19] 1 = Carrier gas inlet; 2 = vapor-jacket surrounding the column; 3 = segmented column packed with activated charcoal; 4 = exit to thermal-conductivity detector and fraction collector. The length of each segment of the column was 10 cm, with the respective diameters of 16, 8, and 2 mm.

this presentation as the first thorough report on the use of GC for the analysis of this group of compounds.

Soon after the Faraday Symposium, in 1950–1951, when Phillips became aware of the work of James and Martin on GLPC (well before their publications) he immediately switched from the displacement technique to elution GC based on partition, going back to his original ideas. Later in a self-critical assessment Phillips stated that[21]

> ...it soon became all too apparent that (GLPC) was going to be the superior analytical chromatographic technique. In hindsight, of course, this should have been obvious to me when I first started in gas chromatography, but I had just not analysed the problem clearly enough to see it.

In the early 1950s Phillips carried out a number of important investigations on both analytical and non-analytical applications of GC, among them the first use of temperature programming,[22] preceding six years the DuPont team of Dal Nogare[23] which is usually credited with the introduction of this technique. Phillips also wrote the first textbook on gas chromatography published in 1956,[24] representing an excellent summary of the early works of the pioneers.

References

1. A. T. James and R. L. M. Martin, *Biochem. J.* **50**, 679–690 (1952).
2. H. B. Haas, *Nat. Petrol. News* **19**, 251–255 (1927).
3. N. C. Turner, *Nat. Petrol. News* **35**, R234–R237 (1943).
4. N. C. Turner, *Petrol. Refiner* **22**(5) 140–144 (1943).
5. S. Claesson, *Ark. Kem. Mineral. Geol.* **23A**(1), 1–133 (1946).
6. G. Hesse, *Adsorptionsmethoden in chemischen Laboratorium, mit besonderer Berücksichtigung der chromatographischen Analyse (Tswett Analyse)* (W. de Gruyter & Co., Berlin, 1943).
7. G. Hesse, H. Eilbracht and F. Reicheneder, *Liebig's Ann. Chem.* **546**, 233–252 (1941).
8. G. Hesse and B. Tschachotin, *Naturwiss.* **30**, 387–392 (1942).
9. *ACHEMA Nachrichten*, Frankfurt/Main, 1961.
10. O. Bobleter, *Chromatographia* **43**, 444–446 (1996).
11. E. Cremer, *Chromatographia* **9**, 363–364 (1976).
12. *Österr. Chem. Ztg.* **50**, 161 (1949).
13. *Z. Elektrochem.* **55**, 65 (1951).
14. E. Cremer and F. Prior, *Z. Elektrochem.* **55**, 66-70 (1951).
15. E. Cremer and R. Müller, *Z. Elektrochem.* **55**, 217–220 (1951).
16. E. Cremer and R. Müller, *Mikrochem./Mikrochim. Acta* **36/37**, 553–560 (1951).
17. A. I. M. Keulemans, *Gas Chromatography* (Chapman & Hall, London, and Reinhold, New York, 1957), 2nd edn., 1959.
18. A. I. M. Keulemans, *Gaschromatographie.* Translated and supplemented by E. Cremer (Verlag Chemie, Weinheim, 1959).
19. D. H. James and C. S. G. Phillips, *J. Chem. Soc.* 1600–1610 (1953).
20. C. S. G. Phillips, *Disc. Faraday Soc.* **7**, 241–248 (1949).

21. C. S. G. Phillips, in *75 Years of Chromatography — A Historical Dialogue*, eds. L. S. Ettre and A. Zlatkis (Elsevier, Amsterdam, 1979), pp. 315–322.
22. J. H. Griffits, D. H. James and C. S. G. Phillips, *Analyst* (*London*) **77**, 897–904 (1952).
23. S. Dal Nogare and C. E. Bennett, *Anal. Chem.* **30**, 1157–1158 (1958).
24. C. S. G. Phillips, *Gas Chromatography* (Butterworths, London, 1956).

Chapter

20

The Janák-Type Gas Chromatographs of the 1950s*

Modern gas chromatographs have a more or less standard design. The column is in a thermostat and helium, hydrogen or nitrogen is used as the carrier gas. The sample is introduced instantaneously upstream of the column and the column effluent is continuously monitored, measuring some characteristic change (for example, its thermal conductivity) when a separated sample component exits the column. This change is usually recorded in the form of a peak, the area (or height) of which is proportional to the amount of solute present. Naturally, our present-day instruments are more sophisticated than those used 50 years ago. They have a more precise, automated control of both pressures and temperatures, permit programming the column temperature during analysis, and include more sensitive detectors; however,

*Based on the articles by L. S. Ettre published in *LCGC* (*North America*) **20**, 866–874 (2002) and in *LCGC Europe* **15**, 799–804 (2002).

these basic principles have not changed since the introduction of the first commercial instruments 53 years ago.

There were, however, some early systems that differed from this basic design, mainly in two aspects: the method of detection and in conjunction with this, in the choice of the carrier gas. One type of such instruments was quite popular in the 1950s, in laboratories where the main analytical problem was the determination of inorganic gases and volatile C_1–C_4 hydrocarbons. They did not use a detector: instead, the separated sample components were collected and their volume measured. These instruments had a very simple construction and could easily be constructed using standard laboratory hardware.

A principal advantage of these instruments was that it was not necessary to consider the problem of detector response factors which, in a thermal conductivity detector, can be quite different for the components of a sample consisting of both inorganic and organic substances. One simply measured the volume of the collected fractions, which permitted the establishment of sample concentrations directly in volume percent.

These systems were based on the work of Jaroslav Janák, in Czechoslovakia, in 1951–1952. Their development was based on good science and evaluation of the literature; it was also necessitated by the political events in Central Europe.

We have to go back to early 1939, a few months after the Munich agreement that gave the Sudetenland, the German-speaking mountainous area on the western and northern part of Czechoslovakia, to Germany. At that time Germany was already preparing for war and an important part of this preparation was the building of factories producing synthetic gasoline from brown coal. Such large coal deposits were located in the northwestern part of the area annexed by Germany and thus, within a few months, the construction of a large plant started in the town of *Most* (its German name was *Brüx*). It utilized a modified Bergius process (named after Friedrich Bergius who, together with Carl Bosch, received the 1931 Chemistry Nobel Prize for its invention), producing gasoline-type fuel by the high temperature and pressure hydrogenation of brown coal. Production started toward the end of 1942 and by the end of the Second World War, the plant provided a significant part of the needs of the German armed forces. After the war, when the territory was again incorporated into Czechoslovakia, the profile of the plant shifted from gasoline production to the chemical utilization of brown coal and coal tar, to the recovery of phenols from waste water, and to the synthesis of some chemicals. At that time the plant in Most was one of the most modern facilities in East-Central Europe and I remember that in 1950–1952, a number of my colleagues in Hungary visited it to learn the new techniques. (In fact, I was also scheduled for such a visit, but meanwhile I changed my job.)

In 1947 Jaroslav Janák (born 1924), a recent graduate of the Technical University in Prague, joined the Most plant and there he was placed in charge of the microchemical as well as the gas and water analysis control laboratories. At that time gas analysis was carried out by the classical Orsat method which, however, did not determine the individual volatile hydrocarbons; on the other hand, the new, more sophisticated chemical processes now required the monitoring of their concentration. Therefore, more advanced analytical methods were needed. The plant had an extensive library containing not only German journals, but also the newest American publications: even although Germany was in war against the United States, these journals (and even some special chemicals) were obtained from the US through neutral Switzerland until the end of 1944. Janák felt that low-temperature fractional distillation would be the best method for their purpose, and so an order was placed for a unit to the Podbielniak Co. (Chicago, IL) which had a virtual world-wide monopoly on such systems.[1] However, due to the Communist take-over in Czechoslovakia in the spring of 1948, the US government imposed trade restrictions and export of this instrument was denied. Therefore, other ways had to be found for the needed analytical work.

Janák was very familiar with the liquid column and paper chromatographic analysis of mono and diphenols and felt that the chromatographic principle could also be applied to gaseous samples. In two autobiographical treatments[2,3] he described in detail the events leading eventually to the new technique. After studying the available literature, among them the papers of Turner[4,5] and Claesson[6] that described

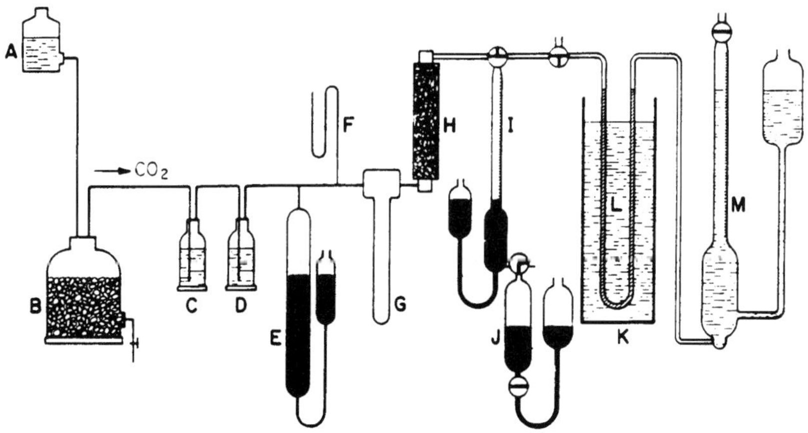

Fig. 20.2. The basic GC system of Janák.[15] A = HCl container; B = marble for the production of CO_2; C = $NaHCO_3$ scrubber; D = conc. H_2SO_4 scrubber; E = mercury pressure regulator; F = manometer; G = flow meter; H = $CaCl_2$ scrubber; I = gas microburet (for sample introduction); J = gas storage unit; K = column thermostat (Dewar flask); L = chromatographic column; M = nitrometer.

a plot of collected gas volume vs. time and read the heights of the individual plateaus. An example is given in Fig. 20.3.

Soon after the development of the new technique Janák started to publish detailed reports on it, its variations and applications. The first paper was published in 1953 and in the next six years, a total of 18 papers were authored and coauthored by him on this subject. All these papers were first published in Czech, in *Chemické Listy*, the journal of the Czechoslovak Chemical Society; subsequently, their German (in two cases, Russian) translations were published in the *Collection of Czechoslovak Chemical Communications*, an international scientific journal founded in 1929 and published ever since. Table 20.1 lists the titles of these papers, indicating the wide variety of samples analyzed.

In his original work Janák used activated charcoal and silica gel as the column packing for the separation of the inorganic gases and gaseous hydrocarbons. Later, zeolites ("molecular sieves") as well as partition-type columns prepared with various liquid phases also were utilized.

Janák's GC almost immediately aroused a lot of interest in Europe, and he was invited to participate in a number of symposia and to visit

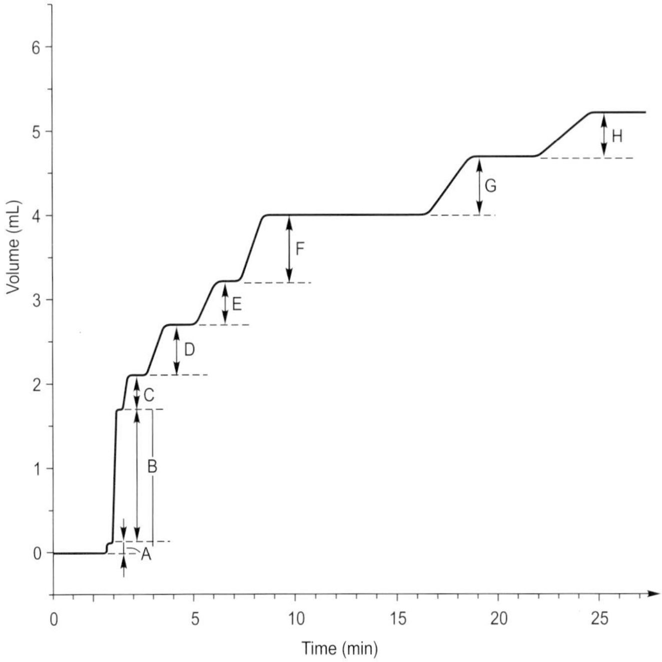

Fig. 20.3. Graphical presentation of the volumes of the collected fractions vs. time. Total sample volume: 5.2 mL. Sample composition (concentrations in vol-%): A = air (1.9%); B = methane (30.8%); C = ethane (7.7%); D = propane (11.5%); E = 2-methylpropane (9.6%); F = *n*-butane (15.4%); G = 2-methylbutane (13.5%); H = *n*-pentane (9.6%). The chromatographic column used contained dimethylsulfolane as the liquid phase. (From the author's work, in 1957–1958.)

major laboratories in Western Europe (Fig. 20.4). A number of laboratories built their own system for the routine analysis of gaseous samples. Some of the companies where self-built Janák-type gas chromatographs were in use are: BASF, Farbwerke Hoechst and Lurgi, in Germany; British Petroleum, in England; Shell Laboratories, in the Netherlands; and the Societé Française du Pétrole, in France; These instruments were so popular that at Hoechst (Frankfurt am Main, Germany), they even created a new verb to describe analysis with this system: *janákieren* ("to Janák"). A few small instrument companies (e.g., Hereus, in Germany, Griffin & George, in the United Kingdom, and Kavalier, in Czechoslovakia) also developed commercial versions.

Table 20.1. Jaroslav Janák's articles on "Chromatographic semi-microanalysis of gases" published in *Chemické Listy* and in the *Collection of Czechoslovak Chemical Communications* (1953–1959).[a]

J. Janák, Chromatographic semi-microanalysis of gases — Preliminary communication, *Chem. Listy* **47**, 464–467 (1953); *Collection* **18**, 798–802 (1953) (*).
J. Janák, Chromatographic semi-microanalysis of gases, I. Theory and method of analysis, *Chem. Listy* **47**, 817–827 (1953).
J. Janák, Chromatographic semi-microanalysis of gases, II. Analysis of natural gas and determination of methane in mine gases, *Chem. Listy* **47**, 828–836 (1953).
J. Janák, Chromatographic semi-microanalysis of gases, III. Analysis of gases rich in hydrogen, *Chem. Listy* **47**, 837–841 (1953).
J. Janák, Chromatographic semi-microanalysis of gases, I–III. Theoretical and practical basis of analysis, *Collection* **19**, 684–699 (1954).
J. Janák, Chromatographic semi-microanalysis of gases, IV. Analysis of gaseous paraffins, *Chem. Listy* **47**, 1184–1189 (1953).
J. Janák and M. Rusek, Chromatographic semi-microanalysis of gases, V. Analysis of unsaturated C_2 and C_3 hydrocarbons, *Chem. Listy* **47**, 1190–1196 (1953).
J. Janák and M. Rusek, Chromatographic semi-microanalysis of gases, IV–V. Separation and analysis of gaseous hydrocarbons, *Collection* **19**, 700–711 (1954).
J. Janák, Chromatographic semi-microanalysis of gases, VI. Analysis of rare gases, *Chem. Listy* **47**, 1348–1353 (1953); *Collection* **19**, 912–924 (1954).
J. Janák and I. Paralova, Chromatographic semi-microanalysis of gases, VII. Analysis of dissolved gases, *Chem. Listy* **47**, 1476–1480 (1953); *Collection* **20**, 336–341 (1955). (*).
J. Janák and M. Rusek, Chromatographic semi-microanalysis of gases, VIII. Separation and analysis of several halogenated hydrocarbons, *Chem. Listy* **48**, 207–211 (1954); *Collection* **20**, 520–524 (1955).
J. Janák and M. Rusek, Chromatographic semi-microanalysis of gases, IX. Determination of nitrous oxide, *Chem. Listy* **48**, 397–400 (1954); *Collection* **20**, 343–347 (1955).
J. Janák and K. Tesarik, Chromatographic semi-microanalysis of gases, X. Determination of small minute amounts of helium, neon and hydrogen in gases, *Chem. Listy* **48**, 1051–1057 (1954); *Collection* **20**, 348–355 (1955).
J. Janák and M. Rusek, Chromatographic semi-microanalysis of gases, XI. Direct determination of individual olefins in gases, *Chem. Listy* **49**, 191–199 (1955); *Collection* **20**, 923–932 (1955).
J. Janák, M. Rusek and A. Lazarev, Chromatographic semi-microanalysis of gases, XII. Separation and analysis of gaseous cycloparaffins, *Chem. Listy* **49**, 700–705 (1955); *Collection* **20**, 1199–1205 (1955).
J. Janák, M. Nederost and V. Bubenikova, Chromatographic semi-microanalysis of gases, XIII. Separation of chlorine, bromine and iodine, *Chem. Listy* **51**, 890–984 (1957); *Collection* **22**, 1799–1804 (1957).

(*Continued*)

Table 20.1. (*Continued*)

J. Janák and J. Novák, Chromatographic semi-microanalysis of gases, XIV. Direct determination of individual gaseous paraffins and olefins in 1,3-butadiene, *Chem. Listy* **51**, 1832–1837 (1957); *Collection* **24**, 384–390 (1959).
J. Janák and K. Tesarik, Chromatographic semi-microanalysis of gases, XV. Automation of the measuring unit of a gas chromatograph, *Chem. Listy* **51**, 2048–2054 (1957); *Collection* **24**, 536–544 (1959).

[a]Parts I–III and IV–V were published as separate publications but in the same issue of *Chemické Listy*, and then republished as combined papers with a somewhat modified title in the *Collection of Czechoslovak Chemical Communications.* From Part VI on, each paper was published individually.
The language of the publications in *Chemické Listy* is always Czech; in the *Collection of Czechoslovak Chemical Communications* it was German except the two marked with an asterisk (*) which were published in Russian, with an English summary. We list here the titles in English translation.
In the listing we abbreviate the title of the *Collection of Czechoslovak Chemical Commmunications* as "*Collection.*"

Fig. 20.4. J. Janák (right) with A. V. Kiselev (Moscow; left) and J. F. K. Huber (at that time in Eindhoven), at the 1962 Hamburg Symposium (author's collection).

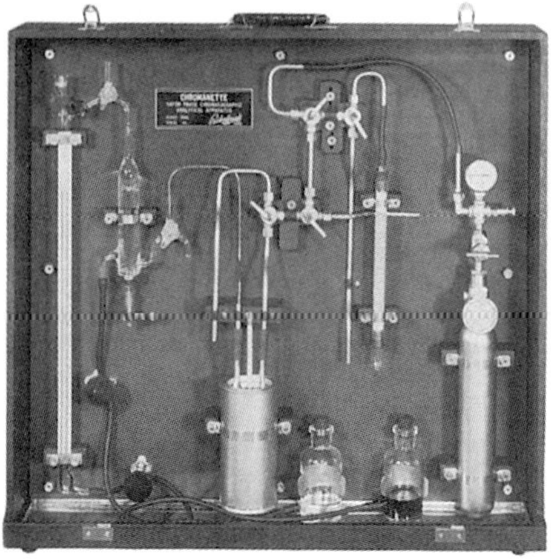

Fig. 20.5. The prototype of the Chromanette 9495 Series Janák-type portable gas chromatograph, developed by the Podbielniak Co. (1956) (courtesy of S. T. Preston III).

Podbielniak Co., in the US built the prototype of a portable commercial unit named the *Chromanette 9495 Series* (Fig. 20.5). It was described in 1956, at a meeting of the California Natural Gasoline Association;[16] however, I have no information that it was actually sold commercially. In 1956–1957 the Hungarian Petroleum and Natural Gas Research Institute (MÁFKI) developed a portable Janák-type gas chromatograph which was also produced commercially. According to the available information about 50 such units were built for various Hungarian laboratories;[17] this is a high number and illustrates the popularity of the Janák-type instruments. However, it was not difficult to construct such a system using standard laboratory hardware, and thus, most of the laboratories built their own instrument. Figure 20.6 shows the author with his self-built Janák gas chromatograph in the laboratories of the LURGI Companies, in Frankfurt am Main, in 1957.

Naturally, the self-built units more or less deviated from Janák's original design, improving it in certain aspects. For example, in the system built by me carbon dioxide was not generated from marble, but a small autoclave was filled with solid CO_2 ("dry ice") and this provided the needed continuous gas flow. When filling the autoclave, some air was also trapped inside; however, this remained in a gaseous

Fig. 20.6. The author with his Janák-type gas chromatograph in the laboratories of the Lurgi Companies, in Frankfurt am Main, in 1957.

state and could be easily blown out by briefly opening the autoclave. This was also essentially the technique of Rouit who, besides Janák, presented the most detailed description of this type of gas chromatograph and its operation.[18] Another modification of my system was the way the gaseous samples were introduced. Janák used gas microburets, pressing with mercury the desired sample volume into the carrier gas flow; in my system I utilized a standard rotary-type gas sampling valve with calibrated sample loops, available by that time from the Perkin-Elmer Corporation.[19,20] Also, I used a standard rotameter for the measurement of the carrier gas flow that could be controlled by the exit valve of the autoclave.

As mentioned the main advantage of the Janák system was the possibility of directly reading the volumes of the separated fractions. Let us not forget that at that time — even after the introduction of the first commercial gas chromatographs — peak area had to be established by carefully measuring its height and width or, for more accurate results, utilizing a hand-held planimeter to obtain the peak area. This was a tedious, time consuming process. Next, one had to normalize the peak area with help of the pre-determined detector response factors and calculate in this way sample concentration. Thus, for the routine analysis of gas samples, the use of the Janák system had major advantages.

In spite of its ease of operation, the Janák system also had a number of limitations. It was essentially restricted to samples that are gaseous at ambient temperature. Since the column was in a liquid thermostat (a Dewar flask), its temperature could be varied from sub-ambient (needed for the separation of rare gases) to about 100°C; however, the nitrometer and the connecting tubes were at ambient temperature, and their heating would have made the system much too complicated. Obviously carbon dioxide, if present in the sample, could not be detected because it was absorbed by the caustic solution. Similarly, sample components with an acidic character (e.g., H_2S) were absorbed in the nitrometer. Finally, compounds that are partially soluble in KOH solution gave incorrect quantitative results: a typical case was acetylene. These problems were discussed in detail during the first gas chromatography symposium ("Gas Kolloquium") held in Germany in November 1956, where Janák served as the keynote speaker.[21]

As seen earlier, the Janák system relied on the direct observation of the volumes of the separated fractions. Toward the end of the 1950s Janák tried to automate the readout of the nitrometer (see No. XV of his publication series) and the Podbielniak instrument also was supposed to provide some automation of the readout. However, these made the system too complicated. By that time general-purpose laboratory gas chromatographs had already been marketed by a number of companies, and these were much more versatile than the Janák-type instruments. Thus, from the end of the 1950s on, these units were gradually replaced by the more modern commercial, general-purpose instruments. Today, the Janák-type gas chromatograph exists only in our memories; but nevertheless, it played an important role in the early evolution of gas chromatography.

References

1. L. S. Ettre, *J. Chromatogr. Sci.* **37**(9), 2A–8A (1999).
2. J. Janák, in *75 Years of Chromatography — a Historical Dialogue*, eds. L. S. Ettre and A. Zlatkis (Elsevier, Amsterdam, 1979), pp. 173–185.
3. J. Janák, in *Chromatography — a Century of Discovery 1900–2000*, eds. C. W. Gehrke, R. L. Wixom and E. Bayer (Elsevier, Amsterdam, 2001), pp. 263–270.
4. N. C. Turner, *Nat. Petrol. News* **35**, R234–237 (1943).
5. N. C. Turner, *Petr. Refiner* **22**(5), 140–144 (1943).
6. S. Claesson, *Ark. Kem. Mineral. Geol.* **23A**(1), 1–133 (1946).
7. A. A. Zhukhovitskii, O. V. Zolotareva, V. A. Sokolov and N. M. Turkel'taub, *Dokl. Akad. Nauk SSSR* **77**, 435–438 (1951).
8. E. Cremer and F. Prior, *Z. Elektrochem.* **55**, 66–70 (1951).
9. E. Cremer and R. Müller, *Z. Elektrochem.* **55**, 217–220 (1951).
10. E. Cremer and R. Müller, *Mikrochem./Mikrochim. Acta* **36–37**, 553–560 (1951).
11. A. J. P. Martin and R. L. M. Synge, *Biochem. J.* **35**, 1358–1368 (1941).
12. A. T. James and A. J. P. Martin, *Biochem. J.* **50**, 679–690 (1952).
13. E. Smolková-Keulemansová, *J. High Resolut. Chromatogr.* **23**, 497–501 (2000).
14. J. Janák, Apparatus for the Quantitative and Qualitative Analysis of Hydrocarbons and Other Gases (Gas Chromatograph) Czechoslovak Patent No. 83,991 (Filed: 20 September 1952; Issued: 1 February 1953).
15. L. S. Ettre, *Chromatographia* **55**, 625–631 (2002).
16. W. J. Podbielniak and S. T. Preston, The Future Possibilities of Gas Chromatography, Lecture at the 11 October 1956, meeting of the California Natural Gasoline Association, Pasadena, CA.
17. L. Szepesy, *A Kromatográfia és Rokon Elválasztási Módszerek Története és Fejlesztése Magyarországon* (The history and development of chromatography and related methods in Hungary) (Hungarian Separations Science Association, Budapest, 2007), p. 147.
18. Ch. Rouit, in *Vapour Phase Chromatography* (*1956 London Symposium*), ed. D. H. Desty (Butterworths, London, 1957), pp. 291–303.

19. E. S. Watson and D. R. Bresky (to Perkin-Elmer Corp.), *U.S. Patent* No. 2,757,541 (Filed: 27 February 1956; Issued: 7 August 1956).
20. H. H. Hausdorff, in *Vapour Phase Chromatography* (*1956 London Symposium*), ed. D. H. Desty (Butterworths, London, 1957), pp. 377–387.
21. *Proceedings of the Gas Kolloquium*, Hamburg, 14–16 November 1956.

Chapter

21

The Beginning of GC Instrumentation*

The interest in gas chromatography was almost instantaneous after Martin's lecture at the First International Congress for Analytical Chemistry, in Oxford, England, in September 1952, and the publication of the fundamental papers of James and Martin, in the same year.[1–4] However, the basic difference between liquid and gas chromatography became apparent: while classical liquid chromatography could be carried out in the laboratory practically by everybody using the usual laboratory setups, GC required components which were not found in standard laboratory equipments. Companies with large research laboratories usually had mechanical shops capable of constructing the required sophisticated systems; however, smaller laboratories did not have the means to construct and build their own gas chromatographs. Therefore, it soon became obvious that the immense potential of gas chromatography could only be exploited fully if proper instruments became available.

*Based on the articles by L. S. Ettre published in *LCGC* (*North America*) **8**, 716–724 (2000) and **23**, 142–148 (2005), and in *LCGC* (*Europe*) **18**, 416–421 (2005). The figures are from the author's collection.

The years after the Second World War saw the birth of the scientific instrument industry. The first real "instruments" developed and built by the instrument companies were the infrared spectrophotometers, but at that time these were used only in a limited number of laboratories and their operation and use needed special skills. The gas chromatographs represented the first truly automated, complex analytical instruments which did not need specially skilled scientists for their operation and could be used practically by every laboratory. Thus it is not surprising that the early availability of commercially produced gas chromatographs had a major role in the rapid expansion of the technique and its use. At the same time this new industrial branch benefited greatly from the growing demand for this instrument. We may even say that some kind of symbiosis existed between GC and the scientific instrument industry: the evolution of the former could not happen without the involvement of the latter. A number of companies prominent today in the scientific instrumentation field started as small companies founded by some enterprising chemists who were starting to build gas chromatographs for the analytical chemists.

A couple of British companies (Griffin & George, in London, and Metropolitan Vickers Electrical Co., in Manchester) tried to provide gas chromatographs soon after the publications of James and Martin. However, they did not have the resources for more fundamental development work and their instruments were not much different from a self-constructed laboratory setup. The field was soon taken over by American companies which had gained experience in electronics and optics and in producing high-precision systems to fulfill the need of the Allied military during the Second World War. The two companies that introduced gas chromatographs almost simultaneously in the spring of 1955 — just three years after the publications of James and Martin — were the Burrell Corporation (Pittsburgh, PA) and the Perkin-Elmer Corporation (Norwalk, CT).

21.1. Burrell's Kromo-Tog

Burrell Technical Supply Corporation was founded after the First World War to supply gas analysis equipment. In 1943 they introduced a system for the determination of the components of natural gas by selective adsorption–desorption. This was the so-called Turner–Burrell Adsorption Fractionator and was based on the work of Nelson C. Turner. It was a fairly complicated, floor-standing machine, and its use was very tedious: one analysis took at least 8 h. In the subsequent years, the instrument was somewhat simplified, reducing the analysis time to 1–3 h: the result was the so-called Fracton introduced in 1953 (see Chapter 20).

Both Burrell instruments utilized what we would call today a combination of frontal and displacement analysis. The large gas sample was adsorbed on the column, and then it was desorbed and pushed upward by a combination of a moving heater and a displacer: liquid mercury in the Turner instrument and tetrachloroethylene vapor in the Fracton. During this slow movement the sample components were more or less separated. Finally the separated sample fractions were purged by hydrogen gas into a thermal conductivity detector.

At the September 1954 National Meeting of the American Chemical Society H. W. Patton of Tennessee Eastman Co., in Kingsport, TN, presented the first American paper on GC.[5] In it he described a self-constructed system using an adsorption column in the elution chromatography mode, an inert carrier gas, and commercially available thermal conductivity cells as the detector. L. V. Guild of Burrell was present at the meeting and he realized the possibility of modifying the Fracton into a "real GC," by using a carrier gas instead of the displacer. The new instrument, the so-called Kromo-Tog Model K-1 (Fig. 21.1) was announced in March 1955. It used an unheated charcoal column and had only a limited use for the analysis of samples which were gas or vapor at room temperature. A fairly complicated bypass device was used for sample introduction and there was no possibility to inject liquid samples.

In the next few years Burrell improved this instrument, adding the possibility of column heating (with a wrap-around nichrome heating wire) and liquid sample injection. However, the company did not have the needed R&D resources to keep up with continuous improvements, and thus by the mid-1960s their production and marketing of gas chromatographs were discontinued.

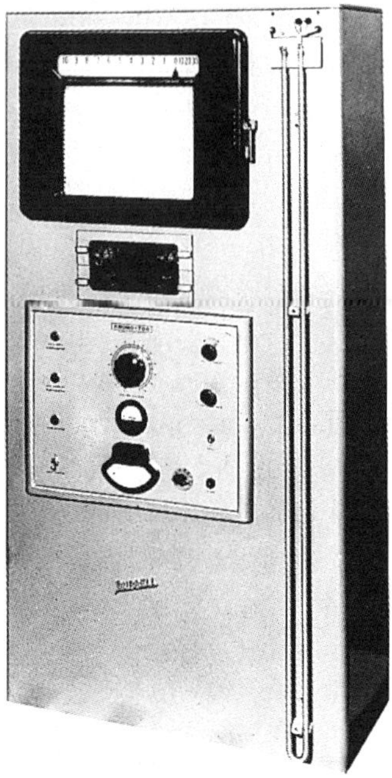

Fig. 21.1. The Burrell Kromo-Tog Model K-1 gas chromatograph introduced in March 1955. The cover over the U-shaped column was removed; it was not thermostatted and at room temperature. Gas samples could be introduced with a multivalve inlet system (behind the small door just below the recorder).

21.2. Perkin-Elmer's Vapor Fractometer

While the Burrell instrument was relatively short-lived, the one introduced by Perkin-Elmer represented the first instrument in a continuous line of gas chromatographs.

The Perkin-Elmer Corporation (PE) was founded in 1937 and established itself during the war as one of the leading manufacturers of precision optics. In mid-1940 the company also became involved in manufacturing infrared (IR) spectrophotometers and through this product line had increasing contacts with leading scientists in the United States and Europe. Through these contacts Perkin-Elmer's representative heard in 1953 about the GC work of C. S. G. Phillips at Oxford University and of James and Martin in the laboratories of the British Medical Council, in London. In 1954 Harry H. Hausdorff, then the head of PE's IR applications laboratory, went to visit them to learn more about the new technique; when returning home he gave an enthusiastic report about its possibilities. Based on these visits and also on information from E. F. Williams at the central research laboratories of American Cyanamid Co., in Stamford, CT, a breadboard model was constructed by the middle of 1954; detailed investigations were carried out on the best instrument design and on the influence of operation parameters on the analytical results. This led to the development of the final version of the instrument which was introduced in May 1955: the so-called Model 154 Vapor Fractometer (Fig. 21.2). Besides Hausdorff, two other PE associates should particularly be mentioned

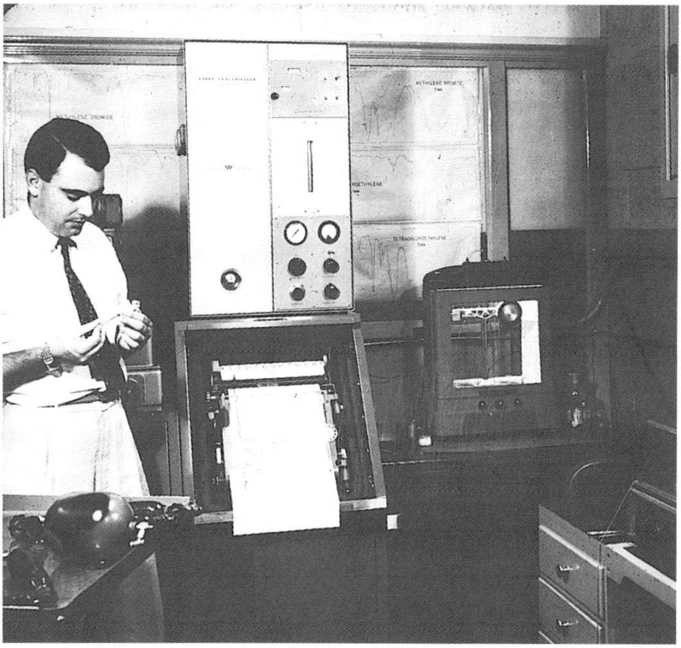

Fig. 21.2. The original version of the Perkin-Elmer Model 154 Vapor Fractometer, with Harry H. Hausdorff who had a major share in its development. The door on the left side covers the air thermostat for the U-shaped column mounted on the thermal conductivity detector block. A flash vaporizer was incorporated into the system, with a rubber septum, permitting syringe injection of liquid samples (small circle at the lower left). Starting with the Model 154-B (introduced in 1956) the potentiometric recorder was housed in a box of the same size as the instrument.

who participated in the development of the instrument: A. Savitsky and E. S. Watson.

The Model 154 became an instant success. This was best demonstrated by the comments of Ralph H. Müller, the legendary columnist of *Analytical Chemistry*, who within one month after the introduction of the instrument characterized it in his monthly column as "a splendid example of automatic analysis." Commenting on the chromatograms that were obtained (using a potentiometric recorder) he stated that "if the reader can tolerate our exuberance, which we hope is contagious, these recordings are a delight to behold".[6]

Three major factors contributed to the success of the Model 154. The first was its versatility and ease of operation. It incorporated a heated thermal-conductivity detector utilizing thermistor beads. The U-shaped columns in the air thermostat were mounted directly on the detector block eliminating unnecessary dead volumes; they were easily interchangeable. Liquid samples were injected with a syringe through a rubber septum into a flash vaporizer connected directly to the column inlet and within a year, a newly developed rotary type gas sampling valve, with interchangeable sample loops of different volumes, became available. PE also supplied standardized columns with a variety of stationary phases.

The second major factor contributing to the success of the instrument was the availability of a simple text from the company that explained the principles of GC and the proper selection of operational parameters. This 31-page long booklet by Hausdorff, published in September 1955,[7] helped the novices (and at that time, *everybody* was a novice in GC!) to master the intricacies of the technique. Ralph Müller characterized this publication as "a compact and very informative summary of the theory, uses, instrumentation, and general practices of gas chromatography".[8]

Finally the third factor in the success of the Model 154 was the help provided by the company to the prospective users in solving their analytical problems through its Application Group, via personal contacts, and by issuing data sheets and notes on some key applications. Even the company's advertisements were directed to show the solution of a practical problem. For example, the chromatogram shown

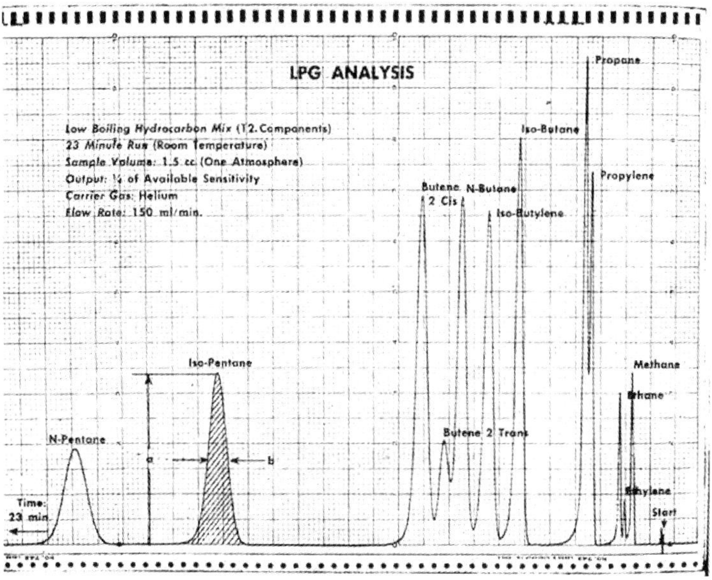

Fig. 21.3. The chromatogram showing the separation of 12 C_1–C_5 hydrocarbons in 23 min, used in the first advertising of Perkin-Elmer's Model 154.[9] A 2-m long column containing 30% triisobutylene on Celite was used at room temperature. The ad also explained the way of quantitative evaluation of the chromatogram.

in Fig. 21.3 was part of the first ad of the Model 154: the analysis of C_1–C_4 hydrocarbons was at that time a crucial application problem, and this illustration also explained the method for quantitative evaluation of the chromatogram. Soon Applications Groups were also formed by the other instrument companies and it is safe to say that without the extensive activities of these groups the meteoric rise of GC would have been impossible.

For over five years Model 154 was the most widely used gas chromatograph (Fig. 21.4) and the instrument was kept up-to-date by adding the newest developments to its construction. Although starting in 1962 PE also introduced new, more advanced, and sophisticated gas chromatographs, the Model 154 continued to be produced until the late 1960s, and many laboratories kept their old instrument for decades. (A few years ago I learned of a unit purchased in 1956 by a Swedish laboratory that was used until 2003!)

Fig. 21.4. The control laboratory of a petroleum company in 1957, showing eight Model 154s. The operator is standing before a box containing a potentiometric recorder. The recorders for the other instruments are on the lower level of the laboratory bench.

21.3. Additional Instruments

Burrell and Perkin-Elmer were soon followed by a number of other companies, both in the United States and overseas, that introduced new gas chromatographs. By 1956 six additional companies were in the market, with seven new models: Podbielniak, Inc. (Chicago, IL), Fisher Scientific Co. (Pittsburgh, PA), Beckman Instruments, Inc. (Fullerton, CA), Consolidated Electrodynamics Co. (Pasadena, CA), Hallaikainen Instruments Co. (Emeryville, CA), and Wilkens Instruments & Research, Inc. (Walnut Creek, CA):

- Podbielniak had been a supplier of low- and high-temperature laboratory distillation apparatuses used mainly in the natural gas and petroleum industry, and had a virtual monopoly for the low-temperature distillation systems. Their first GC, the Chromacon, was introduced in December 1955.
- Fisher Scientific's first GC was introduced in March 1956, at the Pittsburgh Conference. It was based on the development work at

Fig. 21.5. The Fisher-Gulf Model 300 Partitioner introduced in March 1956. The small box at the lower part of the front plate is their micro-dipper for liquid sample introduction.

Gulf Oil Research & Development Co.: it was called the Fisher-Gulf Model 300 Partitioner (Fig. 21.5).

- Beckman Instruments had been in the scientific instrument field since the 1940s and was particularly known for their ultraviolet spectrophotometers. Their first GC, the GC-1, was also introduced at the 1956 Pittsburgh Conference. This instrument had one drawback: while the other GCs available by then could be thermostatted in a wide temperature range, the GC-1 could only be used at a fixed temperature, 40°C. Within a year Beckman's GC-2, incorporating an air thermostat with variable temperature setting was also introduced (Fig. 21.6).
- Consolidated Electrodynamics (CEC) had been in the mass spectrometer business and the development of a GC was a logical extension of their activities. Their Model 26-201 was introduced in the summer of 1956.

Fig. 21.6. The Model GC-1 of Beckman instruments, introduced in March 1956. The Model GC-2 looked the same.

- Hallaikainen Instruments was a small California company that was licensed to utilize the development work of Shell Development Co. Their GC, the so-called Chromagraph, introduced in the spring of 1956 was the first commercial system having a separate refrigerating unit that permitted column temperatures down to 0°C.
- The last company entering the GC market in 1956 was Wilkens Instrument & Research, founded by K. P. Dimick, formerly associated with the Western Regional Research Laboratory of the US Department of Agriculture. Their first instrument, the Model Aerograph A-90, was introduced by the year's end.

By 1962 the number of American companies offering gas chromatographs multiplied: in that year 24 companies were listed offering more than 50 different basic models, most of them available in several versions depending on the detectors used and other additional features. Table 21.1 presents a listing of these companies.[10]

Naturally, in the subsequent years the field underwent many changes. Most of the smaller companies merged with others or discontinued their operations and even some of the major suppliers lost their original identity. Thus, F&M Corporation — that introduced temperature programming and exhibited its first instrument at the

Table 21.1. American companies producing and marketing laboratory gas chromatographs in 1962.[a,10]

Company	Number of models[b]
American Instrument Co., Silver Spring, MD	1
Barber-Colman Co., Rockford, IL	5
Beckman Instruments, Inc., Fullerton, CA	2
Burrell Corp., Pittsburgh, PA	4
Central Scientific Co., Chicago, IL	2
Consolidated Electrodynamics Corp., Pasadena, CA	3
Dohrman Instruments Corp., San Carlos, CA	1
Dynatronic Instruments Corp., Chicago, IL	1
R. L. Faley & Associates, Houston, TX	1
F&M Scientific Corp., Avondale, PA	4
Fisher Scientific Corp., Pittsburgh, PA	1
Glowall Corp., Glenside, PA	1
Greenbrier Instrument Inc., Roncoverte, WV	1
Hallaikainen Instruments Co., Emeryville, CA	1
Jarrell-Ash Co., Newtonville, MA	2
Loe Engineering Co., Altadena, CA	4
Nester/Faust, Newark, DE	2
Micro/Tek Instruments, Baton Rouge, LA	3
Packard instruments Co., LaGrange, IL	1
Perkin-Elmer Corp., Norwalk, CT	5
Podbielniak Inc., Chicago, IL	series
Precision Scientific Co., Chicago, IL	1
Research Specialities Co., Richmond, CA	2
Wilkens Instrument & Research Inc., Walnut Creek, CA	4

[a]Preparative and process instruments are not included in the listing.
[b]Generally each model had several versions depending on the detectors used and other additional features.

1959 Pittsburgh Conference (Fig. 21.7) — was bought in 1965 by Hewlett-Packard Co., which in the 1990s split into two companies: the instrument business is now Agilent Corp. Wilkens Instrument & Research was bought in 1966 by Varian Associates, and it is now part of its instrument business. As of January 2007 when I write these words, only three of the 24 companies listed in Table 21.1 remain in the GC business: Agilent, Varian, and the Life & Analytical Science Division of PerkinElmer Co., the successor of Perkin-Elmer's original Instrument Division.

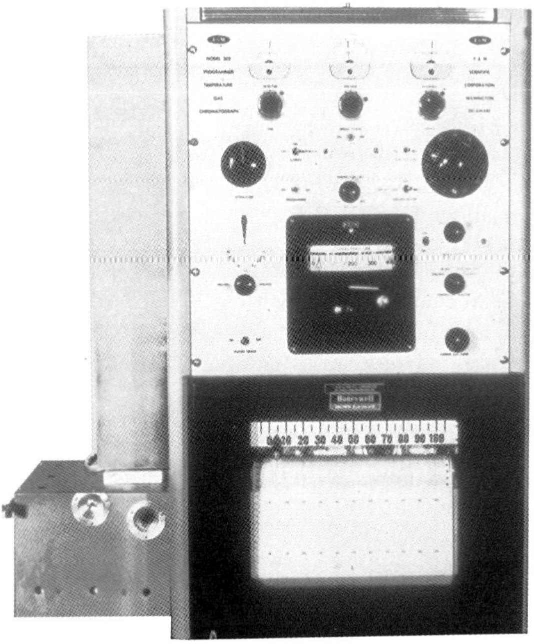

Fig. 21.7. The F&M Model 202 programmed-temperature GC first shown at the Pittsburgh Conference in March 1959. The column is in the "chimney" (an air thermostat) on the left, mounted on the box containing the thermal conductivity detector. The injection port is just below the column on the left.

References

1. A. T. James and A. J. P. Martin, *Biochem. J.* **50**, 679–690 (1952).
2. A. T. James, A. J. P. Martin and G. H. Smith, *Biochem. J.* **52**, 238–242 (1952).
3. A. T. James, *Biochem. J.* **52**, 242–247 (1952).
4. A. T. James and A. J. P. Martin, *Analyst* 77, 915–932 (1952).
5. H. W. Patton, J. L. Levis and W. I. Kaye, 126th National American Chemical Society Meeting, New York, NY, September 12–17, 1954; *Anal. Chem.* **17**, 170–174 (1955).
6. R. H. Müller, *Anal. Chem.* **27**(6), 33A–36A (1955).
7. H. H. Hausdorff: *Vapor Fractometry (Gas Chromatography) — A Powerful New Tool in Chemical Analysis* (The Perkin-Elmer Corp., Norwalk, CT 1955), 31 pp.
8. R. H. Müller, *Anal. Chem.* **27**(12), 33A–35A (1955).
9. *Anal. Chem.* **27**(9), 22A (1955).
10. L. S. Ettre, *J. Chromatogr. Sci.* **15**, 90–110 (1977).

Chapter

22

The Invention, Development, and Triumph of the Flame-Ionization Detector*

22.1. Background

In classical column chromatography the separated compounds were observed as colored rings on the column, and in paper- and thin-layer chromatography the separated spots can be visualized relatively easy. However, the situation is different in gas chromatography (GC): here, some means have to be found for the detection of the separated compounds that are eluted from the column. As we have seen (see Chapter 14) A. T. James and A. J. P. Martin used titration in their first investigations on gas–liquid partition chromatography (GLPC) because their samples consisted of fatty acids[1,2] and amines.[3,4] However, naturally, titration could not be used for neutral compounds; also, the automated titration unit developed by Martin was too complicated for wide-spread use.

*Based on the articles by L. S. Ettre published in *LCGC* (*North America*) **20**, 48–60 (2002), and *LCGC Europe* **15**, 364–373 (2002).

Gas (adsorption) chromatography had already been carried out before James and Martin, by utilizing adsorption columns (see Chapter 19). In these investigations, and also in other measurements involving gas mixtures, thermal-conductivity detectors (TCD) had been utilized. Thus, it was logical for N. H. Ray of the British ICI laboratories — who visited Martin's laboratory to learn about the new technique in 1950–1951 when GLPC was still in the developmental stage — to adapt this detector to his gas chromatograph and use it for the analysis of impurities in ethylene.[5,6]

When, in 1955–1956, the first commercial gas chromatographs were introduced in the United States, they all used thermal-conductivity detectors.[7] These detectors were rugged, reliable, and easy to use. However, they had two limitations. The first particularly restricted their use: for best performance helium was needed as the carrier gas, but at that time (in the second half of the 1950s) helium was practically unavailable outside America, and even if one could get it, its price was often prohibitive. The second limitation was the limited sensitivity of the TCD even when using helium, insufficient to analyze low sample concentrations. For these reasons early gas chromatographs built in England utilized other detectors that did not need helium. Thus, the instruments of Griffin & George utilized the gas density balance developed by Martin in 1954,[8,9] while the GC of Shandon Scientific Co. incorporated the so-called hydrogen flame detector of R. P. W. Scott.[10,11] In this detector the carrier gas contained hydrogen, which was burned at column end, and the temperature of the flame was continuously monitored. Without sample, the carrier gas produced a constant signal; however, when an organic vapor eluted from the column, the temperature of the flame increased, resulting in a response proportional to the amount of the eluting compound. For a short time, there was great hope with these detectors, however, they were difficult to use, and did not yield much improvement in the analytical results. A number of other detectors have also been investigated, but none of these reached beyond the breadboard stage and are completely forgotten today.

Then, in 1958, suddenly two ionization detectors were introduced, which completely changed the situation: the argon-ionization detector and the flame-ionization detector.

The argon-ionization detector (AID) was invented by J. E. Lovelock (see Chapter 23). This detector was almost immediately commercialized by W. G. Pye & Co., introducing the so-called Argon Chromatograph at the Second International GC Symposium held in May 1958, in Amsterdam, The Netherlands. The AID had a high sensitivity and opened GC to the field of biochemistry. However, within a few years it was superseded by the flame-ionization detector, which did not need a radioactive source and had better linearity.

The second ionization detector that was described almost simultaneously with the AID was the flame-ionization detector (FID). Due to its high sensitivity, predictable response, and extended linear range, it became, within a few years, the most widely used GC detector and was included in practically every gas chromatograph.

22.2. Invention

The FID was developed in 1957 in two locations: at the Central Research Laboratories of Imperial Chemical Industries of Australia and New Zealand (ICIANZ) by I. G. McWilliam and R. A. Dewar, and at the Department of Physical Chemistry, University of Pretoria, South Africa by J. Harley, W. Nel, and V. Pretorius. Neither group knew about the activities of the other. They worked almost simultaneously, the ICIANZ group having a slight edge; also, while the Pretoria group did not follow up their original work, the Australian group pursued it to completion.

22.2.1. *Work in Australia*

The laboratories of ICI in Australia and in England had a fairly close contact, and the associates of ICIANZ frequently visited the English laboratories. In 1955, during such a visit, R. A. Dewar (1908–1981), associate research manager of ICIANZ, became familiar with the results of Ray[6] and, upon returning to Australia, he initiated work on GC. However, being aware of the limitations of the TCD, his aim was not just to copy Ray's system (using a TCD), but to improve it. At that time Ian Gordon McWilliam (born 1933), a fresh graduate of Melbourne University, joined ICIANZ, and he was assigned to this

Fig. 22.1. I. G. McWilliam (right) and R. A. Dewar, around 1958. A self-constructed gas chromatograph can be seen in the background. (Courtesy of Dr. McWilliam.)

project (Fig. 22.1). Twenty-five years later, he discussed in details their work and results in two retrospective articles,[12,13] and thus, we have a first-hand account of this exciting story.

McWilliam and Dewar tried to improve the TCD and investigated various modifications, but without any real advances. Meanwhile, they also learned about the hydrogen flame detector of Scott,[10] built one, and investigated it. However, they soon realized its fundamental problem: the continuously burning hydrogen (part of the carrier gas) results in a high background, while the burning of small concentrations of sample components cause only small temperature changes. (It is like establishing the weight of a captain by weighing the ship with and without him.) Therefore, they decided to modify the system: they would measure the ion current in the flame and not its temperature.

These investigations started in early 1957. A 23-gauge hypodermic needle (i.d.: about 0.34 mm) was used as the jet, two metal electrodes were placed on the opposite sites of the flame, and the ion current produced by placing a battery in series with the electrodes was measured. Originally, air for combustion was simply obtained from the

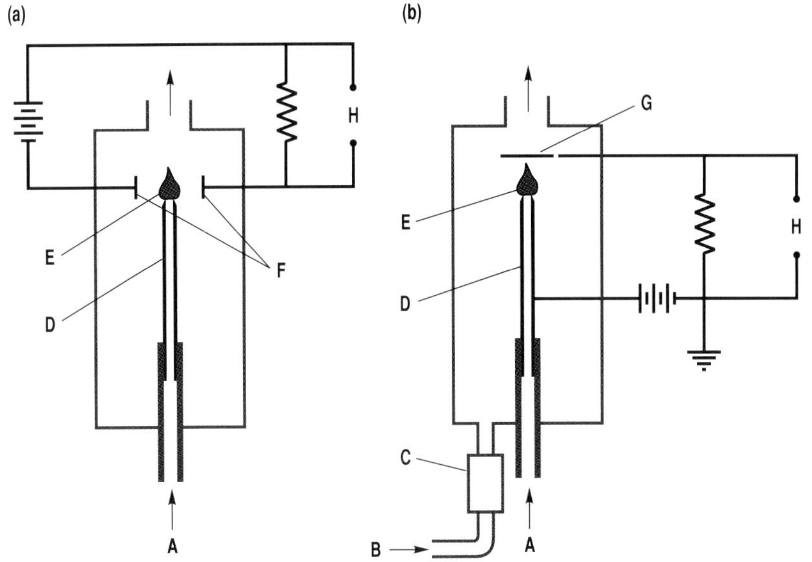

Fig. 22.2. Early FID designs of McWilliam and Dewar. Left: first design with parallel-plate electrodes; right: FID having the jet as one of the electrodes and using filtered air.[13] A = column effluent + hydrogen, B = air, C = filter, D = metal jet, E = flame, F = electrodes, G = wire gauze collector electrode, H = recorder.

surrounding atmosphere; however, it was found that this contained a large amount of dust particles which burned in the flame, creating a noisy background. Therefore, the design was modified, introducing filtered air into the detector housing. Changes were also made in the electrical system, among them using the metal jet as one electrode and wire gauze as the collector electrode (Fig. 22.2). A further modification was to use two FIDs, one at the end of the analytical column and the second at the end of a reference column, measuring the difference between the outputs of the two detectors. In this way, background current disturbance could be offset.

By the summer of 1957 the prototypes of the FID showed excellent performance, high sensitivity, and good linearity. At this time McWilliam and Dewar prepared a short paper describing the principles of the FID (both in the single and dual jet form) for submission to the journal *Nature*, whose editorial offices are in London, and ICIANZ also filed a patent application (see below).

Just at that time McWilliam was scheduled for a two-year stay at the ICI laboratory in Manchester, England. Therefore, in order to speed up the handling of their manuscript, he took it with him and personally delivered it on 18 September 1957 to the editorial offices of *Nature* in London, instead of mailing it from Australia. Soon after arriving in England McWilliam participated on 4 October 1957 at an informal discussion organized by the British Gas Chromatography Discussion Group at Cambridge University, and there he briefly described the newly developed FID. His presentation, which was the first public disclosure of the detector, raised considerable interest.

Meanwhile, the manuscript personally submitted to *Nature* underwent the usual editorial process and was tentatively accepted with the condition that the somewhat lengthy description of the electrical circuit is replaced by a diagram, and the authors were notified by this editorial decision in a letter. However, now the comedy of errors started. The administrators of the journal ignored the fact that McWilliam gave them his address *in England* and sent this letter, dated 6 November 1957, *to Australia* by *sea mail*, which, of course, took many weeks to arrive there. ICIANZ returned the letter to McWilliam who immediately carried out the required modification; however, due to this delay, he could submit the final manuscript to the editorial offices of *Nature* only at the beginning of January 1958.

A few days later McWilliam was shocked to read in the January 18 issue of the journal a short paper on "flame-ionization detector for gas chromatography" (the same title as of his manuscript!) by Harley, Nel, and Pretorius of the University of Pretoria, in South Africa.[14] Upon his inquiries, the editorial office of *Nature* apologized: apparently, the two papers were handled by different editors and since there was no editorial comment to the South African paper, it was published earlier. The situation was even worse since *Nature* does not give the date when a manuscript was submitted and thus, for the reader, the Pretorius paper had the priority. Due to this mix-up the McWilliam–Dewar paper was published only two months later in the 15 March 1958 issue of the journal.[15]

22.2.2. Work in South Africa

Let us now turn our interest to the University of Pretoria, in South Africa, where at that time Victor Pretorius (Fig. 22.3) had just started his independent academic career.

Victor Pretorius (1928–1989) was a descendant of Andries Pretorius, one of the leaders of the Boer pioneers in the first part of the 19th century, after whom the city of Pretoria was named. A brilliant student at the University of Pretoria, Victor was awarded in 1952 a Rhodes Scholarship to Oxford University, in England, where he fully utilized his potential, obtaining a Ph.D. under the Nobel Laureate Sir Cyril Hinshelwood. In Oxford, Pretorius' research subject was the investigation of the products formed during the gas-phase polymerization of olefins. Pretorius heard early about the work of Ray at ICI, and after a visit to his laboratory he also built a gas chromatograph with a TCD.

Upon returning to South Africa at the end of 1954 and rejoining the University of Pretoria, Pretorius continued these investigations. However, just as the ICIANZ group in Australia, he was not satisfied with the sensitivity of the TCD (particularly since helium was unavailable in South Africa), and therefore he tried other means of detection, among them a self-built glow-discharge detector.[16] In the early 1957, at the request of a colleague, he also built a hydrogen

Fig. 22.3. Victor Pretorius (Author's collection).

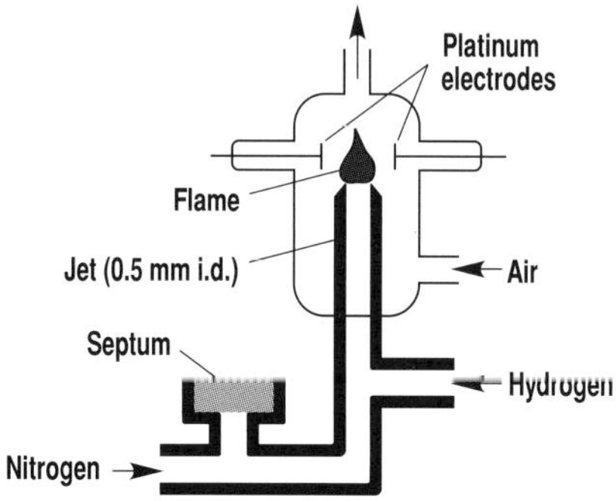

Fig. 22.4. The original all-glass FID constructed by Pretorius in early 1957. In its initial testing, no column was used: the sample was directly injected into the nitrogen stream.[17] The Pt electrodes were 5 mm apart and 2 mm above jet.

flame detector based on Scott's work.[10] As narrated by Pretorius,[17] while testing this detector he remembered reading somewhere that the electrical resistance of a flame changes with the composition of the gases being combusted. He constructed a simple, all-glass detector to test this possibility (Fig. 22.4). In this respect, it is interesting to note that McWilliam, in his 1958 presentation in Amsterdam,[18] specifically advised against using a glass jet, because traces of alkaline ions would provide an abnormally high background. The electrical circuit of the newly built FID was put together by J. Harley, a technical assistant at the university, and already the first tests had shown the high sensitivity of the new detector. After some additional investigations — now in a GC system, with a column — Pretorius submitted a short paper to *Nature*. Besides Pretorius and Harley, W. Nel, an assistant of Pretorius, was listed as the third co-author of this paper, which — as mentioned earlier — was published in the 18 January 1958 issue of the journal,[14] two months before the communication of McWilliam and Dewar.[15]

22.3. Further Developments

Pretorius did not follow up his initial work on the FID; however, the ICIANZ group carried out detailed investigations on detector characteristics, optimum conditions, and the nature of its response. At the Second International GC Symposium held in Amsterdam, The Netherlands, in May 1958, McWilliam presented a detailed report on these results.[18] He also stated in his lecture that the FID is essentially a "carbon counter": its response is proportional to the number of carbon atoms in the compound's molecule. This observation was very important because it facilitated the prediction of detector response.

In Amsterdam a very extensive discussion followed McWilliam's presentation; particularly on the possibility of using the FID in conjunction with capillary columns, which also were introduced at the symposium, where sample sizes and column flow rates are much smaller than with packed columns. Soon after the symposium extensive research was started at a number of places, particularly in conjunction with capillary columns. Probably the first group involved in such investigations that constructed a gas chromatograph incorporating a capillary column and a FID was Desty's at British Petroleum. He first reported publicly about the system and the results obtained at a GC Symposium held in the fall of 1958 in Leipzig, East Germany,[19] followed a few months later by a lecture at an informal symposium of the British GC Discussion Group, in London on 10 April 1959.[20] At that time McWilliam also carried out investigations on the use of the FID with glass capillary columns (Fig. 22.5).

As mentioned earlier, McWilliam already gave a brief informal description of the new detector at a meeting on 4 October 1957 in London, six months before the publication of their paper in *Nature*. Scientists from the Koninklijke/Shell Laboratories in Amsterdam participated at this meeting and apparently immediately picked up the new idea. Thus, at the 1958 Amsterdam Symposium, Hendrik Boer of Shell could already report on some of their results during the discussion of McWilliam's paper.[18] Meanwhile, one of the Shell scientists, A. I. M. Keulemans, became professor and head of the Laboratory of Instrumental Analysis at the new University of Technology at Eindhoven. One of his first graduate students was Leo Ongkiehong, a chemist at the Koninklijke/Shell laboratory, who had already been involved in these preliminary investigations on the FID and thus this subject was selected for his Ph.D. thesis. At that time the laboratories

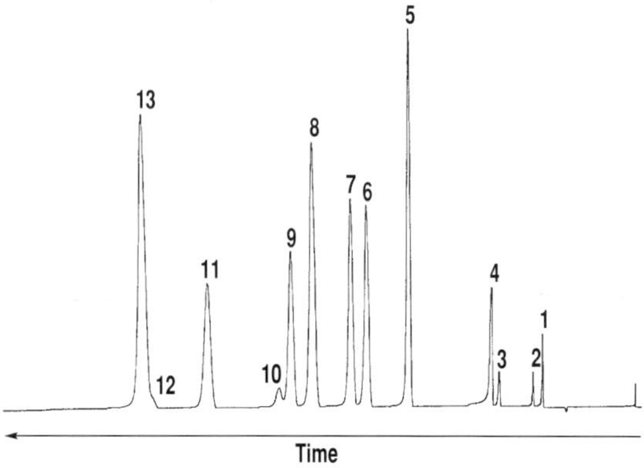

Fig. 22.5. A chromatogram obtained by McWilliam in 1959, using a glass capillary column with an FID. Peaks: 1 = acetone, 2 = methyl acetate, 3 = ethyl acetate, 4 = carbon disulfide, 5 = benzene, 6 = cyclohexane, 7 = cyclohexene, 8 = isoctane, 9 = *n*-heptane, 10 = isobutyl acetate, 11 = methyl cyclohexane, 12 = *n*-butyl acetate, 13 = toluene. (Courtesy of Dr. McWillliam.)

of the new university were not yet ready, and therefore Ongkiehong actually carried out his thesis work at the Koninklijke/Shell laboratories and completed it toward the end of 1959.[21] This thesis represented probably the most detailed investigations on this detector. A summary of Ongkiehong's results was presented at the Third International GC Symposium, held in June 1960, in Edinburgh.[22] At the laboratories of The Perkin-Elmer Corporation (Norwalk, Connecticut) we received a copy of Ongkiehong's thesis from Prof. Keulemans in January of that year, and thus my colleague R. D. Condon was able to further expand these investigations, reporting on additional data at Edinburgh.[23] At the same meeting, D. H. Desty also reported on their investigations on the performance of the FID.[24] We should also mention here the basic paper of J. C. Sternberg, presented one year later, at the International GC Symposium organized by the Instrument Society of America and held in East Lansing, MI.[25] These papers — together with the seminal paper of McWilliam and Dewar[18] as well as two additional papers by them, reporting on further investigations[26,27] — provided the theoretical and practical basis for the rapid and wide-spread application of the FID.

22.4. Instrumentation

The possibility of using the FID in conjunction with capillary columns was already discussed at the Amsterdam Symposium. In fact, the FID was the ideal detector for such applications because it had high sensitivity and zero dead volume, important for the small sample sizes and low flow rates used with capillary columns. According to my best knowledge the first instrument company that investigated the possibility of incorporating the FID into a commercial gas chromatograph was Perkin-Elmer. These investigations started almost immediately after Amsterdam and by October 1958 there was already a prototype detector designed by R. D. Condon. It eventually became part of the Model 154-C gas chromatograph introduced at the Tenth Pittsburgh Conference, in March 1959. This was the first commercial instrument incorporating an FID; it was described at the Conference in a paper by Condon, together with illustrations of its use with capillary columns (Fig. 22.6).[28]

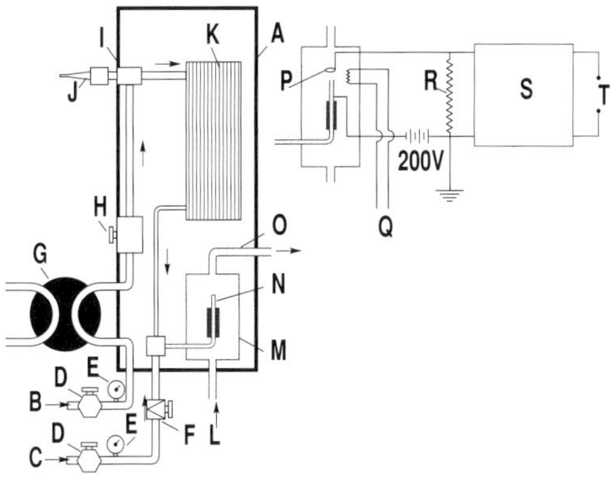

Fig. 22.6. Schematic diagram of the Perkin-Elmer Model 154-C gas chromatograph with an FID, for use with capillary columns, introduced in March 1959.[28] A = GC oven, B = carrier gas, C = hydrogen, D = pressure regulator, E = pressure gauge, F = restriction valve, G = gas sampling valve, H = flash vaporizer for liquid samples, I = split point, J = variable needle restrictor (split vent), K = capillary column, L = filtered air inlet, M = FID housing, N = jet, O = detector vent, P = collector electrode, Q = ignitor, R = high impedance resistor, S = amplifier, T = recorder.

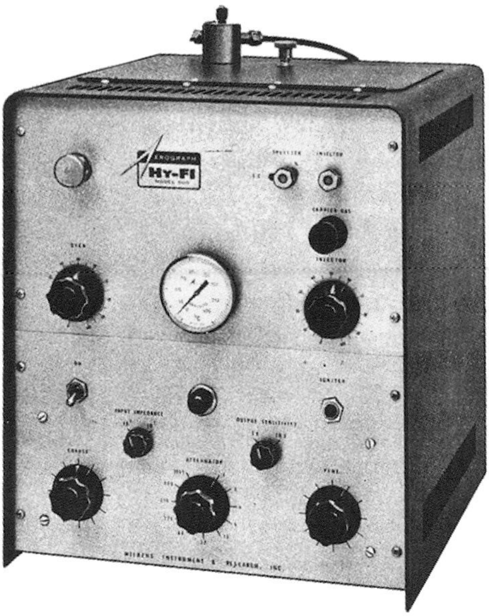

Fig. 22.7. The Aerograph Model 600 Hy-Fi gas chromatograph, introduced in the spring of 1961 by Wilkens Instrument & Research Co. The upper part of the FID can be seen on the top of the instrument.

In the next few years the FID became the most commonly used GC detector and it was incorporated into practically every gas chromatograph manufactured by scores of instrument companies. The FID was particularly important for the growth of Wilkens Instrument & Research Co., the predecessor of the present Varian Chromatography Division. The phenomenal growth of this company in the early 1960s could be mainly attributed to the introduction of their Model 600 gas chromatograph in the spring of 1961 (Fig. 22.7). This was a simple instrument, easy to use, that incorporated an FID. It was commonly called the Aerograph Hy-Fi, and as the story tells, this name was concocted by Mrs. Dimick (the wife of Dr. Keene Dimick, the founder of the company) from the initials of *hy*drogen *f*lame *i*onization.[29,30]

As mentioned above, originally, the ICIANZ group developed two types of FID, with single or dual jets, respectively, the latter configuration aiming to offset background current disturbance. Soon, however, it was found that this is not necessary: the same effect can also be obtained by a proper electrical circuit. Thus, most of the commercial instruments utilized only a single flame. In some instruments two FIDs were installed, however, for a different purpose: for use in the dual column–baseline compensation technique developed in 1961 by Emery and Koerner[31,32] to compensate for the baseline drift in programmed-temperature operation due to the increase in "bleeding" of the liquid stationary phase with temperature. These systems essentially consisted of two separate FIDs, with two separate electrical circuits, and the baseline drift measured from the reference column was automatically subtracted from the output of the FID connected to the outlet of the analytical column. At that time a special instrument was also developed for this purpose: this was the Model 800 gas chromatograph introduced by Perkin-Elmer at the 1962 Pittsburgh Conference. In this instrument the two jets fed one amplifier, providing a single output.[33] In the last two decades the use of systems with two FIDs have lost their importance: present-day stationary phases have very little "bleeding", and if still needed, correction for baseline drift can be carried out by computers.

The introduction of the FID coincided with the growing interest in air pollution research and control. In this respect, an easy and accurate way to measure the total organic content of the atmosphere and automobile exhaust was much sought after. Since the FID was essentially a "carbon counter," it was proposed almost immediately after its introduction that it should be adapted for this purpose. Such portable instruments — the so-called hydrocarbon analyzers (Fig. 22.8) — were developed in 1959 almost simultaneously by a number of companies such as American Cyanamid Co.,[34] Perkin-Elmer,[35,36] and Beckman Instruments.[37] In these instruments no column was used, and the sample gas (for example, atmospheric air or automobile exhaust) was pumped through the detector in lieu of the carrier gas at a constant flow rate. The detector's response was proportional to the total concentration of organics present in the sample gas.

22.5. Patents

In the story of the FID the patent situation is particularly important. On 4 July 1957 ICIANZ filed a patent application in Australia, in

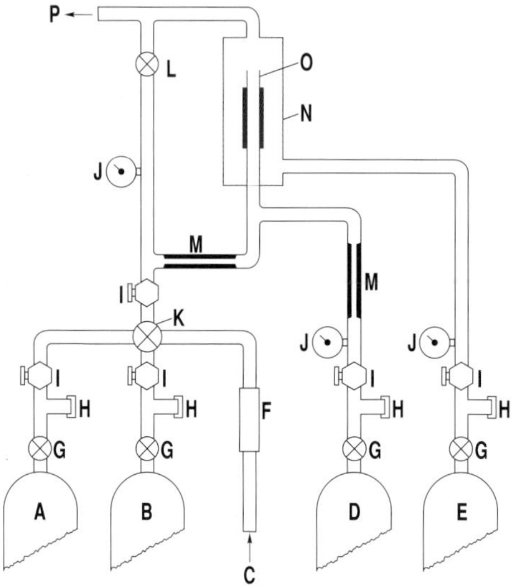

Fig. 22.8. Flow schematic of the Perkin-Elmer Model 213 hydrocarbon analyzer, introduced in the spring of 1960.[35,36] The gases for the portable instrument were supplied from built-in lecture-size gas cylinders (A, B, D, E); however, it was also possible to connect the instrument to outside gas sources via panel fittings (H). A = zero gas, B = test gas, C = sample gas inlet, D = hydrogen, E = air, F = filter, G = valves, H = alternate gas inlets, I = pressure regulators, J = pressure gauges, K = selector valve, L = toggle valve, M = restrictor, N = FID housing, O = detector jet, and P = vent.

McWilliam's name[38] and it was issued on 21 October 1959. Subsequently, applications were filed in a number of countries, among them naturally, also in the United States. Meanwhile, Harley left the University of Pretoria and joined the South African Iron and Steel Corporation, where there was an active patent department which encouraged employees to file patent applications at no cost to the employee. Since Pretorius expressed his disinterest in a patent application, Harley utilized this opportunity and filed a patent (including also in the United States) on their FID in his own name.[17] Thus, there were some complications with the American ICI application. Although there was a clear priority of the Australian application, based on their notebook recordings and patent filing in Australia, ICI decided not to get into a legal fight; rather they made an agreement with Harley, providing him a regular yearly payment. This agreement cleared the way for the issuance of the United States patent to ICIANZ.[39] Subsequently, ICI granted 37 licenses to instrument companies, and their income from these licenses was substantial. As mentioned by McWilliam,[13] at one time the U.S. royalties alone paid for the whole cost of the ICI office in New York City!

One country was inadvertently forgotten by ICI when filing the patent applications: Japan. This was obviously a serious oversight and they tried to correct the situation as soon as the error was recognized, however, without success.[12] Thus, Japanese instrument companies could produce gas chromatographs without requiring any license for the FID.

22.6. Triumph

In a review article on GC detectors,[40] E. R. Adlard, one of the GC pioneers, characterized the FID in the following way:

> The flame ionization detector was first described ...at a crucial moment in the development of gas chromatography, when a high sensitivity, low dead volume detector was required. ...The ease of construction of the FID and its relative insensitivity to variation in operating parameters ensured an instant success.

G. R. Primavesi, another GC pioneer, called it "an almost incredible analytical tool,"[41] and we could continue with quotations praising the importance of the FID in the evolution of GC. Indeed, the FID was the right detector, introduced at the right time, at the moment when the meteoric rise of the use of gas chromatography started, when the technique used up to then only by a limited number of laboratories suddenly became everybody's tool.

Today, there is practically no gas chromatograph without a FID. In the first 15 years of its existence, over 60,000 detectors were manufactured under license from ICI[42] and today, the number of these detectors used in the world can be counted in the hundreds of thousands. Thus, there is probably no other analytical instrument that has made as great contribution to the daily investigations in research and industry, biochemistry and clinical chemistry, and environmental protection, as the little device conceived in 1957. Its triumph is the continuous proof of the foresight of its inventors.

22.7. Personalities

We cannot finish the story of the FID without saying a few words about the continuing activities of the principal players.

In 1960 Victor Pretorius became professor of physical chemistry and director of the Chromatography Research Unit at the University of Pretoria; then, in 1973, he expanded it into an Institute for Chromatography which he lead until his untimely death on 28 December 1989, at the age of 61. He became one of the most original thinkers in chromatography, always full of ideas, mostly ahead of time. However, his major handicap was that he was working thousands of miles away of other chromatography groups and, particularly in the first decade, with very limited means: it was like being out nowhere, on an island, where a ship arrives only once in every six months. Because of this situation, most of his ideas remained unfinished: after trying an idea and illustrating its feasibility by a few experiments, he moved to another subject, with the hope that somebody else will eventually pick up his results.

Ian McWilliam remained with ICIANZ until 1968 when he joined Monash University on a Shell Research Fellowship. Two years later he moved to the Department of Applied Chemistry at Swinburne College (now Institute) of Technology. During his more than three decades at this university, he has been involved in a number of research projects, mainly related to investigations of various aspects of ionization in a flame. Thus, he remained faithful to his "first love" for the rest of his professional career.

References

1. A. T. James and A. J. P. Martin, 20 October 1950 Meeting of the Biochemical Society; *Biochem. J.* 48(1), vii (1951).
2. A. T. James and A. J. P. Martin, *Biochem. J.* 50, 679–690 (1952).
3. A. T. James, A. J. P. Martin and G. H. Smith, *Biochem. J.* 52, 238–242 (1952).

4. A. T. James, *Biochem. J.* **52**, 242–247 (1952).
5. N. H. Ray, in *75 Years of Chromatography — A Historical Dialogue*, eds. L. S. Ettre and A. Zlatkis (Elsevier, Amsterdam, 1979), pp. 345–350.
6. N. H. Ray, *J. Appl. Chem.* **4**, 21–25, 82–85 (1954).
7. L. S. Ettre, *J. Chromatogr. Sci.* **15**, 90–110 (1977).
8. A. T. James and A. J. P. Martin, *Brit. Med. Bull.* **10**, 170–176 (1954).
9. A. J. P. Martin and A. T. James, *Biochem. J.* **63**, 138–142 (1956).
10. R. P. W. Scott, *Nature* (*London*) **176**, 793 (1955).
11. R. Müller, *Anal. Chem.* **29**(3), 55A (1957).
12. I. G. McWilliam, in *Proceedings of the International Conference on Detectors and Chromatography* (*Melbourne, 30 May–3 June 1983*), ed. A. J. C. Nicholson (Australian Scientific Industry Assoc., Melbourne, 1983), pp. 5–20.
13. I. G. McWilliam, *Chromatographia* **17**, 241–243 (1983).
14. J. Harley, W. Nel and V. Pretorius, *Nature* (*London*) **181**, 177–178 (1958).
15. I. G. McWilliam and R. A. Dewar, *Nature* (*London*) **181**, 760 (1958).
16. J. Harley and V. Pretorius, *Nature* (*London*) 178, 1244 (1956).
17. V. Pretorius, in *75 Years of Chromatography — A Historical Dialogue*, eds. L. S. Ettre and A. Zlatkis (Elsevier, Amsterdam, 1979), pp. 333–338.
18. I. G. McWilliam and R. A. Dewar, in *Gas Chromatography 1958* (*Amsterdam Symposium*), ed. D. H. Desty (Butterworths, London, 1958), pp. 142–152.
19. D. H. Desty, in *Gas Chromatographie 1958*, ed. H. P. Angelé (Akademie Verlag, Berlin, 1959), pp. 176–184.
20. D. H. Desty, A. Goldup and B. H. F. Whyman, *J. Inst. Petrol.* **45**, 287–298 (1959).
21. L. Ongkiehong, *The Hydrogen Flame Ionization Detector*, Ph. D. Thesis, Institute of Technology, Eindhoven (1960).
22. L. Ongkiehong, in *Gas Chromatography 1960* (*Edinburgh Symposium*), ed. R. P. W. Scott (Butterworths, London, 1960), pp. 7–15.
23. R. D. Condon, P. R. Scholly and W. Averill, in *Gas Chromatography 1960* (*Edinburgh Symposium*), ed. R. P. W. Scott (Butterworths, London, 1960), pp. 30–45.
24. D. H. Desty, C. J. Geach and A. Goldup, in *Gas Chromatography 1960* (*Edinburgh Symposium*), ed. R. P. W. Scott (Butterworths, London, 1960), pp. 46–64.
25. J. C. Sternberg, W. S. Gallaway and D. T. L. Jones, in *Gas Chromatography* (*1961 Lansing Symposium*), eds. N. Brenner, J. E. Callen and M. D. Weiss (Academic Press, New York, 1962), pp. 231–268.

26. I. G. McWilliam, *J. Chromatogr.* **6**, 110–117 (1961).
27. R. A. Dewar, *J. Chromatogr.* **6**, 312–323 (1961).
28. R. D. Condon, 10th Pittsburgh Conference, 2–6 March 1959; *Anal. Chem.* **31**, 1717–1722 (1959).
29. L. S. Ettre, *J. Chromatogr. Sci.* **15**, 90–110 (1977).
30. K. P. Dimick, *LCGC North America* **8**, 782–786 (1990).
31. E. M. Emery and W. E. Koerner, *Anal. Chem.* **33**, 523–527 (1961).
32. E. M. Emery and W. E. Koerner, *Anal. Chem.* **34**, 1196–1198 (1962).
33. L. S. Ettre, R. D. Condon, F. J. Kabot and E. W. Cieplinski, *J. Chromatogr.* **13**, 305–318 (1964).
34. A. J. Andreach and R. Feinland, 11th Pittsburgh Conference, 29 February–4 March, 1960; *Anal. Chem.* **32**, 1021–1024 (1960).
35. H. N. Claudy and L. S. Ettre, Instrument-Automation Conference, Instrument Society of America, San Francisco, CA, 9–12 May 1960; *ISA Preprint* No. 20-SF-60.
36. L. S. Ettre, 53rd Annual Meeting of the Air Pollution Control Association, Cincinnati, OH, 22–26 May 1960; *J. Air Pollut. Cont. Assoc.* **11**, 34–43 (1961).
37. R. L. Chapman, 53rd Annual Meeting of the Air Pollution Control Association, Cincinnati, OH, 22–26 May 1960; *J. Air Pollut. Cont. Assoc.* **10**, 463–464 (1960).
38. I. G. McWilliam (assigned to ICI of Australia and New Zealand Ltd.), Australian Patent 224, 504 (applied: 4 July 1957; issued: 21 October 1959).
39. I. G. McWilliam (assigned to ICI of Australia and New Zealand Ltd.), US Patent 3,039,856 (issued: 19 June 1962).
40. E. R. Adlard, *Critical Rev. Anal. Chem.* **5**(1), 1–36 (1975).
41. G. R. Primavesi, in *75 Years of Chromatography — A Historical Dialogue*, eds. L. S. Ettre and A. Zlatkis (Elsevier, Amsterdam, 1979), pp. 339–343.
42. Leaflet issued by the Central Research Laboratories of ICI Australia and New Zealand Ltd., on Open Day, 26 June 1986.

Chapter

23

The Development of the Electron-Capture Detector*

In addition to the universal detectors used in gas chromatography (GC), selective detectors also have played an important role in the rapid spread of the technique. Probably the most important selective GC detector is the electron-capture detector (ECD), with a very high sensitivity to organic compounds containing chlorine and fluorine atoms in their molecules. Initially, and continuing through the present, the ECD has played a vital role in environmental protection and control: its use helped to prove the ubiquitous presence of chlorinated pesticides in nature and halocarbons in our atmosphere, and made us aware of the global extent of pollution. It was the ECD that made concentration ranges of parts-per-billion (ppb: $1{:}10^9$) or even parts-per-trillion (ppt: $1{:}10^{12}$) detectable. Today, these terms are routinely used without realizing how formidable such a sensitivity really

*Based on the article by L. S. Ettre and P. T. Morris, published in *LCGC* (*North America*) **25**, 164–178 (2007).

is: one ppb means that a spaceship (or an UFO, depending on one's inclination) could pick up a particular family of six from the whole living population of the Earth, and one ppt means that it could even find one piece of chewing gum in the pocket of one of the children. James Lovelock — the inventor of the ECD — illustrated its superior sensitivity by the following metaphor[1]: if one would pour about one liter of a perfluorocarbon liquid onto a blanket in Japan, and left it out to dry in the air by itself, a few weeks later one could detect on the west coast of England the vapor that had evaporated into the air in Japan from that blanket and carried by the jet stream around the world.

The ECD is an ionization detector and its response is based on the capability of molecules with certain functional groups to capture electrons generated by a radioactive source. The detector chamber contains two electrodes and a radioactive foil as the radiation source. Using an inert carrier gas with no affinity for electrons, the ions formed by the ionizing radiation can be collected, creating a steady standing current in the detector. When molecules of certain electron-absorbing solutes enter the detector chamber, they will capture electrons, resulting in a decrease of the standing current, giving a negative peak. In practice the recorded peaks are made positive by reversing the polarity of the recorder.

Since its invention the design of the ECD underwent a number of changes, but its principles remained the same, as shown in Fig. 23.1.[2] Also, different radioactive sources have been used: in Lovelock's original design the foil contained ^{90}Sr, but soon this was changed to tritium occluded in titanium foil. Today it almost universally contains ^{63}Ni. However, questions regarding the detector's construction are not our subject.

As already mentioned, the ECD is the brainchild of the extraordinary British scientist James E. Lovelock. Lovelock (born 1919) graduated in 1941 as a chemist from Manchester University, and then in 1948 obtained a Ph.D. in Medicine at the London School of Hygiene. After 1941 he was associated with the British Medical Research Council for almost 20 years. In 1958–1959 he was a visiting scientist at Yale University Medical School (New Haven, Connecticut) and then, from

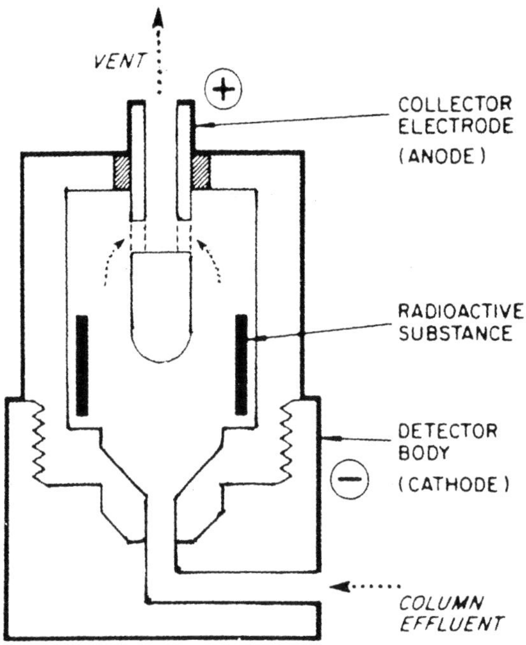

Fig. 23.1. Schematic of a typical early electron-capture detector.[2]

1960 to 1964 he was associated with Baylor College of Medicine in Houston, Texas, and the University of Houston as a professor. Since 1964 he has been a free-lance scientist serving as a consultant to various companies and institutions; among others he also cooperated in NASA's space programs. In the early 1970s Lovelock proposed his theory of Gaia, the living Earth, functioning as a superorganism where the physical environment and the life forms inhabiting the planet interact to maintain a more-or-less steady state. His fundamental contributions to our understanding of the impact of environmental pollution were recognized by three major awards: the Heineken Prize for the Environment of the Royal Dutch Academy of Sciences (1991), the VOLVO Prize (1996), and the Blue Planet Prize (1997), the latter generally considered as the environmental equivalent of the Nobel Prizes.

Lovelock has written a number of autobiographical treatments[1,3,4] in which he dealt in detail with his involvement in GC and the invention and development of the ECD. Our discussion is mainly based on these.

23.1. Inventions

The ECD was actually invented by Lovelock in four stages: around 1948, in 1954–1955, in 1956–1957, and finally in 1959. In the first two cases, the aim of his work was something different; the electron capturing effect he obtained was unexpected and at first not understood. In the third stage, the development of the argon-ionization detector, the crucial phenomenon was not an electron capturing effect, and only the fourth stage was aimed directly at the development of the ECD. These represent a fascinating story that illustrates how a scientist gradually perfects his work.

23.1.1. *First Stage: An Anemometer*

At the British Medical Research Council, Lovelock was seconded to its Common Cold Research Unit. This unit aimed to find reasons for the common cold, and Lovelock's task was to investigate the validity of the common belief that draughts of cold air can cause the illness. Scientists always try to express variables in quantitative terms, and thus air movements are to be expressed in terms of velocity. However, air movement due to draughts in a closed room is so slight that the then existing anemometers could not detect it, let alone measure it. Therefore, Lovelock invented a new anemometer, based on the disturbance of the slow flow of positive ions produced by the radiation from radium by even a very slight air flow. He obtained radium by scraping the dial paint from some gauges taken from the flight deck of Second World War military aircraft, ashing the scrapings, suspending the ash in a lacquer, and then painting with it the anemometer's ion source, which was surrounded by three wire circles in an open sphere and served as the collector of ions. This device worked well as an anemometer; in fact, it was sensitive not only to small air flows but even to the slightest cigarette smoke in the atmosphere. Lovelock also investigated whether other gases cause similar disturbance and found among others that very small concentrations of halocarbons also result in a similar response. In other words, his anemometer acted as an electron-capture detector although the reason for this behavior was not clear. At that time, nobody was interested in this observation and the device was too sensitive for use as an anemometer. Thus, while Lovelock described it in a short paper,[5] no further work was done with it.

23.1.2. Second Stage: Search for a High-Sensitivity Detector

In 1951 Lovelock was transferred to the National Institute for Medical Research (NIMR) of the Council, in London. There he was involved in another type of research where he needed to determine the composition of the lipids in the membrane of a living cell. Close to Lovelock's laboratory was the laboratory of A. J. P. Martin and A. T. James who were working on the adaptation of partition chromatography — invented in 1941 by Martin and R. L. M. Synge, see Chapter 14[6] — to systems permitting the separation of volatile compounds using an inert gas as the mobile phase, in other words gas-liquid partition chromatography, and who illustrated its possibilities through the separation of volatile fatty acids.[7] Lovelock asked their help and soon, a close collaboration was established between them.

Originally, James and Martin used titration to measure the eluted fatty acids and then Martin developed a gas density balance to serve as a general-purpose GC detector. However, he soon found out that it was not sensitive enough and was too complicated to be operated by an average chemist. (A formal description of the detector[8] was published only a few years after its actual development, although by then a few laboratories had self-constructed systems.) Therefore, Martin suggested to Lovelock that he invent a more sensitive detector. While he was working on this problem, Lovelock heard about a new GC detector described by the associates of two Shell research laboratories: Otvos and coworkers at the laboratories in Emeryville, California,[9,10] and Boer at the Koninklijke/Shell lab in Amsterdam.[11] This detector was based on the ionization of molecules by β-ray using ^{90}Sr as the radiation source. Lovelock immediately tried to build one, but he had problem with the carrier gas. Both Otvos and Boer used the light gases hydrogen and helium, but these two could not be used at the NIMR: helium was prohibitive due to its high price, and hydrogen was unacceptable for safety reasons. Thus, Lovelock tried to work with nitrogen, but his first results were discouraging: the sensitivity of the device did not meet their requirements. At that time Lovelock remembered his experience with the anemometer the sensitivity of which was very dependent on the applied potential and thus, tried to use different ranges with the new detector to find out whether the results can be improved. Indeed, when using low potentials he suddenly obtained dozen of large peaks when analyzing allegedly pure substances, but their retention times differed from the values the compound's peak should have had. In fact, what he now had was an electron-capture detector and the large peaks were due to trace quantities of electron-absorbing impurities in the sample.

In his recollections[4] Lovelock describes an interesting and most annoying observation during his investigations. One of the samples he tried out on the new detector was dissolved in carbon tetrachloride which was considered a non-reactive solvent. However, upon injection the ion current fell to zero and no further peaks could be observed. In fact it remained there for weeks and all attempts to restore the detector's operation failed. Only much later did he realize that CCl_4 is one of the most intensely electron-absorbing substances. Part of its vapors eluting from the column became adsorbed on the silicone rubber seal between the column and the detector in his home-made GC: this amount of CCl_4, slowly desorbing into the column effluent, became an almost permanent source of its vapor, making the detector useless for weeks. As he said, they had no use for such a temperamental detector: at that time it seemed to be useless for their purpose, and thus it was shelved and no results were published.

It is difficult to set exactly the dates of these developments. Martin certainly had developed the gas-density balance detector by 1954, and thus Lovelock had to start looking for a more sensitive detector soon after this time. Most likely he tried to use the modified version of the Shell detector around 1955 and had the argon incident, mentioned below, in 1956, leading to the argon-ionization detector (AID).

23.1.3. *Third State: The AID*

The development of this detector is part of the events leading to the electron-capture detector, and therefore should be briefly summarized here. While Lovelock was continuing his investigations targeting the construction of a universal, sensitive detector that utilized the principles of ionization, an unexpected event occurred: the Institute's store was temporarily out of nitrogen, but a flask of argon was available. Not wanting to wait until a new shipment of nitrogen arrived, Lovelock tried to use argon as the carrier gas. Applying a fairly high potential suitable to ionize the solute molecules, he indeed obtained the desired results: large peaks for the sample components and good, noise-free baseline. When the new nitrogen shipment arrived, he tried the same conditions with that carrier gas, but the results were poor, showing only unsatisfactory low sensitivity.

Further studies revealed that the reason for the unexpected results was the so-called Penning effect. In 1934 F. M. Penning of Philips (Eindhoven) discovered that when rare gas (argon) atoms are exposed to radiation, the resulting metastable atoms have a relatively long life and their concentration approaches that of the ions during steady irradiation. If traces of the vapors of some other gases are present, the metastable argon atoms transfer their energy on collision with these molecules, as long as the ionization potential of the other molecule is less than the energy level of the metastable argon atoms.[12] The ions so formed yield an increased cell current related to the concentration of the other vapor in the detector chamber. This brief summary explains how the argon-ionization detector (AID) was invented due to a lucky coincidence.

The AID was first discussed publicly by Lovelock in 1957 at an informal meeting of the newly formed Gas Chromatography Discussion Group, in Oxford, and then described in two publications.[13,14] It was commercialized almost immediately by the British instrument company W.G. Pye & Co., which introduced the so-called argon chromatograph at the Second International GC Symposium held in May 1958, in Amsterdam, The Netherlands (Fig. 23.2).

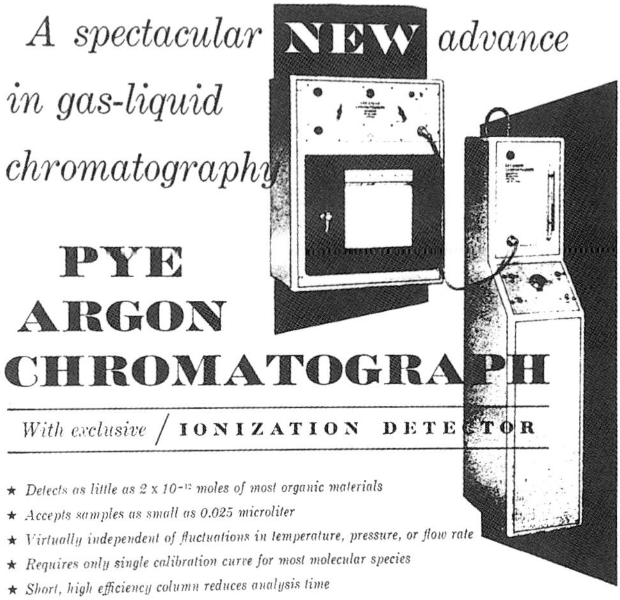

Fig. 23.2. The advertisement of the Pye Argon Chromatograph, the gas chromatograph equipped with an argon-ionization detector.

In the spring of 1958 Lovelock was invited to present a paper on the AID at the symposium Analysis of Mixtures of Volatile Substances organized 10–11 April in New York City by the New York Academy of Sciences.[15] There he met S. R. Lipsky, professor at Yale University Medical School in New Haven, Connecticut. Lipsky already built an ionization detector based on the Shell design (mentioned above), but had problems with it, and after Lovelock's lecture asked his advice. As a conclusion of their discussion, Lipsky invited Lovelock to stay for several months as a visiting professor at Yale.

Lovelock arrived at Yale soon after the Second International GC Symposium, held in May 1958, in Amsterdam, The Netherlands, where M. J. E. Golay presented his fundamental paper on open-tubular (capillary) columns (see Chapter 24),[16] and Lipsky, who at that time had been carrying out investigations trying to separate saturated and unsaturated fatty acids (involved in the development of arteriosclerotic heart disease), immediately wanted to use these columns with their superior separation capability. However, due to the columns' small sample capacity, the GC system would have needed a high-sensitivity detector. Lovelock's AID was an obvious choice, except that its original construction had a too large volume for the low flow rates used with these columns. Therefore, Lovelock redesigned it, making a small-volume version.[17] Using the low-volume AID and high-resolution capillary columns they soon were able to accomplish important separations of long-chain saturated and unsaturated fatty acids (as their methyl esters). After a presentation at the 134th National American Chemical Society Meeting held in September 1958 in Chicago, IL, they submitted two papers reporting on their results, showing, among others, for the first time the separation of methyl oleate from its *trans* isomer methyl elaidate.[18,19]

Toward the year's end Lovelock also visited A. Zlatkis at the University of Houston, Texas. Lovelock took with him the low-volume AID, and they used it with a capillary column for the separation of petroleum products. Their first chromatogram was obtained on 6 November 1958, in the presence of a number of visitors who signed it (Fig. 23.3). A more complex chromatogram containing the peaks of 25 C_5–C_8 paraffins and cycloparaffins was included in a short paper submitted within one month to *Analytical Chemistry*.[21]

The chromatograms included in these publications by Lovelock, Lipsky, and Zlatkis were the first showing the superior performance of capillary columns after the fundamental paper of Golay presented at the Amsterdam Symposium.

Soon after, W. G. Pye & Co. other instrument companies also added the AID to their GCs, most notably Barber-Colman in the United States (see Fig. 23.4). For a few years this detector had been used widely, mainly among biochemists, however, soon it was replaced by the flame-ionization detector (FID), which was first described almost simultaneously with the AID.[22,23] The FID proved itself to be simpler to operate and had a wider linear range. Today, the AID is almost forgotten.

23.1.4. *Fourth State: The Invention of the ECD*

In New Haven, Lovelock finally had time to return to his observations related to the electron-capturing effect of some molecules containing

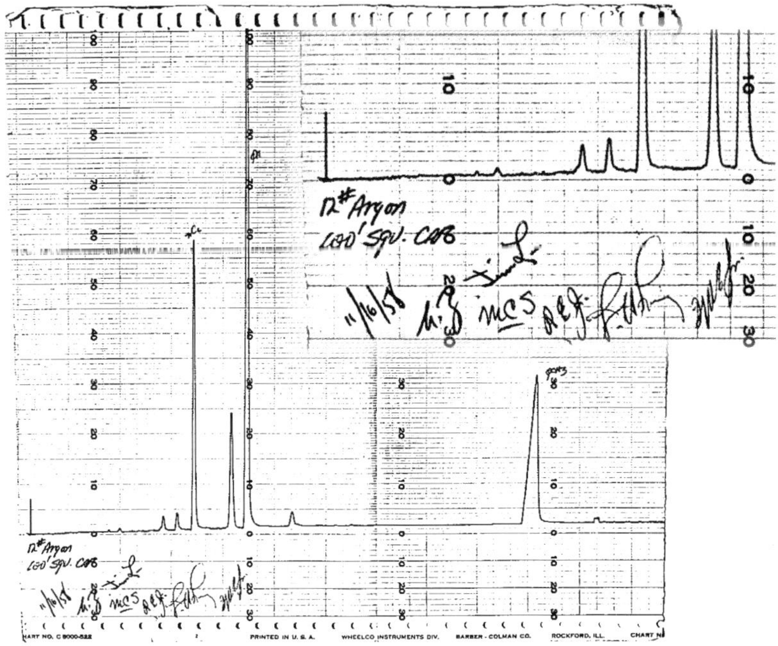

Fig. 23.3. The first chromatogram obtained at the University of Houston on 18 November 1958, using a capillary column with the small-volume argon-ionization detector of Lovelock. The last three peaks are *n*-hexane, benzene, and toluene. Identified signatures are those of A. Zlatkis, J. E. Lovelock, M. C. Simmons (Shell), R. E. Johnson and H. Lilly (both with Barber-Colman).[20] (Collection of L. S. Ettre)

certain functional groups and — together with Lipsky — reduced the design of the ECD to practice. He was also helped by Ken McAffee of Bell Telephone Laboratories, who suggested an improvement to the original electrical system of the detector. Their manuscript dealing with the principles and construction of the detector was finally submitted on 14 May 1959 to the *Journal of the American Chemical Society* and published in January 1960.[24] This paper generally is cited by everyone when speaking about the ECD and its applications, however it is most likely that very few people actually read it: namely it does not mention the ECD's particular sensitivity to compounds with certain functional groups. The paper suggests its use as a device for qualitative analysis instead, for the identification of certain compounds or compound groups. The suggested technique was

Fig. 23.4. The Barber–Colman Model 10 gas chromatograph, equipped with a glass column and an argon-ionization detector (the first instrument). Various other gas chromatographs can be seen in the background. (Collection of L. S. Ettre.)

to gradually increase the applied potential, decreasing in this way the electron capturing effect of individual compounds, until a saturation current was reached; at that point electron capturing by the particular compound or compound group becomes virtually non-existent, and the corresponding peak(s) disappear(s). Then, if the potential is increased further, the response becomes positive. As pointed out in the paper, the potential at which this transition occurs can be used for the characterization of the major classes of organic compounds.

23.2. Commercial Realization of the ECD

Instrument companies started to supply ECDs around 1961–1962. While in Houston, Lovelock also partnered with Al Zlatkis to form the small company Ionics Research, producing such detectors for commercial use (Fig. 23.5). For a few years they supplied the detectors for the instruments of the Perkin-Elmer Corporation. In general the early detectors were fairly delicate to use and I remember that we often had to call them in Houston to straighten out some problems. The requirement for a special license to use the detectors, which had radioactive material in them, represented a general complication for customers.

Fig. 23.5. J. E. Lovelock (left) and A. Zlatkis, in Houston, in 1961. Zlatkis is holding an electron-capture detector. (Collection of L. S. Ettre.)

Also, at the very beginning the field of application of the detector was not clear: it became only gradually the most important device in the environmental movement, and such applications first had to be demonstrated. In this respect Keene Dimick, the founder of Wilkens Instrument and Research Co. (the present-day Chromatography Division of Varian) and the editor of the company's quarterly publication called *Aerograph Research Notes*, was particularly successful. The title page of its Summer 1962 issue was particularly striking: it showed a single peak of a pesticide corresponding to 10^{-12} g, with the headline "Have you ever seen a picogram?" printed in large, bold-faced characters. Figure 23.6 shows a similar publicity chromatogram from Perkin-Elmer, demonstrating the analysis of the extract of an earthworm that indicated the presence of five pesticides.[25]

23.3. The Electron Capture Detector and the Environmental Movement

In general the start of the environmental movement is identified with the publication of the book *Silent Spring* by Rachel Carson, in 1962,[26] documenting the detrimental influence of pesticides to the environment. Its title refers to the result of their indiscriminate use: these

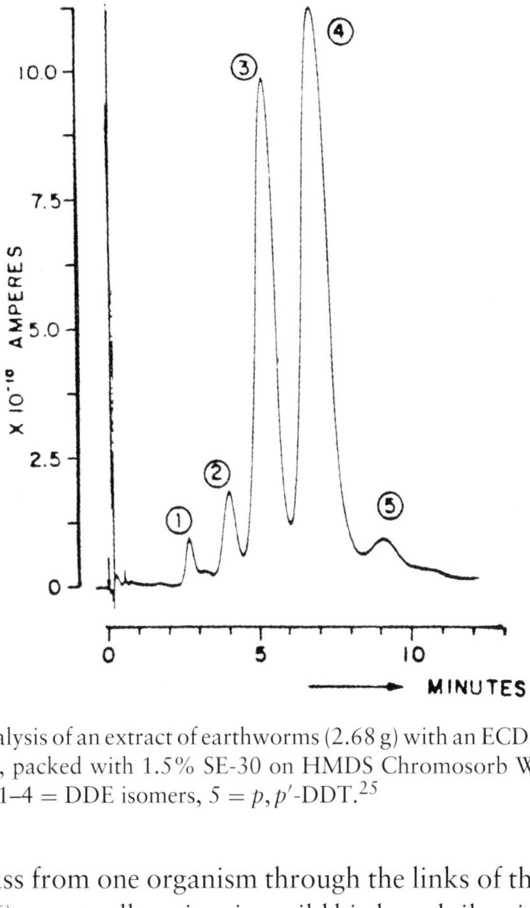

Fig. 23.6. Analysis of an extract of earthworms (2.68 g) with an ECD. Column: 6 ft × 2 mm i.d. glass, packed with 1.5% SE-30 on HMDS Chromosorb W 80/100 mesh; 175°C. Peaks: 1–4 = DDE isomers, 5 = p,p'-DDT.[25]

chemicals pass from one organism through the links of the food chain (see Fig. 23.6), eventually poisoning wild birds and silencing the forests and meadows.

In the literature one can often find implications that Carson's book was made possible by the use of the ECD: even a relatively recent article in *The New York Times* expressed this opinion, saying that Lovelock's invention was "providing a foundation for the work of Rachel Carson."[27] This is, however, not true: *Silent Spring* was based on investigations in the 1950s using other analytical techniques, and this is clear if one checks the more than 500 literature references given in the book. However, it is true that the use of the ECD provided the infallible proof of the correctness of Carson's conclusions. Lovelock cites the first two papers reporting on the use of the ECD for the determination of pesticide residues. The first was a presentation by British scientists from Shell at the 18th International Congress on Pure and Applied Chemistry, in August 1961, in Montreal, Canada,[28] and the other by associates of the U.S. Food & Drug Administration (FDA) presented at the 75th Annual Meeting of the Association of Official Agricultural Chemists, held in October 1961 in Washington, DC.[29]

23.3.1. *The Chlorofluorocarbon Problem*

There is one major field where Lovelock's personal activities have served as the foundation of our awareness of the detrimental effect of air pollution and eventually led to banning the use of halocarbons. We should finish our discussion with a brief summary of these activities.

In 1966 Lovelock spent the summer with his family on the westernmost coast of Ireland and was surprised to see haze often over the Bay, depending on the wind. Next summer, upon returning to Ireland he brought with him a portable gas chromatograph and regularly measured the concentration of chlorofluorocarbon Freon F11 (trichlorofluoromethane) — the most widely used halocarbon — during and after the outbreak of hazy air. His measurements clearly established a halocarbon concentration of around 150 ppt in the case of a haze. However, it also demonstrated that even on clear days there is a small steady background concentration of about 50 ppt. The high concentration during haze could easily be attributed to pollution brought by winds from the European continent; however, there was no explanation for the steady background concentration on clear days when the winds came from the Atlantic Ocean: did this mean that the air is polluted there? To investigate this, in November 1971 Lovelock joined the Research Ship Shackleton of the British Natural Environment Research Council, for a six-month voyage traveling from England to Antarctica, and he carried out regular measurements of atmospheric halocarbon concentration over the Atlantic Ocean. The results of these measurements[30] clearly indicated the accumulation of Freon 11 and other halocarbons (used in aerosol cans and as refrigerant) in the Earth's atmosphere, serving as the source of the steady background concentration he observed on the Irish coast. Lovelock's data collected during this voyage have served as the basis on which F. S. Rowland and M. J. Molina were able to develop their theory on the decomposition of the halocarbons in the stratosphere, releasing chlorine that in turn, is catalyzing the depletion of stratospheric ozone.[31] For their work Roland and Molina (together with P. Crutzen of The Netherlands) received the 1995 Chemistry Nobel Prize.

For a long time the ECD was the most sensitive GC detector, with its unique selectivity. Recently, improvements in GC–mass spectrometer systems make these a rival to it. However, GC–ECD systems still remain the workhorse instruments for routine pesticide determinations in water and soil, PCBs in the transformer oils, and halocarbons in air.

References

1. J. E. Lovelock, *Homage to Gaia* (Oxford University Press, Oxford, 2000), pp. 181–190.
2. L. S. Ettre, *J. Chromatogr. Sci.* **16**, 396–417 (1978).
3. J. E. Lovelock, in *75 Years of Chromatography — A Historical Dialogue*, eds. L. S. Ettre and A. Zlatkis (Elsevier, Amsterdam, 1979), pp. 277–284.
4. J. E. Lovelock, in *Electron Capture Theory and Practice in Chromatography*, eds. A. Zlatkis and C. F. Poole (Elsevier, Amsterdam, 1981), pp. 1–11.
5. J. E. Lovelock and E. M. Wasilewska, *J. Sci. Instrum.* **26**, 367–370 (1949).
6. A. J. P. Martin and R. L. M. Synge, *Biochem. J.* **35**, 1358–1368 (1941).
7. A. T. James and A. J. P. Martin, *Biochem. J.* **50**, 679–690 (1952).
8. A. J. P. Martin and A. T. James, *Biochem. J.* **63**, 138–142 (1956).
9. J. W. Otvos and D. P. Stevenson, *J. Am. Chem. Soc.* **78**, 546–551 (1956).
10. C. H. Deal, J. W. Otvos, W. N. Smith and P. S. Zucco, *Anal. Chem.* **28**, 1958–1964 (1956).
11. H. Boer, in *Vapour Phase Chromatography (1956 London Symposium)*, ed. D. H. Desty (Butterworths, London, 1957), pp. 169–184.
12. F. M. Penning, *Physica (Amsterdam)* **1**, 1028–1044 (1934).
13. J. E. Lovelock, *Nature (London)* **181**, 1460–1462 (1958).
14. J. E. Lovelock, *J. Chromatogr.* **1**, 35–46 (1958).
15. J. E. Lovelock, A. T. James and E. A. Piper, *Ann. N.Y. Acad. Sci.* **72**, 720–730 (1959).

16. M. J. E. Golay, in *Gas Chromatography 1958* (*Amsterdam Symposium*), ed. D. H. Desty (Butterworths, London, 1958), pp. 36–55.
17. J. E. Lovelock, *Nature* (*London*), **182**, 1663–1664 (1958).
18. S. R. Lipsky, R. A. Landowne and J. E. Lovelock, *Anal. Chem.* **31**, 852–856 (1959).
19. S. R. Lipsky, J. E. Lovelock and R. A. Landowne, *J. Am. Chem. Soc.* **81**, 1010 (1959).
20. L. S. Ettre, *Anal. Chem.* **57**, 1419A–1438A (1985).
21. A. Zlatkis and J. E. Lovelock, *Anal. Chem.* **31**, 620–621 (1959).
22. I. G. McWilliam and R. A. Dewar, *Nature* (*London*) **181**, 760 (1958).
23. I. G. McWilliam and R. A. Dewar, in *Gas Chromatography 1958* (*Amsterdam Symposium*), ed. D. H. Desty (Butterworths, London, 1958), pp. 142–152.
24. J. E. Lovelock and S. R. Lipsky, *J. Am. Chem. Soc.* **82**, 431–433 (1960).
25. E. W. Cieplinski, *Instrument News* **15**(2), 7–8 (1964).
26. Rachel Carson, *Silent Spring* (Houghton Mifflin Co., New York, 1962); 7th printing: 1994.
27. A. C. Revkin, *The New York Times Science Section*, 12 September 2006.
28. E. S. Goodwin, R. Golden and J. G. Reynolds, *Analyst* **86**, 697–709 (1961); **87**, 169 (1962).
29. J. O. Watts and A. K. Klein, *J. Assoc. Off. Agr. Chem.* **45**, 102–198 (1962).
30. J. E. Lovelock, R. J. Maggs and R. J. Waade, *Nature* (*London*) **241**, 194–196 (1973).
31. F. S. Rowland and M. J. Molina, *Rev. Geophys. Space Phys.* **13**, 1–36 (1975).

Chapter

24

Evolution of Open-Tubular (Capillary) Columns for Gas Chromatography*

In this chapter we shall summarize the events in the evolution of open-tubular (capillary) columns for gas chromatography, explain the key problems the pioneers were facing, and point out individual achievements.

To start, we have to go back over 50 years, to the beginnings of gas chromatography. At that time, Marcel J. E. Golay (Fig. 24.1) had joined the Perkin-Elmer Corporation as a consultant, after a 25-year distinguished career at the US Signal Corps Engineering Laboratories, in Fort Monmouth, NJ. He was originally trained as an electrical engineer and mathematician at the Federal Technical University (ETH) in Zurich, Switzerland, at that time the world's most prestigious technical school, and he received his PhD in nuclear physics from the University of Chicago. Golay had a very broad range of interest and had worked in a number of fields. His connection with Perkin-Elmer was mainly

*Based on the article by L. S. Ettre published in *LCGC* (*North America*) **19**, 48–59 (2001).

Fig. 24.1. M. J. E. Golay, around 1960.

due to his involvement in the development of an IR detector, originally conceived as an aircraft detecting device, and of a multi-slit IR spectrometer.

24.1. Invention

When Golay joined Perkin-Elmer, everybody were excited by the versatility and separation power of gas chromatography and inevitably, he also became involved in various discussions on the new technique, for him a totally unknown field. He became intrigued by the mathematics of the separation process and being an electrical engineer by training and experience, he tried to interpret it with the help of the "telegrapher's equation" used to describe the process in transmission lines. He presented this unique comparison at the GC symposium organized in conjunction with the Spring 1956 National Meeting of the American Chemical Society[1] (see Chapter 31). In the subsequent months Golay continued to investigate — at first theoretically — the separation process taking place in the (packed) chromatographic column. In order to simplify the system he constructed in his mind a model, consisting of a bundle of capillary tubes, each corresponding to a passage through the column packing. These ideal capillaries would not be restricted by the geometry of the packing or by the randomness of the passages through it, which are beyond our control; therefore, they should behave close to the theoretical possibilities. Golay's considerations were outlined in a number of internal reports from which the one dated September 5, 1956, was most important.[2] In this he suggested some experiments with a capillary tube, 0.5–1 mm in diameter, wetted with a suitable stationary phase (corresponding to one of these passages), to determine the correctness of his assumptions.

These experiments were carried out in the fall of 1956. A serious handicap was represented by the large volume of the then existing (thermal conductivity) detectors and of the injection block. Therefore, Golay designed a micro-thermal conductivity detector and used that for the investigations. The results were indeed striking: Figure 24.2 presents Golay's first chromatograms obtained on a 12 ft × 0.055 in. (1.37 mm) i.d. column, showing the analysis of a Phillips 37 hydrocarbon mixture and of a mixture of isomeric pentanes.

In the next 16 months Golay carried out a very intensive theoretical study, in addition to some experimental work. He presented an interim report at the GC symposium organized by the Instrument Society of America, held in August 1957 in East Lansing, MI[3] (see Chapter 31) and he also discussed the new concept and theory with a number of scientists, most notably with A. J. P. Martin, the inventor of gas–liquid partition chromatography, and A. I. M. Keulemans, one of the most widely known scientists in this field at that time. His final report including the full theory of open-tubular (capillary) columns — which is still valid today, 50 years later — was presented at the next international GC symposium held in May 1958 in Amsterdam, The Netherlands.[4]

24.2. Realization

Golay's presentation at the Amsterdam Symposium with its 93 equations was impressive enough in itself. Still, in the original form published in the preprints of the lectures, it probably would have had little impact: it sounded too theoretical. However, in his actual presentation at the meeting he showed two chromatograms obtained just a couple of days earlier by Richard D. Condon, his young associate at Perkin-Elmer, showing the separation of C_8 hydrocarbons and the

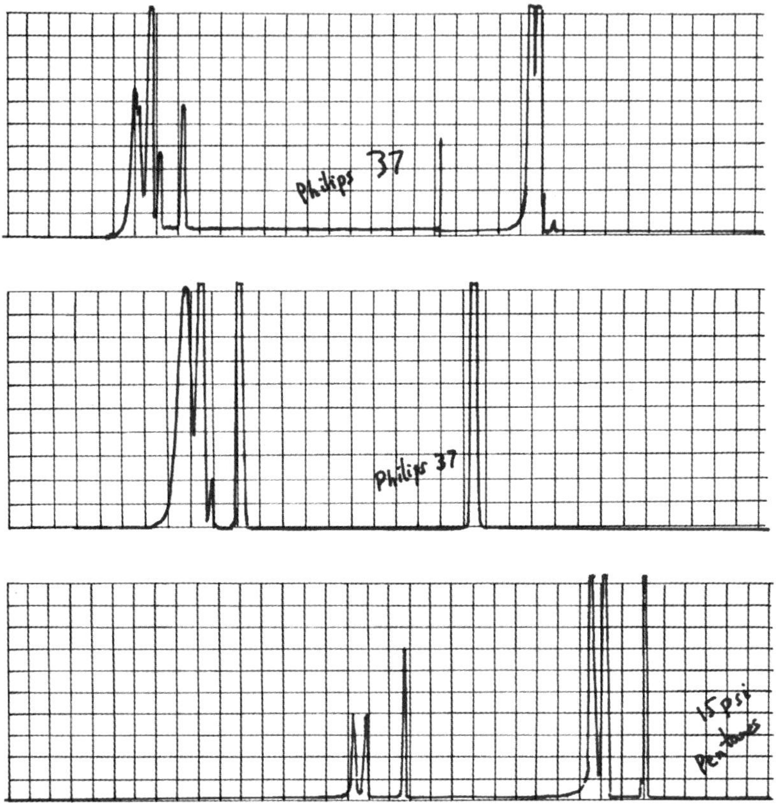

Fig. 24.2. Golay's first chromatograms on capillary columns showing the analysis of a Phillips 37 mixture and isomeric pentanes. Column: 12 ft (366 cm) × 0.055 in. (1.37 mm) i.d. coated with Carbowax 1540 poly(ethylene glycol). Room temperature. Sanborn high-speed recording galvanometer and a specially built microthermal conductivity detector.

xylene isomers, on a 150 ft × 0.010 in. (0.25 mm) i.d. stainless-steel column, coated with diisodecyl phthalate. Thirty years later, Dennis H. Desty, the organizer of the Symposium (who in the subsequent years had a major role in the general use of capillary columns), still remembered the excitement caused by these chromatograms, demonstrating the exceptional separation power of capillary columns, which was up to then impossible to achieve even on the best packed column[5]:

> I well remember the gasp of astonishment from the audience at this fantastic performance that was to change the whole technology of

gas chromatography over the next decade. We were all enraptured by the elegant simplicity of Marcel's concept and I could not wait to dash off to my laboratory to start experiments with the wonderful new tool.

As mentioned earlier, Golay had to build a special micro-thermal conductivity detector for his investigations because the existing detectors had much too large volume and not enough sensitivity for the small sample sizes and low carrier gas flow rates needed. Fortunately, at the same Amsterdam Symposium I. G. McWilliam and R. A. Dewar described in detail the flame-ionization detector (FID),[6] and within a short time James E. Lovelock modified his argon-ionization detector making it suitable for the capillary column to work[7] (see Chapters 22 and 23). Thus, within a short time, all the necessary basic ingredients were available for the practical use of the capillary columns.

Most likely, the first who utilized a capillary column — FID system was Desty at British Petroleum Co. Ltd in England. According to his personal recollections[8] they put together a crude setup practically days after returning from Amsterdam. It consisted of a breadboard model of the FID and a 250-ft long stainless steel tube coated with squalane using the dynamic coating technique described by G. Dijkstra and J. De Goey at the Amsterdam Symposium;[9] their system also included a (crude) split injection system. Within a few weeks Desty's group consisted of B. H. F. Whyman, A. Goldup, and W. T. Swanton constructed a complete apparatus for operation up to 250°C and explored the separation of a wide variety of samples using columns made of stainless steel and copper tubes. Desty first reported on this system at a symposium held October 9–11, 1958, in Leipzig, East Germany,[10] followed by a detailed presentation at the meeting of the (British) Gas Chromatography Discussion Group held on April 10, 1959, in London.[11]

At Perkin-Elmer prototypes of the FID were also constructed soon after the Amsterdam Symposium, and by the fall of 1958 R. D. Condon was already obtaining one excellent chromatogram after the other on a working prototype of a gas chromatograph with open-tubular (capillary) columns and an FID. This instrument (the Model 154-C) was then introduced at the Pittsburgh Conference, in March 1959, where Condon had a major presentation describing this system and illustrating the wide range of applications of capillary columns.[12] Parallel to but independently of this work, Albert Zlatkis, at the University of Houston[13] and S. D. Lipsky, at Yale University Medical School,[14,15] both with the help of J. E. Lovelock, explored the use of capillary columns, and reported on their results in the first months of 1959 (see Chapter 23).

The evolution of capillary columns from this beginning to the present universal use went through a number of steps; below we shall deal with the most important key developments which finally made these columns everybody's tool.

24.3. Columns Made of Metal

The very first "capillary column" investigated by Golay was actually a (uncoated) 10 m × 3 mm i.d. Teflon tube; however, soon he switched to glass and then to stainless steel tubes of two internal diameters: 0.010 in. (0.25 mm) and 0.020 in. (0.51 mm). Such tubes could be easily obtained: as mentioned by Golay during the discussion of his Amsterdam paper, "you buy them by weight." Capillary columns made of stainless steel (and to a lesser extent, of copper) have been in general use for well over a decade starting in 1958. Retrospectively, such columns had definite limitations due to the relative unevenness of the inside tube surface which necessitated a relatively thick stationary phase film coating, and to the activity of the metal surface. In spite of this, however, properly coated metal capillary columns — with both nonpolar and polar phases — have been successfully used for the analysis of a wide variety of samples: the two chromatograms shown in Figs. 24.3 and 24.4 (both obtained in 1963) illustrate their performance. With respect to the lifetime of the columns, it is sufficient to quote the statement of Halász during a discussion at the 1961 Lansing Symposium[16]:

> With a copper column coated with squalane, we worked for about 7 months, eight to ten, sometimes for 24 hours daily.

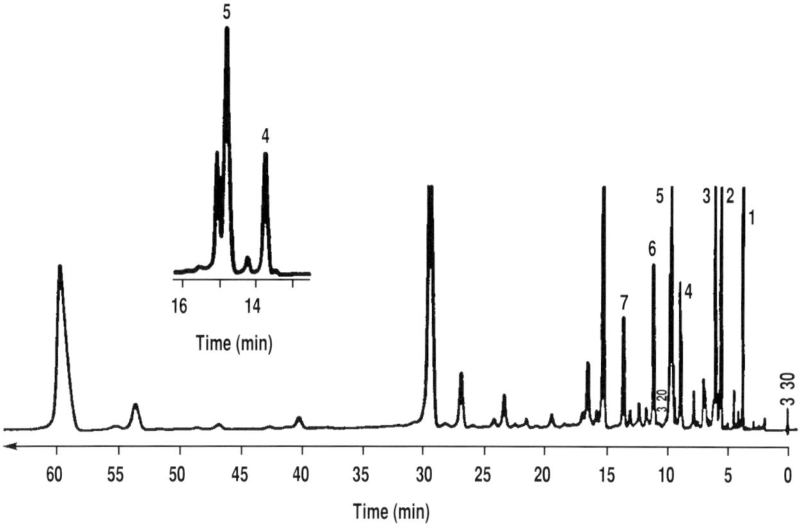

Fig. 24.3. Analysis of a fatty acid methyl ester mixture obtained from menhaden oil. (F. J. Kabot, Perkin-Elmer, 1963). Column: 150 ft (45.7 m) × 0.010 in. (0.25 mm) i.d. capillary, coated with butanediol succinate. Carrier gas: nitrogen for the full cromatogram (inlet pressure: 40 psig) and helium for the cut-out chromatogram (inlet pressure: 24 psig). Column temperature: 185°C. Split injection. Flame-ionization detector. Peaks: methyl esters of (1) myristic, (2) palmitic, (3) palmitoleic, (4) stearic, (5) oleic, (6) linoleic, and (7) linolenic acids.

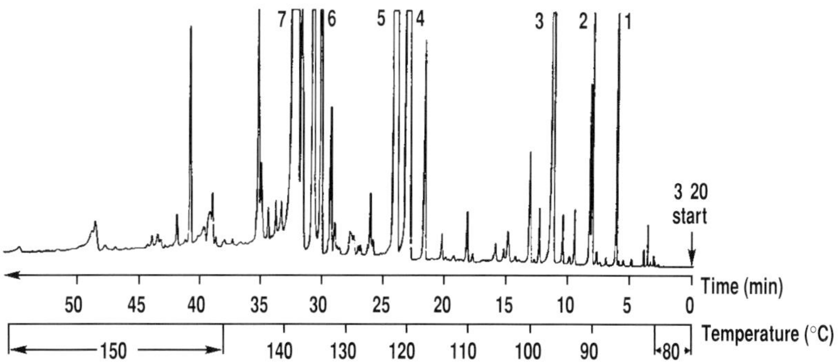

Fig. 24.4. Analysis of a peppermint oil sample from Yakima Valley, Washington (E. W. Cieplinski and W. Averill, Perkin-Elmer, 1963). Column: 150 ft (45.7 m) × 0.010 in. (0.25 mm) i.d. capillary, coated with Ucon Oil 50 HB 200 poly(propylene glycol). Carrier gas: helium with 20 psig inlet pressure. Column temperature: programmed, at 2°C/min. Split injection. Flame-ionization detector. Peaks: (1) α-pinene, (2) β-pinene, (3) eucalyptol, (4) menthone, (5) menthofuran, (6) menthyl acetate, (7) menthol.

The early stainless steel columns were made of tubing with a thick wall (about 1.6 mm) and these were heavy and bulky (see Fig. 24.1). However, by 1962, improved stainless steel tubing with a thin (0.12–0.15 mm) wall became available. This tubing was much more flexible and had a better, more inert, and smooth inner surface.

For the reduction of the activity of the inner tube surface Warren Averill proposed in 1961 the addition of a small amount (about 1%–2%) of a surface-active agent to the stationary phase.[17] A typical such additive is Atpet 80, which is chemically sorbitan monooleate. The polar hydroxy groups at one end of its molecules are permanently adsorbed by the tube wall, thereby deactivating its active sites; at the same time, the long hydrocarbon chain remains free and results in a velvet-like structure that spans the interface between the tube wall and the coated stationary phase film. As pointed out by F. Farré-Rius *et al.*,[18] such additives also reduce the surface tension and thus facilitate the spreading of the stationary phase.

24.4. Coating Technique

With regard to the coating technique practically everybody adapted the dynamic procedure originally described by G. Dijkstra and J. de Goey.[9] In this, a plug of the stationary phase solution is slowly forced through the tubing with the aid of a dry inert gas, wetting in this way the inside wall of the tube with the solution. Subsequently the solvent is evaporated by blowing dry gas through the column for a few hours. This technique has been discussed in detail in the literature[19,20] and if carried out skillfully, it resulted in columns with good performance and long life. A major advantage of the dynamic method was that it did not need any complicated setup; however, its shortcoming was that the thickness of the coated film depended on the coating conditions, and it could not be readily established but only estimated. In spite of this, the technique had been in general use for well over a decade and was replaced only slowly in the 1970s by the static coating technique, in the form as described by J. Bouche and M. Verzele in 1968.[21] In this the tube is fully filled with the stationary phase solution, one of its end is closed and then the solvent is slowly evaporated through the open end under reduced pressure, at a temperature below the solvent's boiling point. (Actually a similar technique had already been used in 1961–1962 by Cs. Horváth in the preparation of the support-coated open tubular columns.) The major advantage of this technique is that from the concentration of the coating solution the thickness of the coated film can be readily established.[22]

24.5. Columns Made of Plastic Tubing

In spite of the convenience of metal (stainless steel) columns, it was obvious from the beginning that a more inert tube material would be needed. Plastic tubing was proposed by R. P. W. Scott[23] but there were obvious disadvantages with such materials: temperature limitations, poor coatability, and short life due to plasticizer migration. Thus, except for some early works such columns never gained ground.

24.6. The Era of Glass Capillary Columns

An obvious choice of the tube material would have been glass and this possibility was explored by Golay in his early work: in fact, in the publicity photo made around 1960 (Fig. 24.1) he was holding a glass capillary tube in his hands! At that time, a number of chromatographers tried to prepare capillary columns made of glass. However, this was not as simple as it sounds.

Usually the "era of glass" is considered to have been started with the development of an ingenious device to prepare glass capillary tubes by Desty and his co-workers, in 1960;[24] a similar device was also described at that time in France by A. Kreyenbuhl.[25] In 1960–1961 Desty and his associates published a number of papers in which they used glass capillary columns, the most famous being the analysis of a Ponca Crude petroleum sample on a 263-m long, 0.14 mm i.d. column, in 3½ h.[26] Then, in the second part of the 1960s Desty's machine became commercially available. With them capillary tubes of various lengths and diameters could be prepared using both soda-lime and borosilicate (Pyrex) glass tubes. The capillary tubes produced in these machines had a thick wall, and their final form was that of rigid coils, typically with a coil diameter of 13–15 cm and column internal diameter of 0.23–0.27 mm, with a wall thickness of about 0.20–0.25 mm. Naturally, glass capillary columns were fragile; in the hands of skilled operators, however, very little damage was done. Thus, one may ask, why did glass tubing replace metal only in the early 1970s, over ten years after the description of the glass drawing machine?

The problem arose from the poor coatability and short life of glass capillary columns prepared in this period. The situation was well characterized by Halász in his (already quoted) remark at the 1961 Lansing Symposium.[16] While emphasizing the long life and good performance of metal columns, he continued by saying that

> with glass capillary columns ...coated with squalane, we were unable to work longer than 2 or 3 days. On glass columns coated with squalane, you can see with your eyes after two days that your film is not in one place.

Although squalane was a bad example — it was found that the forming of a stable squalane film on glass is very difficult, even with the best column pretreatment and coating technique — this statement illustrates the state of the art in the first part of the 1960s. It took years until the reasons for this problem were understood: it was due to the strong cohesive forces of liquids on the glass surface. These forces are characterized by the surface tension which in turn can be characterized by the contact angle of a drop on the solid surface: the higher the contact angle the poorer is the spreading of the liquid. The extent of this phenomenon was first investigated in 1962, by Farré-Rius and co-workers who measured the contact angles of liquid phases on various potential column tube materials.[18] Because of this problem the inside surface of the glass tube has to be treated in some way prior to coating, in order to increase its wettability.

The breakthrough came in 1965–1968 through the work of K. Grob[27] describing a way to deposit a carbon layer and M. Novotný and K. Tesařik[28,29] who etched the internal surface of the tube with dry HCl or HF. In the decade that followed scores of different ways were developed for the treatment of the inside surface of the glass capillary tube. Such treatment was needed not only to improve the coatability, but also to make the tube more inert: metal or other ions in its composition could be detrimental and had to be eliminated. A particular problem was boron, present in fairly high concentration (13% as B_2O_3) in borosilicate glass.[30]

The decade of the 1970s was the most exciting period in the evolution of capillary columns. This was the time when these columns really started to become everybody's tool, and chromatographers even created a special acronym to describe their field, calling it $(GC)^2$ for glass capillary gas chromatography. These columns were made in different lengths and diameters and coated with a wide variety of stationary phases, having a wide range of film thickness. One of the most impressive chromatograms from that period is shown in Fig. 24.5, obtained

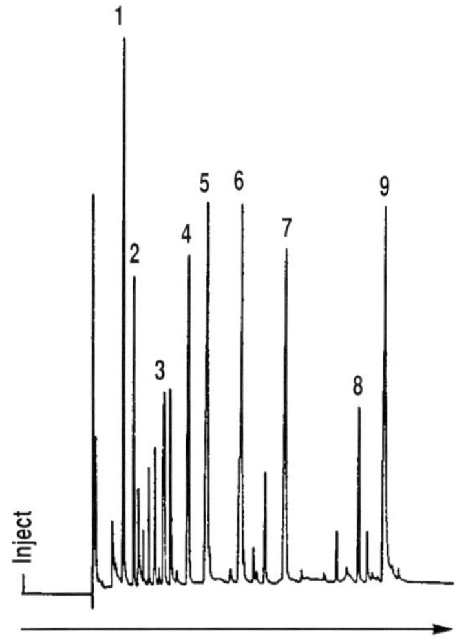

Fig. 24.5. Analysis of the amino acids in ribonuclease hydrolyzate, in the form of the n-propyl, N-acetyl derivatives. Column: 50 m × 0.27 mm i.d. glass capillary, coated with a 1:1 mixture of Carbowax 20M and Silar CP. Film thickness: $<0.1\ \mu m$. Carrier gas: helium with a flow rate of 0.75 mL/min. Column temperature programmed at 8°C/min from 110°C to 190°C and then ballasted to 250°C. Split injection. Thermionic detector sensitive for nitrogen-containing compounds. Peaks: 1 = alanine, 2 = valine, 3 = norleucine (internal standard), 4 = threonine, 5 = serine, 6 = aspartic acid, 7 = glutamic acid, 8 = tyrosine, 9 = lysine. The retention time of the last peak is less than 30 min.[31]

in a column having a coated film thickness of less than 0.1 μm: it demonstrates the analysis of amino acid derivatives from a ribonuclease hydrolyzate.[31]

In the 1970s supply houses already manufacturing packed columns started to also provide capillary columns for chromatographers. However, at the end of the decade the situation suddenly changed almost from one day to the other, by the introduction of the capillary columns made of flexible fused-silica tubing.

24.7. Fused-Silica Columns

Back in 1960, when the glass drawing machine was developed by Desty's group, its use to make capillary columns from quartz was also mentioned.[32] However, at that time no further work was done along this line, because for such use the burner softening the tube material would have to be modified to facilitate much higher temperature. Later Desty went back to this possibility and built a modification of the capillary drawing machine for this application. In a paper presented at the 1975 Capillary GC Symposium held in Hindelang, Bavaria, he briefly described this modified capillary drawing machine, showing its photograph,[33] but gave no data on actual column manufacturing. It should be noted that the system developed by Desty would have produced rigid, thick-walled quartz (fused-silica) columns and his major problem at that time was to find suitable platinum tubes which can be heated to the needed 1250°C–1350°C for the formation of the coiled capillary tubes.

It is interesting to note that in the mid-1970s Grob was also considering to use quartz as the column tube material. He recognized that it is more inert than glass; however, he believed that his methods developed to modify the inner surface of glass capillaries were satisfactory and saw no need to change to a new tube material.[34]

Because of this lack of previous work, the paper of R. D. Dandeneau and E. H. Zerenner of Hewlett-Packard, presented at the Third Hindelang Symposium (April 29–May 3, 1979), was a complete surprise to the participants.[35] They described the production and use of *thin-walled*, *flexible* fused-silica columns; this tubing was an adaptation of the production of fiber optics already carried out at

Hewlett-Packard's Palo Alto, California, facilities. They also realized that in the thin wall of the tubing cracks can develop, eventually leading to breakage. In order to prevent this, the outside of the tubing was coated immediately after drawing, first with silicone rubber, but soon this was changed to polyimide.[36,37]

There is no question that the introduction of fused-silica columns changed not only capillary gas chromatography, but the whole field of separation techniques. Within a few months column supply houses had already started to offer fused-silica capillary columns to the users and very soon these made the glass columns obsolete. A good comparison of the characteristics of glass and fused-silica capillary columns was given soon after their introduction in a small monograph by Jennings.[38]

Above, the words "quartz" and "fused silica" are used interchangeably. In practice, however, the first term is always used for the natural material, while the latter for the synthetic product prepared from silicon tetrachloride. The difference between the two is in the amount of impurities present: natural quartz may contain trace amounts of metals up to a total concentration of about 50–60 ppm, while the total amount of metallic impurities in the synthetic fused silica is in the order of 0.08–0.5 ppm.[39]

Following the presentation of Dandeneau and Zerenner, other scientists also investigated in detail the various questions associated with fused-silica capillary columns. In this respect the activities of S. R. Lipsky should be particularly emphasized.[39,40] A few years later he also pioneered in the development of columns which could be used up to about 400°C–450°C, having a bonded stationary phase and a very thin aluminum outer coating instead of polyimide.[41]

Today, fused-silica capillary columns are used universally in gas chromatography. They are manufactured by a number of companies providing columns with specified parameters and performance, also illustrating their application fields.

24.8. Immobilized and Bonded Stationary Phases

One cannot finish the discussion of the evolution and continuous improvements of capillary gas chromatography without mentioning one additional subject: improvements in the stationary phases.

In the first two decades of the evolution of GC most of the stationary phases used in the columns were taken "off the shelf": chemicals readily available in the laboratory were utilized. Except for a few these were compounds with fairly low boiling and low molecular weight, with significant vapor pressures even at moderate temperatures. This fact restricted the temperature range of GC, and experience has shown that the upper temperature limit of capillary columns was actually lower than that of packed columns. In fact, even by the middle of the 1970s, one could rarely find a capillary chromatogram in which the column was heated above 200°C.

In the 1970s the situation changed drastically with the introduction of silicone (polysiloxane) phases, custom-made for GC and particularly for capillary column use. These were high-molecular-weight polymers with molecular weights in thousands or tens of thousands range, and a few even exceeding 100,000. In contrast, squalane, one of the most common phases of the 1960s, has a molecular weight of 423, and the average molecular weight of Carbowax 1540 poly(ethylene glycol) — another popular phase of the period — is 1540. These new phases can be coated well from their solution on the inner surface of the glass and also fused-silica tubing; they provide excellent chemical and thermal stability, with low bleeding, permitting the extension of the column's upper temperature limit.

The molecular weight of the stationary phase used for coating the capillary columns is restricted by the need that it must be soluble, since the capillary tube is coated using a solution of the phase, and in general, the higher the molecular weight, the more difficult it is to dissolve the substance. Therefore, ways had to be found to overcome this limitation. This was accomplished by an additional step: a secondary polymerization in the column, resulting in a coated stationary phase film with very high molecular weight. The products of this process were called *immobilized phases*. Additionally a chemical bond may also be formed between the stationary phase molecules and the surface silanol groups on the inside surface of the fused-silica tubing: in this way, the so-called *bonded phases* are created.

Besides providing capillary columns which can be safely used at higher temperatures, immobilization and/or chemical bonding of the stationary phase result in two additional advantages. Such a phase can tolerate the injection of large volumes of a solvent without dissolution in it. In addition, it is possible to prepare stable capillary columns with a wide range of film thickness, even with a very thick film. Without this treatment, such columns would soon lose most of their original coating through bleeding.

The basic work on immobilization and cross-linking was carried out in the decade between 1976 and 1986 by a number of groups such as those of C. Madani in France, L. Blomberg in Sweden, K. Grob in Switzerland, P. Sandra in Belgium, G. Schomburg in Germany, V. Pretorius in South Africa, as well as S. R. Lipsky and M. L. Lee in the United States. A good summary of the questions associated with this revolution in column technology was provided by E. F. Barry, giving also the pertinent references.[42] From the mid-1980s on, these techniques are part of the routine column manufacturing technology.

References

1. M. J. E. Golay, *Anal. Chem.* **29**, 928–932 (1957).
2. M. J. E. Golay, Discussion of the Results with Three Experimental Gas-Liquid Chromatographic Columns, *Engineering Report No. 523* (The Perkin-Elmer Corp., Norwalk, CT, September 5, 1956).
3. M. J. E. Golay, in *Gas Chromatography (1957 Lansing Symposium)*, eds. V. J. Coates, H. J. Noebels and I. S. Fagerson (Academic Press, New York, 1958), pp. 1–13.
4. M. J. E. Golay, in *Gas Chromatography 1958 (Amsterdam Symposium)*, ed. D. H. Desty (Butterworths, London, 1958), pp. 36–55.
5. D. H. Desty, *Chromatographia* **23**, 319 (1987).
6. I. G. McWilliam and R. A. Dewar, in *Gas Chromatography 1958 (Amsterdam Symposium)*, ed. D. H. Desty (Butterworths, London, 1958), pp. 142–152.
7. J. E. Lovelock, *Nature (London)* **182**, 1663–1664 (1958).
8. D. H. Desty, in *75 Years of Chromatography — A Historical Dialogue*, eds. L. S. Ettre and A. Zlatkis (Elsevier, Amsterdam, 1979), pp. 31–42.
9. G. Dijkstra and J. De Goey, in *Gas Chromatography 1958 (Amsterdam Symposium)*, ed. D. H. Desty, (Butterworths, London, 1958), pp. 56–68.
10. D. H. Desty, in *Gas-Chromatographie 1958*, ed. H. P. Angelé (Akademie Verlag, Berlin, 1959), pp. 176–184.

11. D. H. Desty, A. Goldup and B. H. F. Whyman, *J. Inst. Petrol.* **45**, 287–298 (1959).
12. R. D. Condon, *Anal. Chem.* **31**, 1717–1722 (1959).
13. A. Zlatkis and J. E. Lovelock, *Anal. Chem.* **31**, 620–621 (1959).
14. S. R. Lipsky, R. A. Landowne and J. E. Lovelock, *Anal. Chem.* **31**, 852–856 (1959).
15. S. R. Lipsky, J. E. Lovelock and R. A. Landowne, *J. Am. Chem. Soc.* **81**, 1010 (1959).
16. I. Halász, in *Gas Chromatography (1961 Lansing Symposium)*, eds. N. Brenner, J. E. Callen and M. D. Weiss (Academic Press, New York, 1962), p. 560.
17. R. Kaiser, *Chromatography in the Gas Phase, Vol. II: Capillary Gas Chromatography* (Original German edition; Bibliographisches Institut, Mannheim, 1961; English translation: Butterworths, London, 1963).
18. L. S. Ettre, *Open Tubular Columns in Gas Chromatography* (Plenum Press, New York, 1965).
19. J. Bouche and M. Verzele, *J. Gas Chromatogr.* **6**, 501–505 (1968).
20. L. S. Ettre, *Chromatographia* **34**, 613–528 (1992).
21. W. Averill, in *Gas Chromatography (1961 Lansing Symposium)*, eds. N. Brenner, J. E. Callen and M. D. Weiss (Academic Press, New York, 1962), pp. 1–6.
22. F. Farré-Rius, J. Henniker and G. Guiochon, *Nature (London)* **196**, 63–64 (1962).
23. R. P. W. Scott, *Nature (London)* **183**, 1753–1754 (1959).
24. D. H. Desty, J. N. Haresnape and B. H. F. Whyman, *Anal. Chem.* **32**, 302–304 (1960).
25. A. Kreyenbuhl, *Bull. Soc. Chim. France* **1960**, 2125–2127.
26. D. H. Desty, A. Goldup and W. T. Swanton, in *Gas Chromatography (1961 Lansing Symposium)*, eds. N. Brenner, J. E. Callen and M. D. Weiss (Academic Press, New York, 1962), pp. 105–135.
27. K. Grob, *Helv. Chim. Acta* **48**, 1362–1370 (1965); **51**, 718–737 (1968).
28. M. Novotný and K. Tesařik, *Chromatographia* **1**, 332–333 (1968).
29. K. Tesařik and M. Novotný, in *Gas-Chromatographie 1968 (Berlin Symposium)*, ed. H. G. Struppe (Akademie Verlag, Berlin, 1968), pp. 575–584.
30. M. L. Lee, F. J. Yang and K. B. Bartle, *Open-Tubular Column Gas Chromatography: Theory and Practice* (Wiley, New York, 1984), p. 53.
31. R. F. Adams, F. L. Vandemark and G. L. Schmidt, *J. Chromatogr. Sci.* **15**, 63–68 (1977).

32. D. H. Desty, J. N. Haresnape and B. H. F. Whyman (assigned to British Petroleum Co.), British Patent No. 899,909 (Applied: April 9, 1959; Issued: June 27, 1962).
33. D. H. Desty, *Chromatographia* **8**, 452–455 (1975).
34. K. Grob and G. Grob, *Wissenschaftl. Zeitschr. der Karl-Marx-Univ., Math.-Naturwiss. Reihe* **26**(4), 379–384 (1977).
35. R. D. Dandeneau and E. H. Zerenner, *J. High Resol. Chromatogr.* **2**, 351–356 (1979).
36. R. D. Dandeneau and E. H. Zerenner, *LCGC* (*North America*) **8**, 908–912 (1990).
37. R. D. Dandeneau, quoted by R. Stevenson, *Amer. Lab.* **30**(5), 30–34 (1998).
38. W. G. Jennings, *Comparison of Fused-Silica and Other Glass Columns in Gas Chromatography* (Huethig Verlag, Heidelberg, 1981).
39. S. R. Lipsky, W. J. McMurray, M. Hernandez, J. E. Purcell and K. A. Billeb, *J. Chromatogr. Sci.* **18**, 1–9 (1980).
40. S. R. Lipsky and W. J. McMurray, *J. Chromatogr.* **217**, 3–17 (1981).
41. S. R. Lipsky and M. L. Duffy, *J. High Resolut. Chromatogr.* **9**, 376–382, 725–730 (1986).
42. E. F. Barry, in *Modern Practice of Gas Chromatography*, ed. R. L. Grob, 3rd edn. (Wiley, New York, 1995), pp. 198–203.

Chapter

25

The Beginnings of Headspace Analysis*

Since the beginning of gas chromatography analysts have encountered the need to analyze the volatile constituents of an essentially nonvolatile sample. If such a liquid sample is introduced into a gas chromatograph, the volatile components evaporate and their vapor is carried through the column by the mobile phase (the carrier gas). However, the nonvolatile matrix will remain in the injector and contaminate it. The investigation of the volatile components present in a solid sample is even more complicated. Such a sample obviously cannot be introduced into the instrument: this requires an elaborate sample preparation procedure consisting among others, extraction of the volatile components. Furthermore, some samples have a matrix that is somewhat volatile, but with a higher boiling point; thus, it eventually will evaporate in the injector but not together with the components of interest. The presence of a very high amount of a higher boiling constituent requires much too long a time for the analysis. A separate

*Based on the article by L. S. Ettre published in *LCGC* (North America) **20**, 1120–1129 (2002), and on discussions with Prof. G. Machata.

problem is often encountered — e.g., in physicochemical measurements — when one is interested in the composition of the vapors in equilibrium with a liquid or solid matrix. In such a case, none of the above procedures could be used, because they automatically destroy the equilibrium conditions.

Headspace (HS) analysis evolved out of necessity in order to solve these problems. It is essentially a sampling method that permits the analyst to take an aliquot of the gas (vapor) phase in equilibrium with a liquid or solid phase. In equilibrium the distribution of the analytes between the two phases depends on their partition coefficients; thus, from the analytical results of this aliquot the composition of the original (liquid or solid) sample can be established. Sampling of the headspace is done mostly under equilibrium conditions, in a static system. This version is called *static headspace analysis*.

In another technique, the so-called *purge-and-trap method*, the (liquid) sample is purged continuously with an inert gas until all the volatile components are removed from it. The gas effluent leaving the sample vessel is conducted through a trap that is either cooled to low temperature or contains an adsorbent. This trap retards the volatile analytes purged from the sample. When the gas extraction is completed, the collected analytes are released by rapid heating of the trap and their amount determined. This technique is also considered a version of headspace sampling and is called therefore *dynamic headspace analysis*.

In the past decade another version of sample handling was developed: *solid-phase microextraction (SPME)*. Here, a fused-silica fiber coated on its surface with a stationary phase and mounted in a modified syringe is exposed to the headspace above the sample. After equilibrium is reached, the fiber is removed; the analytes collected in its stationary phase coating are thermally desorbed in the injector of the gas chromatograph, and transported by the carrier gas into the column. However, SPME is not limited to headspace sampling: it can also be carried out by immersing the coated fiber into the sample solution.

This chapter will survey the beginnings of static headspace sampling, refer to its first applications, and then discuss the steps leading to present-day instrumental systems. We shall not deal with either purge-and-trap or solid-phase microextraction: further details on the techniques can be found in the book of Kolb and Ettre.[1]

25.1. First Uses of Headspace Sampling

Today headspace sampling automatically is considered as a technique directly connected with GC; however, it is not restricted to it. It can be used with practically any analytical technique and in fact, the first applications of headspace sampling were not in gas chromatography.

The first report on headspace analysis that I could find in the literature is from 1939. It is the abstract of a paper presented by R. N. Harger, E. G. Bridwell, and B. B. Raney of the Department of Biochemistry and Pharmacology of Indiana University School of Medicine, that dealt with an "aerometric method" for the rapid determination of alcohol in water and body fluids.[2] Evidently, the paper described a method representing a combination of static and dynamic sampling: the headspace over the liquid sample was conducted through a sulfuric acid–permanganate reagent, permitting the quantitative determination of its alcohol content. According to the abstract the authors also determined the partition coefficient of alcohol for the air–water system in the temperature range of 0°C–40°C, which values were then used to calculate the alcohol concentration of the original sample from the amount present in the gas phase.

Headspace analysis was also used in Hungary in the first part of the 1950s by E. Schulek, E. Pungor, and J. Trompler at the University of Budapest, for various physicochemical measurements of aqueous solutions. When, in 1955, the political situation in that country somewhat relaxed, Professor Schulek was permitted to present a paper on his group's work in Vienna which then was published in 1956 in an Austrian journal,[3] followed by seven additional publications in the same journal in 1956–1960. They used a self-constructed, all-glass apparatus and investigated changes in the tension of aqueous solutions of alcohols and phenol, in which various nonvolatile substances were dissolved. The concentrations of the analytes in the headspace were determined by classical analytical techniques.

The first report on the combination of headspace sampling with subsequent GC analysis was presented at the 1958 Amsterdam Symposium by Bovijn and co-workers from a power plant in Schelle, Belgium.[4] They used the technique to monitor trace concentrations of hydrogen (at the 1 ppb level) in water present in the high-pressure boilers. Aliquots of the equilibrium gas phase were conducted to a gas chromatograph equipped with a thermal conductivity detector. As mentioned by the authors, the unit already had been in operation at the power plant for more than one year.

Here we should stop for a moment and address the question of nomenclature. In the early days there was no specific expression in English to characterize the technique. As mentioned, Harger spoke about an "aerometric method." In their paper written in German, Schulek and co-workers used the expression *Dampfraumanalyse*, and in the English summary of their paper they directly translated it as "vapor space analytical procedure." Bovijn did not use any specific term; he only spoke about the "gaseous phase in equilibrium with a liquid phase." The terms "headspace" and "headspace analysis" were first used in 1960, in the paper of Stahl (see below); most likely the expression was adapted from the food packaging industry where it had been used to characterize the gas layer above the food in sealed containers.

As mentioned, the German expressions "Dampfraumanalyse" and "Dampfphase" were first used in the papers of Schulek and have remained in use ever since.

25.2. Investigation of Food Volatiles

When GC started its meteoric rise in the second part of the 1950s, it was often compared to our nose: it can "smell" volatile compounds, even if present in small concentrations. Thus it was obvious to apply the technique for the investigation of foods and flavors. However, analysts faced the problem of how to introduce the sample, usually an essentially nonvolatile material, into the gas chromatograph. The possibility of "headspace analysis" was realized practically simultaneously in a number of laboratories. In this respect I would like to mention my own involvement, which was typical of the problems encountered at that time.

In the winter of 1958–1959 I was an applications engineer at The Perkin-Elmer Corporation. One day a chemist from a potato chip company called, and he wanted to find out whether GC could be used to detect the degree of rancidity of the chips during storage. I stored some potato chip bags in a warm room and a few days later opened these, as well as some of the properly stored bags: the difference in the smell could be observed clearly. However, my problem was how to sample the atmosphere within the bags.

For the injection of liquid samples we used at that time fairly large medical syringes of 0.5–1 mL volume (microsyringes did not exist as yet). I pierced the bag with such a syringe, withdrew an aliquot of the gas within the bag, and injected it into the gas chromatograph. Indeed I received different chromatograms depending on the status of the bags. This was, of course, headspace analysis.

According to my best knowledge, the first report on headspace sampling for GC analysis was published in the January 1960 issue of *Food Technology*[5] by Stahl and co-workers at McCormick & Co., in Baltimore, MD. Its subject was the determination of gases (mainly oxygen) in the headspace of cans. A specially constructed piercing device was attached to the top of the can; a 0.5–1 mL aliquot of the headspace gas was withdrawn by a hypodermic syringe and then injected into the gas chromatograph.

Evidently Stahl's work gave the impetus to Beckman Co. (Fullerton, CA) to introduce a device for the investigation of the headspace gas of cans and other sealed containers.[6] The device had a puncturing tool connected to a small closed volume that could be evacuated; in this way gas was drawn into this sampling volume after piercing the container. The principal use of the device was the determination of the presence of oxygen in cans, and for such measurements it could be connected directly to a polarographic oxygen sensor that was coupled to a direct readout.[7] (It is interesting to mention that this oxygen sensor was also used in the Project Mercury space capsules.) In addition, gas samples could also be withdrawn from the Beckman device with a syringe through a rubber septum on the side of the device for

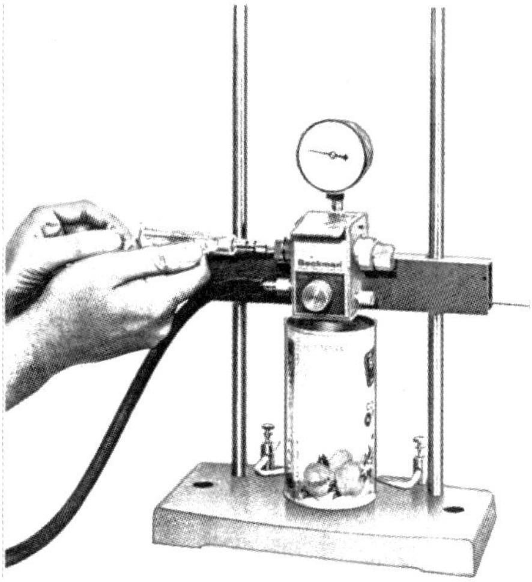

Fig. 25.1. The Beckman Head Space Sampler.[6] The operator is withdrawing a gas sample by a syringe for introduction into another analytical instrument. (Courtesy: Beckman Instruments.)

subsequent GC analysis. The photograph of this unit, now completely forgotten, is shown in Fig. 25.1.

The use of headspace-gas chromatography (HSGC) for the investigation of volatile organic compounds was accelerated by the introduction of the flame and argon ionization detectors toward the end of the 1950s (see Chapters 22 and 23). These had a much greater sensitivity than the thermal conductivity detector used in the first years of GC and with these detectors trace quantities of the odorous compounds could be detected. The reports from this period describe HSGC investigation of a wide variety of samples, such as raspberries, banana, pears, carrots, onions, peppermint oil, coffee, etc. utilizing gas chromatographs with ionization detectors.[8–14] In all these investigations the headspace over the liquid or solid sample confined in a closed container was sampled with a gas-tight syringe, and then the withdrawn aliquot was injected directly into the gas chromatograph. These investigations were mainly qualitative: the researcher was interested in knowing the identity of the large number of compounds present. Almost every report described something new, characterizing the volatile constituents of a large variety of substances. A good example is the paper of Dörrscheidt and Friedrich who demonstrated the presence of 31 volatile compounds in the headspace of honey samples and could show the differences between honeys of different origin.[13]

Manual headspace sample transfer with a syringe has some inherent problems. Even if the samples can be thermostatted at elevated temperatures, the syringe is at room temperature and thus higher boiling but still volatile compounds could condense in it. The second main problem is related to the pressures. When the closed sample vial is thermostatted at an elevated temperature, the pressure in its headspace will be higher than the atmospheric. This gas is sampled with the syringe. However, the internal volume of the syringe is open to atmosphere through the needle and therefore, after withdrawal from the vial, the pressurized gas will expand to atmosphere. In this way, part of the sample is lost and its actual volume in the syringe will depend on the atmospheric pressure. These problems can be partly overcome by automation and special design, but some still remain. Therefore, more precise techniques were sought.

Probably the most important HSGC application, where exact quantitative results are needed, is the determination of alcohol in blood: the results of such analyses must be acceptable to courts of law. Thus, it is not surprising that the more sophisticated, quantitative instrumental methods for HSGC originated from the need in forensic analysis.

25.3. Determination of Alcohol in Blood

The determination of the degree of intoxication of automobile drivers is a major forensic problem. This is generally done by the analysis of blood, urine or breath for its alcohol content. In this respect the regulations in the United States and Europe differ. In the United States, measurement from urine or breath samples was developed because suspects could not be forced to permit drawing a blood sample. (In fact, the investigations of Harger *et al.* mentioned earlier[2] aimed to establish correlation between the alcohol concentration in blood and in either breath or urine.) On the other hand, in Europe scientific opinion opposed testing of breath or urine, because it was felt that the relation of the alcohol concentration between blood and breath (or urine) is not constant and can vary widely. Therefore, in Europe blood testing had always been the preferred method.

The first widely used procedure, the so-called Widmark method, was developed around 1920 by E. M. R. Widmark, professor at the University of Lund (Sweden). A bichromate solution was placed into a special Erlenmeyer-type flask, with a small bowl hanging into the gas phase, into which blood samples were added. The flask was closed and thermostatted at 60°C for 1 h. During this heating the low-boiling volatile compounds present in the blood evaporated and reacted with the bichromate solution, reducing Cr^{VI} to Cr^{III}. The remaining bichromate was determined by titration and the amount of organic volatile compounds reacting with the bichromate calculated by difference. In this way acetone[15] and ethanol[16] could be determined. The method was commonly called "isothermal distillation." Although widely used for the determination of blood alcohol, the technique was not specific; therefore a second independent test was also sought. For this purpose an enzymatic method utilizing alcohol dehydrogenase (ADH) was developed;[17] however, it used expensive reagents which are stable only for a limited time. Therefore, there was a constant search for better, more specific, and simpler methods.

Soon after GC was established as an excellent analytical technique for the analysis of volatile compounds, forensic chemists started to investigate the possibility of using it for blood alcohol determination. This was at the time when the flame-ionization detector was adapted for the instruments and it was particularly attractive for ethanol analysis because it had no response for water: the large water peak does not interfere with the peak of the small amount of ethanol that might be present. However, the essential problem was that blood contains a large amount of nonvolatile substances. Thus, direct injection of blood into a gas chromatograph contaminated the injector block, requiring constant and time-consuming cleaning. A way to overcome this problem was developed in the early 1960s by Dr G. Machata at the Forensic Institute of the University of Vienna, Austria by the use of a precolumn.[18] The blood sample was injected into this small column which contained glass wool or some inert packing material.

After injection, the precolumn was rapidly heated and flushed with the carrier gas, which transported the evaporated volatiles into the chromatographic column while the nonvolatile part of the sample remained adsorbed by the packing of the precolumn. Periodically the packing had to be replaced, but this was a simple procedure. Such systems have been in use in the petroleum industry for the determination of dissolved gases in petroleum products, and the use of the precolumn for blood determination represented a typical parallel development. Based on Machata's work, soon a commercial version of such a system also became available.[19,20] For some time this unit was quite popular in Europe and its use for blood alcohol analysis had also been accepted in Switzerland as the official method. A typical chromatogram is shown in Fig. 25.2;[18] here, acetone was used as the internal standard.

Meanwhile, a brief report in the literature indicated the possibility of using headspace analysis for the determination of the alcohol content in blood,[21] and Machata also started to systematically investigate this possibility. First he used a manual method: placing the blood

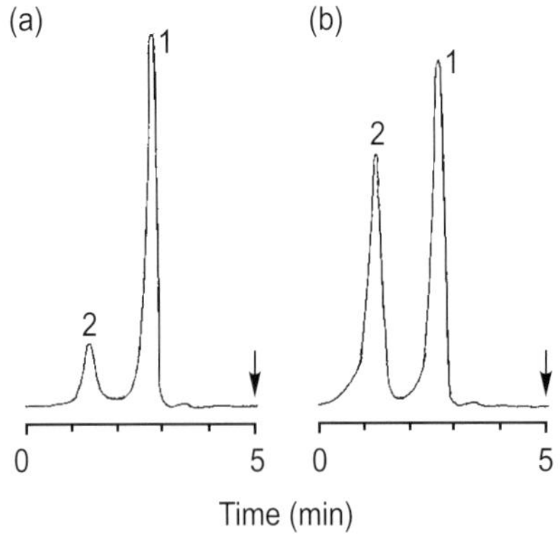

Fig. 25.2. Determination of alcohol in blood using a precolumn.[19] Precolumn: 17 cm × 4 mm i.d., filled with glass beads. Separation column: 2 m × 1/4 in. o.d., containing 15% poly(ethylene glycol) on Celite 545. Column temperature: 100°C. Injected blood volume: 1.2 μL. Flame-ionization detector. Ethanol concentration: (a) 0.022 g/100 mL blood (0.22‰); (b) 0.11 g/100 mL blood (1.10‰). Peaks: (1) acetone (internal standard), (2) ethanol.

sample with the added internal standard into a serum vessel and thermostatting it at 60°C, 1–2 mL of the headspace was injected with a heated syringe into the gas chromatograph.[20] This method was in use in his laboratory for routine analysis starting in 1963. However, due to the manual sample preparation and injection, the number of samples that could be analyzed was limited. Therefore, he began considering automation of the procedure. Cooperation between Machata and Bodenseewerk Perkin-Elmer & Co. (Überlingen, Germany) finally resulted in the first automated HSGC system, introduced in 1967.

25.4. Automated and Integrated HSGC Systems

The main problem in the automation of headspace sampling was the way the sample should be taken from the headspace of the closed sample vials and introduced into the gas chromatograph. Here, it was particularly important to assure precise control and reproducibility. The method finally selected was to pressurize the closed and thermostatted sample vial with an inert gas (usually the carrier gas) and then let the pressurized headspace gas expand for a given time directly into the column. By controlling pressure and time, the volume of the sample aliquot taken from the vial's headspace can be controlled very accurately. This technique was called balanced pressure sampling, and Fig. 25.3 shows the functional schematic of the system. As seen, it consists of three steps: equilibration, when the closed vial is thermostatted; pressurization of the headspace with the carrier gas; and sample introduction. Equilibration requires the longest time of these steps; therefore, in the developed instrumental system the simultaneous thermostatting of a large number of samples was possible. Subsequently, the automated sampling device pressurized the vial and next, controlled aliquots of the pressurized headspace were permitted to expand into the column.

The Model F-40, the first automated and integrated HSGC system based on these principles, was introduced in the spring of 1967 at the ACHEMA Exhibition, in Frankfurt am Main, Germany, by Bodenseewerk Perkin-Elmer & Co. It was capable of simultaneously thermostatting 30 sample vials in a carousel, the temperature of which

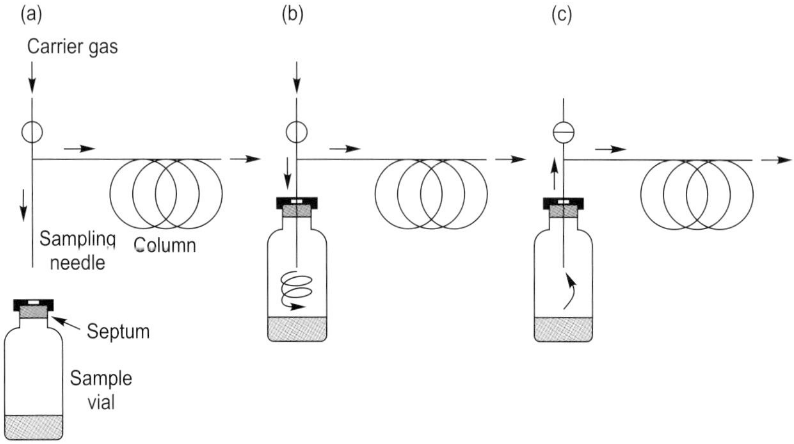

Fig. 25.3. Functional schematic of the balanced-pressure HSGC system; (a) thermostatting (standby), (b) pressurization, (c) sample transfer. During sample transfer, the carrier gas flow to the column is temporarily interrupted and the pressurized headspace gas will expand into the column at a pre-selected time.

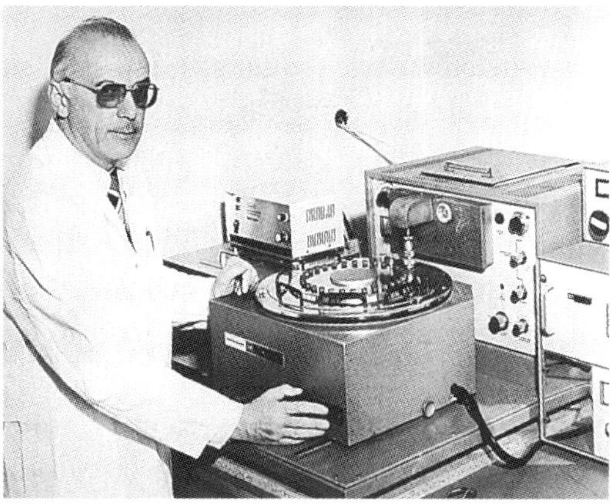

Fig. 25.4. No.1 production unit of the F-40 headspace analyzer, with G. Machata (Courtesy: Prof. Machata).

could be precisely adjusted.[22] Figure 25.4 shows the first production instrument, with Professor Machata. By that time Machata also published the detailed procedure to be used for routine blood alcohol determination.[23]

The availability of an automated instrument for HSGC revolutionized blood alcohol determination: Machata's method and the instrumentation were adopted by many countries both in Europe and America as the standard official method for these measurements. Today millions of such measurements are carried out routinely by forensic laboratories around the world. The method has been improved continuously, by using different internal standards (tert. butanol, n-propanol, or acetonitrile), as well as different columns (packed or capillary) and conditions. As a result of these continuous improvements, the time of analysis has been reduced. Figure 25.5 shows two typical chromatograms from an American forensic laboratory.

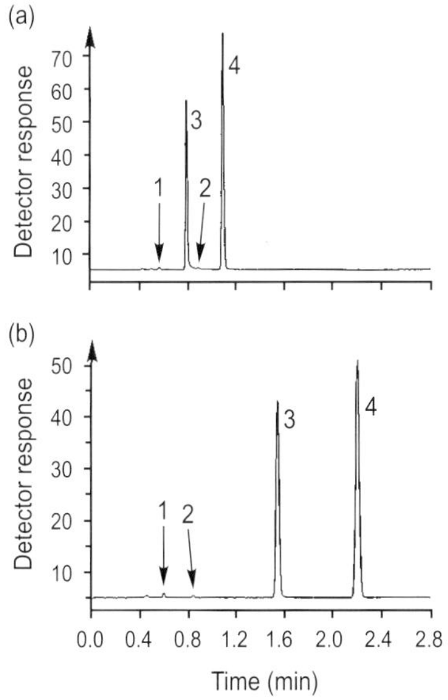

Fig. 25.5. Simultaneous HSGC analysis of a blood sample for ethanol on two columns with stationary phases of different polarities. Concentration in the blood: 0.175 g ethanol and 0.096 g acetonitrile (internal standard) per 100 mL blood (Courtesy: Mesa Police Crime Laboratory, Mesa, AZ, 2002). HS conditions: thermostatting at 60°C for 18 min; pressurization with helium to 23 psig; sampling time: 0.2 min. Chromatographic conditions: 10 m × 0.18 mm i.d. fused-silica capillary column; stationary phase: (a) ALC-2, 0.63 μm film thickness; (b) ALC-3, 0.3 μm film thickness. Column temperature: 40°C. PerkinElmer AutoSystem XL GC with TurboMatrix headspace sampler. Peaks: (1) acetaldehyde, (2) acetone, (3) ethanol, (4) acetonitrile (internal standard).

Naturally, instrument development continued after the introduction of the first instrument, and in the past 40 years a number of other HSGC instruments also have been introduced. Most of these are still based on the balanced pressure approach, but with some modifications: e.g., in some instruments the pressurized headspace sample is not conducted directly into the column, but it first expands into the loop of a gas sampling valve. By rotating the valve, the content of the loop is swept into the column by the carrier gas. Systems using automated syringe injection are also available. Today, HSGC systems are used widely in the industry and research for a wide variety of samples,

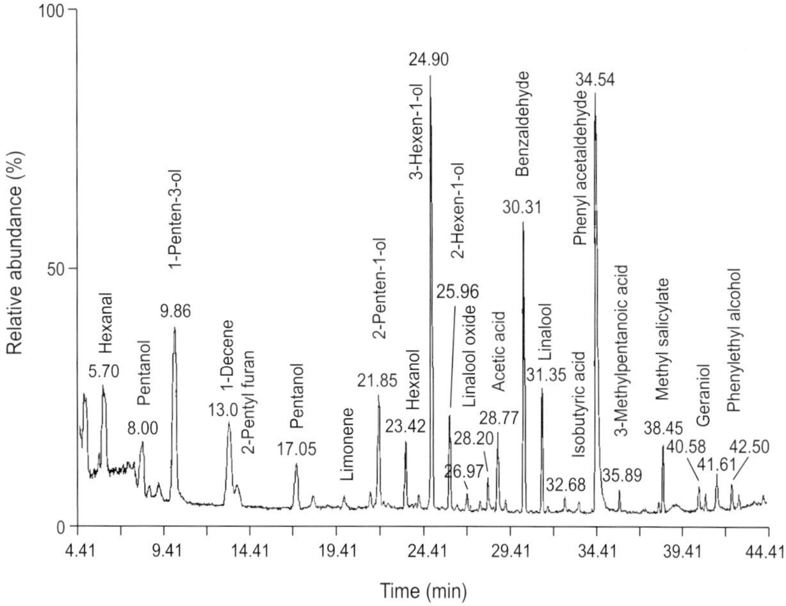

Fig. 25.6. Analysis of an Assam tea sample by HSGC-MS.[25] Sample: 2 g tea leaves + 1 mL water, Headspace conditions: thermostatted at 95°C for 20 min; pressurization with helium to 35 psig; sampling time: 0.5 min. Chromatographic conditions: 15 m × 0.53 mm i.d. fused-silica capillary column coated with poly(ethylene glycol); film thickness: 0.5 μL. Column temperature: isothermal at 50°C for 15 min, then programmed at 4°C/min to 200°C. PerkinElmer AutoSystem XL GC with mass spectrometric detector; total ion chromatogram; Turbomatrix headspace sampler. The numbers above the peaks indicate the retention times (in minutes).

and are indispensable for the handling of samples containing a non-volatile matrix. The capabilities of the technique are best illustrated by Fig. 25.6, showing the investigation of the volatile compounds in an Assam tea sample, by a combined HSGC–mass spectrometer system.[24]

References

1. B. Kolb and L. S. Ettre, *Static Headspace Gas Chromatography: Theory and Practice* (Wiley, VCH, New York, Weinheim; first edn: 1997, second edn: 2006).
2. R. N. Harger, E. G. Bridwell and B. B. Raney, *J. Biol. Chem.* **128**, xxxviii–xxxix (1939).
3. E. Schulek, E. Pungor and J. Trompler, *Mikrochim. Acta* **1956**, 1005–1022.
4. L. Bovijn, J. Pirotte and A. Berger, in *Gas Chromatography 1958 (Amsterdam Symposium)*, ed. D. H. Desty (Butterworths, London, 1958), pp. 310–320.
5. W. H. Stahl, W. A. Voelker and J. H. Sullivan, *Food Technol.* **14**, 14–16 (1960).
6. *Beckman Headspace Sampler. Bulletin No. 7012* (Beckman Scientific and Process Instruments Division, Fullerton, CA, September 1962).
7. *Beckman Laboratory Oxygen Sensor. Bulletin No. 7013* (Beckman Scientific and Process Instruments Division, Fullerton, CA, October 1962).
8. C. Weurman, *J. Food Sci.* **26**, 670–672 (1961).
9. C. Weurman, *Food. Technol.* **15**, 531–536 (1961).
10. D. A. M. Mackay, D. A. Lang and M. Berdick, *Anal. Chem.* **33**, 1369–1374 (1961).
11. R. G. Buttery and R. Teranishi, *Anal. Chem.* **33**, 1439–1441 (1961).
12. S. D. Bailey, D. M. Mitchell, N. L. Bazinet and C. Weurman, *J. Food Sci.* **27**, 165–170 (1962).
13. W. Dörrscheidt and K. Friedrich, *J. Chromatogr.* **7**, 13–18 (1962).
14. R. Bassette, S. Özeris and C. Whitnah, *Anal. Chem.* **34**, 1540–1543 (1962).
15. E. M. R. Widmark, *Biochem. J.* **13**, 432–445 (1919).
16. E. M. R. Widmark, *Biochem. Z.* **131**, 473–484 (1922).
17. R. K. Bonichsen and H. Theorell, *Scand. J. Clin. Lab. Invest.* **3**, 58–62 (1951).
18. G. Machata, *Mikrochim. Acta* **1962**, 691–700.

19. *Zubehör für Gaschromatographie: Vorsäule* (Bodenseewerk Perkin-Elmer & Co., Überlingen, Germany).
20. G. Machata, *Mikrochim. Acta* **1964**, 262-271.
21. A. C. Curry, G. Hurst, N. R. Kent and H. Powell, *Nature* (*London*) **195**, 603–604 (1962).
22. D. Jentzsch, H. Krüger, G. Lebrecht, G. Dencks and J. Gut, *Z. Anal. Chem.* **236** 96–118 (1968).
23. G. Machata, *Blutalkohol* 1 252 260 (1967).
24. S. Tillu and S. T. Kumar, Perkin-Elmer Instruments (Indian Branch, Mumbai, India), unpublished results (2002).

Part Eight

Modern Liquid Chromatography

Chapter

26

The Evolution of Modern Liquid Chromatography*

This chapter deals with the transition from classical liquid chromatography to modern high-performance liquid chromatography (HPLC) and summarizes the merit of the pioneers who, in the 1960s, were instrumental in the development of HPLC. In the following 15–20 years, HPLC experienced an exponential development unparalleled in the history of analytical chemistry — even when compared to the evolution of gas chromatography (GC) a decade earlier.

Liquid chromatography (LC) existed long before HPLC, and for this reason the new name is somewhat misleading. Characterizing present-day LC as high performance may be misunderstood to imply that the LC of earlier days did not exhibit high performance. This is of course not true. By investigating the work of E. Lederer, A. Winterstein, and R. Kuhn (Heidelberg, Germany), and L. Zechmeister and L. Cholnoky (Pécs, Hungary) in the 1930s, A. J. P Martin and

*Based on the paper of L. S. Ettre, published in *LC Magazine* 1, 408–410 (1983).

R. L. M. Synge (Leeds and London, UK) in the 1940s, or S. Moore and W. H. Stein (New York City, USA) in the 1950s, we can realize that the results of these pioneers were indeed characterized by very high performance: they carried out separations up to then considered impossible. Still there are some fundamental differences between LC as carried out earlier, and HPLC as developed in the 1960s. Therefore, when using the new term we do not reduce the importance of the activities of the pioneers; we simply express in this way the numerous improvements in the technique which eventually fundamentally changed it.

26.1. From LC to HPLC

In the first 40 years LC was essentially unchanged: separation was carried out using small glass tubes packed with an adsorbent and simply using gravity to direct the flow of the mobile phase (the eluent). Then, in 1941 Martin and Synge described liquid–liquid partition chromatography (the work for which they received the 1952 Chemistry Nobel Prize; see Chapter 14) and in their paper, they clearly predicted that in order to further improve the column's separation power, very small particles and high pressures would be needed.[1] However, they also stated that due to technical problems, these two goals had to be abandoned, at least temporarily. Then, within a couple of years Martin, with his associates, described paper chromatography[2] (see Chapter 15); this new version of liquid partition chromatography provided such a breakthrough in separation that nobody bothered to further improve liquid column chromatography.

Gas chromatography (GC) was started in the 1940s by a few researchers using adsorbents as the stationary phase (see Chapter 20). In their paper on liquid–liquid partition chromatography Martin and Synge predicted that gas could also be used as the mobile phase, however, nobody followed their suggestion until finally Martin, now with A. T. James, proved the validity of this prediction and in 1952, described gas–liquid partition chromatography.[3] In the next decade GC underwent a meteoric development and over a dozen companies marketed instruments permitting the use of GC as a routine analytical tool. Due to the rapid spreading of its application to the widest possible fields, in 1961 *Chemical & Engineering News* could express the opinion that "if ever analytical chemistry is dominated by a single technique, GC is well on its way to becoming this technique".[4] If we delete the adjective "gas" this prediction, now referring to chromatography in general, became true: in the next two decades LC became an equal partner and eventually even surpassed GC.

The work of a few scientists in ion-exchange chromatography may be considered as the beginning of modern LC. The amino acid analyzer of S. Moore, W. H. Stein and D. H. Spackman, first described in 1958 and introduced as a commercial instrument soon after, actually represented the first modern liquid chromatograph (Chapter 18). Another field that influenced the start of modern LC was size-exclusion (gel-permeation) chromatography, using the polystyrene gels developed by J. S. Moore during the years 1960–1962. The gel-permeation chromatographs introduced in the mid-sixties may also be considered as early liquid chromatographs (Chapter 28). While fully recognizing these achievements, however, the major role in the development of modern LC can be credited to former gas chromatographers. As stated in a paper discussing the foundations of LC[5]

> the conflux of motions and methodology which had continuously evolved in liquid chromatography and the sophistication of gas chromatography in terms of theory and instrumentation had finally precipitated a groundwork from which a rapid development of modern liquid chromatography [would begin] in the sixties.

At the beginning of the 1950s, the theory of GC was fully developed. It was also recognized that the fundamental relationships are the same in both GC and LC; it was thus logical for scientists well versed in the theory of GC to apply their knowledge to the improvement of liquid column chromatography. By generalizing the theory of separation developed for GC, it became apparent that the limitation in LC is the diffusion in the liquid mobile phase that is about three orders of magnitude slower than in a gas. Thus, to obtain speed and efficiency in LC comparable to GC, this handicap had to be overcome by other means. This possibility was formulated in 1963 by J. C. Giddings in his lecture at the First International Symposium on Advances in Gas Chromatography in Houston, Texas[6] and in a follow-up paper,[4] concluding that in order to have column performance in LC approach the performance of GC, much smaller particles (about 2–20 μm in diameter), as well as high pressures along the column would be needed. This was, of course, the same conclusion derived 20 years earlier by Martin and Synge. However, just as stated by them at that time, the proper materials and systems were still not available. Their development took place in the 1960s, in a remarkably short time.

26.2. The Basics of HPLC

26.2.1. *Name*

Originally the modified technique was referred to as *high-pressure* LC because at first glance, the change from atmospheric conditions and gravity flow to pumping systems and pressures in the 100–300 atm range seemed to be the most significant difference. It was soon recognized, however, that high pressure is just one of the characteristics of this new variant of the technique. Therefore, while preserving the acronym HPLC, Csaba Horváth changed the name to high-*performance* LC.

26.2.2. *Differences*

What are then the differences between HPLC and classical LC? Four basic areas can be distinguished in which present-day LC's performance is superior to that obtained by our predecessors.

The first difference is *speed*. What took a couple of hours to separate 60 years ago, we can do today in minutes. Speed became obvious as an important factor at the end of the 1960s, when HPLC began to distinguish itself as the technique we know today. Recent work demonstrates even more the potential of HPLC in this respect.

The second basic difference is in the *system*. Present day HPLC is more precise, and the conditions and results are much more reproducible. The column itself is a significant improvement compared to those used 50 years ago. It contains smaller and more uniform particles, and it is used repeatedly rather than being discarded as was done previously. The mobile phase flow is controlled, and its rate is easily reproduced; it is no longer left to the mercy of gravity. Finally, today we have a variety of detectors that provide very high sensitivity.

The third difference between the old and the new techniques is in the *sample amounts* required. Classical LC was essentially a kind of semi-preparative laboratory method, while HPLC is a micro-technique that significantly reduces the amount of sample needed.

Finally, the fourth difference between the classical LC and the present-day HPLC is that while classical LC was essentially an empirical technique, HPLC has a strong *theoretical basis*. In fact, as mentioned above, its development was based on theoretical predictions that led to improvements in methodology.

26.3. Pioneers in HPLC

We have already mentioned Giddings, who in 1963 first deduced on a theoretical basis the changes needed for LC to be able to approach the performance of GC. His studies were followed in 1966–1967 by the theoretical work of V. Pretorius and co-workers in South Africa,[8,9] J. H. Knox in Scotland,[10] and D. C. Locke in New York City.[11,12] Parallel to these studies a number of scientists initiated the development of modern HPLC systems and columns. The work of five pioneers is particularly worthy of mention: these are Csaba Horváth, J. F. K. Huber, J. J. Kirkland, L. R. Snyder and R. P. W. Scott.

Csaba Horváth (1930–2004) started the development of a system for modern LC in 1964 at Yale University Medical School, in the laboratories of S. R. Lipsky, and by the middle of the year 1965, he had a working system.[13a,14a] In addition to the development of the first working HPLC system, Horváth's major achievements were the development of pellicular (surface-porous) packings, which for over a decade represented the workhorse columns in most laboratories, and one decade later the development of the solvophobic theory which then initiated the renaissance of reversed-phase LC. Horváth's contributions to modern LC are discussed in the next chapter.

J. F. K. Huber (1925–2000) started his activities in LC at the Institute of Technology, Eindhoven, the Netherlands, in 1963–1964, and continued his work at the University of Amsterdam, and from 1974 on, at the University of Vienna, Austria.[13b,14b] His work was theoretical in nature and focused on the study of column performance under different conditions. His first reports were given in a lecture in 1965 at Liverpool,[15] and later, in more detailed presentations, at two international symposia held in 1967 and 1968.[16,17]

J. J. Kirkland (born 1925) was initiated into modern LC during a visit to Huber's laboratory in Eindhoven in 1964.[13c,14c] Upon returning home to E.I. DuPont's Experimental Station, in Wilmington, Delaware, he first concentrated on developing column packings and on liquid–liquid chromatography (LLC). His first report in 1968 described a new LC spectroscopic detector that was used in his investigations, and he illustrated actual separations by LLC.[18] The column packings developed by him were commercialized by Du Pont at the end of the sixties.[19,20]

The fourth pioneer in HPLC is Lloyd Snyder (born 1931).[13d,14d] His activities in this field actually began at the end of the 1950s when he first demonstrated the possibility of liquid–solid elution chromatography with linear isotherms.[21–23] These results were then applied to a detailed investigation of petroleum products and led to his study of modern liquid adsorption chromatography,[24] which he summarized in a major book published in 1968.[25]

And last but not least, the pioneering activities of R. P. W. Scott (born 1924) should be mentioned.[13e,14e] Scott changed his research interests from GC to LC around 1966 and shortly afterwards published detailed reports on the factors affecting the liquid chromatographic system.[26,27]

The activities of these researchers and of the other scientists who had been drawn into the field culminated in 1969 at the Fifth International Symposium on Advances in Chromatography held in Las Vegas, Nevada, January 1969, where a number of landmark papers were presented and lengthy discussions were held on modern LC. Four years later a symposium held in Switzerland showed that the technique found its practical applications in both research and control laboratories (see Chapter 32). From then on the rapid evolution of HPLC continued, eventually surpassing gas chromatography and becoming the most important laboratory investigation technique.

26.4. Bonded Phases

The discussion of the evolution of modern LC would not be complete without some coverage of the development of bonded phases because, without these, HPLC would not have progressed to its present level of efficiency.

In the early days of modern LC, researchers used either solid particles (adsorbents or ion-exchange materials), both fully porous and pellicular, or support particles with controlled-porosity surfaces coated with a liquid stationary phase, just like the column packings used in GC. Coated support particles, however, created serious problems in LC because of the solubility of the stationary phases in the liquid mobile phase. Fortunately, in the early 1970s the bonded-phase packings changed this situation.

The first bonded phase was actually prepared by Martin and Howard[28] in conjunction with the development of reversed-phase partition chromatography. A hydrophobic support was created by treating Kieselguhr (diatomaceous earth) with dichlorodimethylsilane. These silylated particles were then coated with octane or paraffin saturated with aqueous methanol (the mobile phase), while in turn, this mobile phase was saturated with the liquid phase. This technique of treating with a silane had been reintroduced at the end of the 1950s, this time for the preparation of inert supports for GC. In all these reactions, an ether-type bond is formed and the organic groups (in this case, methyl) are bonded to the silicon atoms.

In 1960, C. Rossi and co-workers made use of a different type of reaction, esterifying the silica particles with benzyl or lauryl alcohol;[29] the length of the organic chain depended on the alcohol used. This work was improved by I. Halász and I. Sebestian[30,31] who esterified the silica with 3-hydroxypropionitrile. The "brush-type" phases so obtained showed good performance in GC; however, it was soon found out that such ester phases are only of limited value in LC because of the possibility of hydrolysis of the ester bond.

Stable bonded phases with longer alkyl chains based on ether bonding were first prepared in 1966 by F. H. Pollard's group[32] by reacting hexadecyltrichlorosilane with Celite. Although originally described for GC, this type of bonded phase became most important in LC. Pollard's work was soon followed by H. N. M. Stewart and S. G. Perry [33] who used a silane with a C_{18} chain, and by others. J. J. Kirkland and J. J. de Stefano, in 1970,[34] prepared a variety of bonded-phase packings with non-polar and polar groups, opening the way for the commercial production of such phases. Today, bonded phases are used overwhelmingly in HPLC.

The availability of bonded phases containing alkyl chains made finally the revival of reversed-phase LC possible and today reversed-phase chromatography is probably the most widely used variant of LC — thanks to the theoretical work of Horváth and the development of stable bonded stationary phases.

References

1. A. J. P. Martin and R. L. M. Synge, *Biochem. J.* **35**, 1358–1368 (1941).
2. R. Consden, A. H. Gordon and A. J. P. Martin, *Biochem. J.* **38**, 224–232 (1944).
3. A. T. James and A. J. P. Martin, *Biochem. J.* **50**, 679–690 (1952).
4. *Chem. Eng. News* **39**, 76 (July 3, 1961).
5. L. S. Ettre and C. Horváth, *Anal. Chem.* **47**(4), 422A–444A (1975).
6. J. C. Giddings, *Anal. Chem.* **35**, 439–449 (1963).
7. J. C. Giddings, *Anal. Chem.* **35**, 2215–2216 (1963).
8. V. Pretorius and T. W. Smuts, *Anal. Chem.* **38**, 274–281 (1966).
9. T. W. Smuts, F. A. Van Niekerk, and V. Pretorius, *J. Gas Chromatogr.* **5**, 190–196 (1967).
10. J. H. Knox, *Anal. Chem.* **38**, 253–261 (1966).
11. D. C. Locke, *J. Gas. Chromatogr.* **5**, 202–210 (1967).
12. D. C. Locke and D. E. Martire, *Anal. Chem.* **39**, 921–925 (1967).
13. L. S. Ettre and A. Zlatkis, eds., *75 Years of Chromatography — A Historical Dialogue* (Elsevier, Amsterdam, 1979) (a) C. Horváth, pp. 151–158; (b) J. F. K. Huber, pp. 159–166; (c) J. J. Kirkland, pp. 209–218; (d) L. R. Snyder. pp. 419–424; (e) R. P. W. Scott, pp. 397–400.
14. C. W. Gehrke, R. Wixom and E. Bayer, eds., *Chromatography — A Century of Discovery, 1900–2000* (Elsevier, Amsterdam, 2001).

(a) C. Horváth, pp. 237–248; (b) J. F. K. Huber, pp. 248–250; (c) J. J. Kirkland, pp. 316–329; (d) L. R. Snyder, pp. 552–560; (e) R. P. W. Scott, pp. 506–511.
15. J. F. K. Huber, High Efficiency Columns in Liquid Chromatography, Symposium on Physical Separation Methods in Chemical Analysis, College of Technology, Liverpool, 13–14 May 1965.
16. J. F. K. Huber and J. A. R. J. Hulsman, Symposium on Physical Separation Methods in Chemical Analysis, Amsterdam, 10–14 April 1967.
17. J. F. K. Huber, in *Gas Chromatography 1968 (Copenhagen Symposium)*, eds. C. L. A Harbourn and R. Stock (Institute of Petroleum, London, 1969), pp. 425–427.
18. J. J. Kirkland, *Anal. Chem.* **40**, 391–396 (1968).
19. J. J. Kirkland, *Anal. Chem.* **41**, 218–220 (1969).
20. J. J. Kirkland, *J. Chromatogr. Sci.* 7, 361–365 (1969).
21. L. R. Snyder, *J. Chromatogr.* 5, 430–441 (1961).
22. L. R. Snyder, *J. Chromatogr.* 6, 22–52 (1961).
23. L. R. Snyder, *Anal. Chem.* **33**, 1527–1543 (1961).
24. L. R. Snyder, *Anal. Chem.* **39**, 698, 705–709 (1967).
25. L. R. Snyder, *Principles of Adsorption Chromatography* (M. Dekker. Inc., New York, 1968).
26. R. P. W. Scott, W. J. Blackburn and T. Wilkins, *J. Gas Chromatogr.* 5, 183–189 (1967).
27. R. P. W. Scott and J. G. Lawrence, *J. Chromatogr. Sci.* 7, 65–71 (1969).
28. G. A. Howard and A. J. P. Martin, *Biochem. J.* **46**, 532–538 (1950).
29. C. Rossi, S. Munari, C. Cengarie and G. F. Tealdo, *Chim Ind. (Milan)* **42**, 724–727 (1960).
30. I. Halász and I. Sebestian, Fifth International Symposium on Advances in Chromatography, Las Vegas, Nevada, January 1969; *J. Chromatogr. Sci.* **12**, 161–172 (1974).
31. I. Halász and I. Sebestian, *Angew. Chem Intern. Ed.* **8**, 453–454 (1969).
32. E. W. Abel, F. H. Pollard, P. C. Uden and G. Nickless, *J. Chromatogr.* **22**, 23–28 (1966).
33. H. N. M. Stewart and S. G. Perry, *J. Chromatogr.* **37**, 97–98 (1968).
34. J. J. Kirkland and J. J. de Stefano, *J. Chromatogr. Sci.* **8**, 309–314 (1970).

Chapter

27

The Development of the First High-Pressure Liquid Chromatograph at Yale University*

The development of modern high-pressure, high-performance liquid chromatography (HPLC) is closely connected with the activities of Csaba Horváth and Seymour Lipsky at Yale University (Fig. 27.1), who carried out their fundamental work in 1964–1966. It is primarily their merit to transform classical liquid column chromatography into a modern instrumental separation technique, that within about two decades became the most widely used laboratory method in chemistry and biochemistry.

27.1. Personalities

Seymour R. ("Sandy") Lipsky (1924–1986) was born in San Francisco, California, and studied medicine at New York University. He served

*Based on the article by L. S. Ettre published in *LCGC* (North America) 23, 486–496 (2005).

Fig. 27.1. (Left to right) S. R. Lipsky, Csaba Horváth, and L. S. Ettre (mid-1970s).

his internship in California and first planned to be a practicing physician, but in 1954 he received a fellowship to Yale University Medical School, in New Haven, CT, and from then on was associated with this school in a research capacity. He gradually advanced in the School, and finally in 1966 was appointed a full professor and the director of the School's Physical Sciences Section. In the 1950s he pioneered in the use of GC for the analysis of long-chain saturated and unsaturated fatty acids associated with the triglycerides in humans, and had a role in the development of the electron-capture detector (See Chapter 24). Sandy had an extraordinary sense to pick up new ideas which had the potential to succeed and change the course of science. Thus, in the months after the paper of Giddings, he started to consider the feasibility of improving LC. As mentioned in his autobiography,[1] Lipsky was constantly reminded by his colleagues in the Medical School of all the biological substances that could not be separated by GC. In the mid-1960s he was also selected by NASA as one of the principal investigators for the detection of organic substances which may be present in the Moon rocks eventually brought back from the Moon. This gave him an opportunity to assemble a team and it could be foreseen that they would have plenty of time for other investigations before the rocks arrive. He planned to use this time to study liquid chromatography and was looking for associates.

Csaba Horváth (1930–2004) graduated in 1952 at the Technical University of Budapest, Hungary, as a chemical engineer. For four years he did research at the university; then, after the Hungarian uprising in the fall of 1956 he left Hungary and settled in Germany where he became associated with one of the large chemical companies as a process engineer. In 1961 he decided to get back to a university and earn a PhD degree. He joined at the University of Frankfurt the group of István Halász, who at that time was involved in GC research. Horváth's thesis work dealt with improvements of capillary columns by increasing the inside coated surface of the tubing without a major increase in its diameter and coating this increased surface with a very thin stationary phase film. Horváth was well experienced in surface chemistry and this knowledge came very handy in being able to develop a way to coat a porous layer onto the inside surface of the capillary tubes. The result of his research was the development of the support-coated open-tubular columns which, for almost a decade, enjoyed wide popularity.[2] In addition, he used the same technology to coat glass beads with a thin layer of porous support or adsorbent and used these particles in packed columns; he named this material as *pellicular* particles.[3–5] This column packing was not important for GC; however, as discussed below, they later became important in his early work in LC.

After receiving his doctorate in the spring of 1963, Csaba decided not to get back to industry (as originally planned) but rather to stay in university research, and he accepted a postdoctoral position at the Physics Research Laboratory of Harvard Medical School, in Boston. He became involved in studying the products formed in the irradiation of cholesterol and was horrified upon realizing how primitive the liquid chromatographic techniques used in this analysis were. As an experienced gas chromatographer he was logically thinking about the desirability of having a liquid chromatograph with full control of the operational parameters. He even submitted a proposal to his superiors to develop such an instrument, but he was turned down: nobody at Harvard Medical School had any interest in such a project. Learning about Lipsky's plans, I suggested that he invite Horváth to join him at Yale. As a result, he moved in the fall of 1964 to New Haven, as a research associate of Yale's Medical School.

27.2. The Development of the First High-Pressure Liquid Chromatograph

The next year was most remarkable in Horváth's professional life. His first task was to build a system, an instrument for liquid chromatography. Knowing the existing literature he immediately realized that high pressures and very small stationary phase particles would be needed. Within a couple of months he assembled a high-pressure system for LC, but had problem with finding the needed column packing. At that time particle technology was not advanced enough, and uniformly sieved, very small diameter particles which would be stable at the high pressures were simply not available. Here his previous thesis work suddenly became very useful. The pellicular particles with a thin porous adsorbent layer fulfilled the criteria set over 20 years earlier by Martin and Synge: the very thin porous layer coating on the surface of the glass beads provided the short diffusion paths needed to overcome the slowness of diffusion. Horváth was also able to pack this material into long (1–2 m), narrow-bore (about 1 mm i.d.) columns and successfully used these in the reversed-phase mode. Figure 27.2 shows

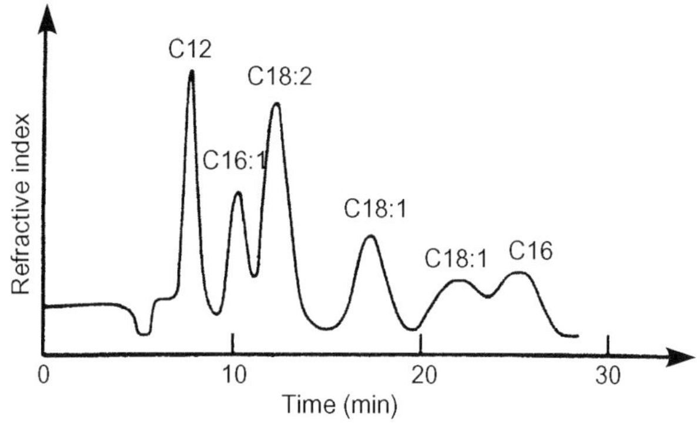

Fig. 27.2. Chromatogram of fatty acids on a 1 m × 1 mm i.d. column packed with pellicular graphitized carbon black (winter 1964/65). Mobile phase: ethanol — 10^{-4} M aqueous NaOH mixture.

one of his early chromatograms obtained in the winter of 1964–1965. Here a mixture of free fatty acids was analyzed on a column packed with such pellicular particles coated with a thin graphitized carbon black layer.

A particular problem inherent to working with such small-diameter columns and with low mobile phase flow rates was extra-column band broadening due to dead volumes, particularly in the detector. Horváth's work was greatly helped by an experimental, small-volume refractive index detector with a 10 μL flow cell provided to him by Perkin-Elmer. (Unfortunately it was never produced and marketed by the company.)

In the subsequent months Horváth further improved his setup, expanding it into a full-fledged scientific instrument. (Fig. 27.3). The eluent (mobile phase) was contained in a reservoir in a thermostatted air bath that could be heated or cooled. The columns were made of 1 mm i.d., 1/16 in. o.d. stainless steel tubes packed with the pellicular material and coiled onto a thin-walled perforated cylinder. A short precolumn (one coil of the tube) was located prior to the injector, which consisted of a 10 μL sliding valve and could be filled with a syringe. In the final system a Hitachi-Perkin-Elmer Model 139 UV–Vis

Fig. 27.3. The high-performance liquid chromatograph developed by Csaba Horváth (Summer, 1965).

spectrophotometer was used as the detector, with a 5 μL flow cell. This cell was made by Horváth of a Swagelock® GC fitting, with quartz windows, and it was an engineering marvel. It may be interesting to note that the first high-performance UV detector specially built for use in modern LC and described in 1968 by Kirkland[6] had a flow-through cell of only 7.4 μL volume. The eluent was transported through the system by a Milton Roy Minipump. For gradient elution a second pump could also be added to the system. This development work was essentially finished by the summer of 1965.

The preliminary investigations with the pellicular graphitized carbon black packing were promising; however, Horváth was looking for another column packing with wider application possibilities and he selected ion-exchange resins. First he coated pellicular support particles with a liquid ion exchanger; however, such columns were not very stable. Therefore, he decided to prepare pellicular particles coated with various porous solid ion-exchange resin layers synthesized *in situ* on the particles' core. He was also considering finding some biologically important problem to demonstrate the feasibility and usefulness of the new technique; after all he was associated with a Medical School! The analysis of nucleic acid derivatives was a logical choice: their separation by classical ion-exchange chromatography had been shown first by Waldo Cohn at Oak Ridge National Laboratory but the analysis time was very long, up to 100 h.[7,8] Thus, while summarizing their work with the liquid ion exchangers in a short paper[9] Horváth and Lipsky started systematic investigations on the analysis of nucleic acids.

Originally they planned to present a report on their work at the Sixth International Symposium on Gas Chromatography that was forthcoming, scheduled for September 1966, in Rome, Italy. However, although they finished the development of the instrument, work on nucleic acid analysis was still in progress, and being in an academic environment they wanted to present a full report on their work. Therefore, it was felt that instead of contributing a formal paper, Horváth should present an interim report during the discussion section dealing with LC (this was the first international symposium where such a session was included in the program). Without any doubt he dominated the discussion and had to answer scores of questions raised by the standing-room-only audience. The report on the discussion published in the proceedings of the symposium gave a brief summary of his presentation and emphasized that "there is no doubt that analysts will soon see a breakthrough in liquid chromatography".[10]

Next spring Horváth and Lipsky — now together with a young colleague, Ben Preiss, who was well experienced in nucleic acid analysis — submitted a detailed report on their work to *Analytical Chemistry.*[11] This fundamental paper consisted of three parts. First a detailed theoretical discussion was given on the possibility of fast liquid chromatography with both small-diameter packed and open-tubular columns, with the conclusion that the latter-type columns are impractical in such systems. Based on the van Deemter equation (originally developed for GC) this treatment included investigations of column characteristics and separation parameters, including temperature, and their influence on the analytical results. Next, the developed high-pressure liquid chromatograph and the method of preparation of the pellicular column packings were described. Finally a detailed report was given on the successful separation of ribonucleosides in less than 1 h. As the conclusion the paper stated that it is "possible to achieve fast separation of complex mixtures by a liquid chromatographic technique similar in speed, resolution and quantitative range to gas chromatography." Without any doubt this paper represents the start of modern LC.

The investigations of Horváth and Lipsky on high-pressure, fast liquid chromatography and nucleic acid analysis did not end with this publication. In a paper submitted on October 20, 1968, to the *Journal of Chromatographic Science* and presented in January 1969 at the Fifth International Symposium on Advances in Chromatography in Las Vegas, NV (see Chapter 32), they further elaborated on column design.[12] This paper can serve as a model on how all aspects of column parameters should be explored. Further improvements in ribonucleoside analysis were reported in a parallel paper submitted on February 24, 1969, to *Analytical Chemistry*[13] demonstrating the separation of the four ribonucleosides in 5–6 min. Figure 27.4 shows two chromatograms from this paper. It may be interesting to point out

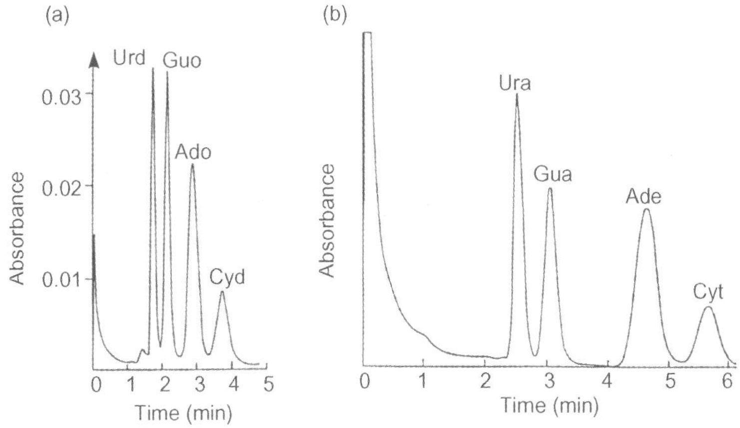

Fig. 27.4. Fast analysis of (a) ribonucleotides and (b) nucleic acid bases. From Ref. 13, reprinted with permission by *Analytical Chemistry*. Column: 1 mm i.d. packed with pellicular cation-exchange resin, sieve fraction #270/325; column length: (a) 151.7 cm, (b) 300.7 cm. Eluent: 0.02 M $NH_4H_2PO_4$ buffer (pH 5.5). Inlet pressure and flow rate: (a) 1925 psi, 35.5 mL/h, (b) 2380 psi, 33.4 mL/h. Temperature: (a) 39°C, (b) 68°C. Sample size: (a) 300 picomoles, (b) 800 picomoles of each component. UV detector at 254 nm; full scale response: (a) 0.04 A.U., (b) 0.02 A.U. Peaks: *Urd* = uridine, *Guo* = Guanosine, *Ado* = adenosine, *Cyd* = cytidine, *Ura* = urasil, *Gua* = guanine, *Ada* = adenine, *Cyt* = Cytosine.

that these analyses were carried out at elevated temperatures. Until recently most LC analyses have been performed at room temperature. Horváth showed already at the beginning the advantages of considering temperature as one of the variables.

27.3. The Rapid Spreading of HPLC

Within a short time after the publication of the first paper by the Yale group the now defunct Picker Nuclear Co. (White Plains, NY) built a commercial instrument based on their work. Unfortunately, however, their interest was fairly narrow, and the company felt that their customers would shy away from a "liquid chromatograph." Therefore they restricted the instrument, calling it a Nucleic Acid Analyzer (Model LCS 1000). However, other instrument companies soon started to pick up the development and by the end of the 1960s several instruments were commercially available (see Chapter 28).

By 1969 pellicular, controlled-porosity support materials had also been developed by Kirkland at DuPont and introduced commercially under the trade name Zipax.[14,15] These support particles could be coated with a liquid stationary phase, just like the column packings used in GC. Coated support particles, however, created problems in LC because of the solubility of the stationary phases in the liquid mobile phase. Therefore, attempts were made to chemically bond the stationary phase to the support particle surface. In their paper[12] Horváth and Lipsky already had pointed out this possibility. The breakthrough in this field occurred around 1970, when column packings with a variety of functional groups were developed by a number of researchers (see Chapter 26). The availability of stable column material with covalently bonded long alkyl chains also made the revival of reversed-phase liquid chromatography possible. In this field Horváth eventually made the fundamental contribution by laying down the theoretical basis of the technique, based on the solvophobic theory. His 1976 paper on this subject[16] — coauthored with Wayne Melander and Imre Molnár — is the most widely cited paper in our field.[17]

27.4. Nomenclature

This discussion would be incomplete without mentioning Horváth's contributions to the vernacular of LC. In their original paper,[11] they spoke about "fast liquid chromatography" and this characterization of the new technique was also used in papers by other authors published soon afterward. By 1969, Horváth and Lipsky changed the name of the technique to "high-pressure liquid chromatography." The acronym HPLC (then meaning high-*pressure* LC) was first used in the paper of Horváth and Lipsky, presented at the 1969 Las Vegas Symposium,[12] and became widely used after Horváth's lecture was presented at the 1970 Pittsburgh Conference[18] where it was used on his slides. That paper also made a change in the meaning of the acronym: from then on it meant "high-*performance* liquid chromatography." Another term introduced in that paper was "isocratic" for constant-strength mobile phase, and it also became universally used within a short time.

27.5. Postscript

Finally a few words about the later activities of Lipsky and Horváth. As already mentioned, Lipsky became involved in the study of moon rocks brought to the earth by the Apollo missions. In the 1970s he became more and more involved in the development of glass, and later fused-silica open-tubular (capillary) columns; he carried out some pioneering work in this field and in the application of capillary GC in biological studies. He also founded Quadrex Corporation to produce such columns and, after retirement from the Medical School in 1984, he fully devoted his time to the improvements of capillary columns. Meanwhile his health started to deteriorate and he died in 1986.

In addition to being associated with the Medical School Horváth also obtained in 1967 an appointment to Yale's Engineering Section and switched over completely in 1972. He became a full professor of chemical engineering in 1979 and served twice (1985–1993 and 1994–1995) as the chairman of the Chemical Engineering Department. He had a most distinguished career at Yale, building up an internationally recognized group and introducing bioengineering among the disciplines of the Chemical Engineering Department, carrying out pioneering research in the separation of proteins and nucleotides. He died on April 13, 2004.

In a paper discussing Horváth's contributions to science, Tyge Greibokk of the University of Oslo, Norway, stated that "it would be difficult to find another person who had a greater impact on liquid chromatography than Csaba Horváth".[19] There is nothing to add to this statement.

References

1. S. R. Lipsky, in *75 Years of Chromatography — A Historical Dialogue*, eds. L. S. Ettre and A. Zlatkis (Elsevier Amsterdam, 1979), pp. 265–276.
2. I. Halász and Cs. Horváth, *Anal. Chem.* **35**, 499–505 (1963).
3. I. Halász and Cs. Horváth, *Nature (London)* **197**, 71–72 (1963).
4. I. Halász and Cs. Horváth, *Anal. Chem.* **36**, 1178–1186 (1964).
5. I. Halász and Cs. Horváth, *Anal. Chem.* **36**, 2226–2229 (1964).
6. J. J. Kirkland, *Anal. Chem.* **40**, 391–396 (1968).

7. W. E. Cohn, *Science* **109**, 377–378 (1949).
8. W. E. Cohn, *J. Am. Chem. Soc.* **72**, 1471–1478 (1950).
9. Cs. Horváth and S. R. Lipsky, *Nature (London)* **322**, 748–749 (1966).
10. J. V. Mortimer, Liquid Chromatography Discussion Section, in *Gas Chromatography 1966* (*Rome Symposium*), ed. A. B. Littlewood (Inst. of Petroleum, London, 1967), pp. 414–418.
11. Cs. Horváth, B. A. Preiss and S. R. Lipsky, *Anal. Chem.* **39**, 1422–1428 (1967).
12. Cs. Horváth and S. R. Lipsky, *J. Chromatogr. Sci.* **7**, 109–116 (1969).
13. Cs. Horváth and S. R. Lipsky, *Anal. Chem.* **41**, 1227–1234 (1969).
14. J. J. Kirkland, *Anal. Chem.* **41**, 218–220 (1969).
15. J. J. Kirkland, *J. Chromatogr. Sci.* **7**, 7–12 (1969).
16. Cs. Horváth, W. Melander and I. Molnár, *J. Chromatogr.* **215**, 129–156 (1976).
17. G. Guiochon and L. A. Beaver, *J. Chromatogr. A* **1043**, 123–126 (2004).
18. Cs. Horváth, Paper presented at the 21st Pittsburgh Conference, Cleveland, OH, 2–6 March 1970.
19. T. Greibokk, *J. Separ. Sci.* **27**, 1249–1254 (2004).

Chapter

28

The Development of GPC and the First Commercial HPLC Instruments*

The previous chapter discussed the activities of Csaba Horváth in the 1960s, leading to modern, high-pressure liquid chromatography. As mentioned, others were also active at that time to convert liquid chromatography from a purely empirical, manual laboratory technique into a more controlled, automated instrumental method.

Csaba Horváth's instrument was, without any question, the first modern, high-pressure liquid chromatograph and as such, it essentially represented the start of those instruments one can find today worldwide in almost every laboratory. However, by the middle of the 1960s instrumentation already existed in two other branches of liquid chromatography: in amino acid analysis by ion-exchange chromatography (see Chapter 18) and in the investigation of molecular-weight

*Based on the article by L. S. Ettre published in *LCGC* (*North America*) 23, 752–761 (2005). The help of Dr. Patrick D. McDonald, Senior Fellow of Waters Corporation, particularly in providing the figures used in this chapter, is greatly acknowledged.

distribution of polymers by the technique of size-exclusion (gel-permeation, GPC) chromatography. In fact, these instruments can be considered as the first automated liquid chromatographs.

The development of the first gel-permeation chromatograph is closely connected with James L. Waters, the founder and for over 20-years the leader of Waters Associates. However, his merit is not restricted to this single product line. By the end of the 1960s he succeeded in becoming a major player in the meteorically growing field of liquid chromatography, transferring his company to "The Liquid Chromatography People." Both achievements can be characterized as true milestones in the evolution of chromatography. This chapter is devoted to his early career, his involvement in making GPC a standard laboratory tool, and the start of Waters Associates' involvement in liquid chromatography.

28.1. Early Activities

The ancestors of the Waters family settled in 1640, in Salem, Massachusetts, and then, with the expansion of the United States, gradually moved westward, eventually reaching Nebraska in 1885. James Logan Waters was born there on 7 October 1925, but in 1942 the family moved back to Massachusetts. After Pearl Harbor he joined the Armed Services, enrolling in the Navy's officer's training program. He graduated in 1946 from Columbia University with a BS degree in physics and with a commission as an ensign in the US Navy. After a short period as a naval officer, he joined Baird Associates, a small company in the Boston area making spectrographs. At that time the US government issued a number of reports on science and technology in Germany during the war and Jim read the one on instrumentation. This inspired him to set up his own company, J. L Waters, Inc., with the aim to develop scientific instruments based on descriptions in this report.

The post-war period can be characterized as the start of the transformation of chemistry from "wet" methods of classical analysis to the use of instruments, based on some physical measurement. This transformation was best characterized in 1962 by H. A. Liebhafsky, with his famous sentence: "Like it or not, the chemistry is going out of analytical chemistry".[1] Waters established some connection with Mine Safety Appliance Co. (MSA), a company in Pittsburgh, Pennsylvania, and designed and built for them an infrared gas analyzer. In 1955 he sold his small company and the design of this instrument to MSA, receiving enough money to continue as an independent instrument developer. Then, in 1958, he formed a new company, Waters Associates.

In the first years the new company was small, with only five employees and operating in the rental basement of the former police station in Framingham, Massachusetts. Waters aim was to develop an instrument for anybody who had the need to measure something. In this period he developed a hydrometer for the US Air Force to be used in balloon flights, a flame photometer for detecting toxic gases and another flame photometer specially built for the Indian Point Nuclear Power Plant to check possible salt contamination in the water used in the boilers which would corrode the heat exchangers; he could also install one photometer in the plant of Baltimore Gas & Electric Co. He also developed a differential refractometer, with a 1-mL cell which found limited use in process control in chemical plants.

28.2. The Breakthrough: GPC

As discussed in earlier chapters, at that time liquid chromatography (LC) was still using small glass columns and gravity flow of the eluent liquid. The only essential difference as compared to the technique of Tswett 60 years earlier was that by now the separated individual fractions did not remain on the column packing but were washed out of the column, the column effluent was collected in small fractions and the content of each fraction determined. However, slowly the method first described in the early 1940s by Arne Tiselius in Uppsala, Sweden, also gained in use: the column effluent was continuously monitored, measuring its refractive index which changed due to the presence and amount of the individual sample components.[2] Learning about this new trend Waters considered modifying his refractometer for LC use. Then, one day in 1961, he was contacted by John C.

Moore, a young scientist at Dow Chemical Co. in Freeport, Texas, who wanted a refractometer with a very small, 0.1 mL cell, to be used at room temperature, with aqueous flow. Jim Waters built one for Dow without knowing its actual use. Nine months later Moore had a new requirement: he now wanted to have a refractometer that could be used with an organic solvent (*o*-dichlorobenzene) instead of water flow, and at 135°C instead of room temperature. Again, Waters developed the detector, still without knowing about its planned use. Only after Dow filed a patent application[3] did he learn that Moore needed the refractometer as a detector for a new method to establish the molecular weight distribution of high polymers: gel-permeation chromatography (GPC). The reason for the two different specifications was that at first, Moore investigated polyglycols, in aqueous solution and at room temperature, followed by the study of polystyrene samples which could only be dissolved in organic solvents like *o*-dichlorobenzene, and the solutions had to be kept at elevated temperatures.

After some negotiations Dow granted exclusive license to Waters Associates on the technique. The task the company faced was great: in addition to designing and building an instrument they also had to master the synthesis of the rigid cross-linked polystyrene–divinylbenzene (PS–DVB) gel used as the column packing. Up to then Moore only produced this in smaller batches but now, Jim Waters had to scale up its production, to be able to supply columns with the instruments. For this they converted the female detention area of the former police station in Framingham into a laboratory (Fig. 28.1). After some experimental work he was able to upscale the original laboratory process to 55 gallon drums and set up regular production of the material with various pore sizes, sold under the trade name Styragel.®

It should be noted that Moore's technique for size-exclusion chromatography had its antecedent: *gel-filtration chromatography* of Porath and Flodin, in Uppsala, Sweden, using soft, compressible, hydrophilic cross-linked dextran gels with the brand name Sephadex®.[4] However, these gels only permitted a slow flow of aqueous solution through relatively short columns, and they could not be used for the separation of high-molecular-weight macromolecules. Usually, these columns were used in the classical way, collecting

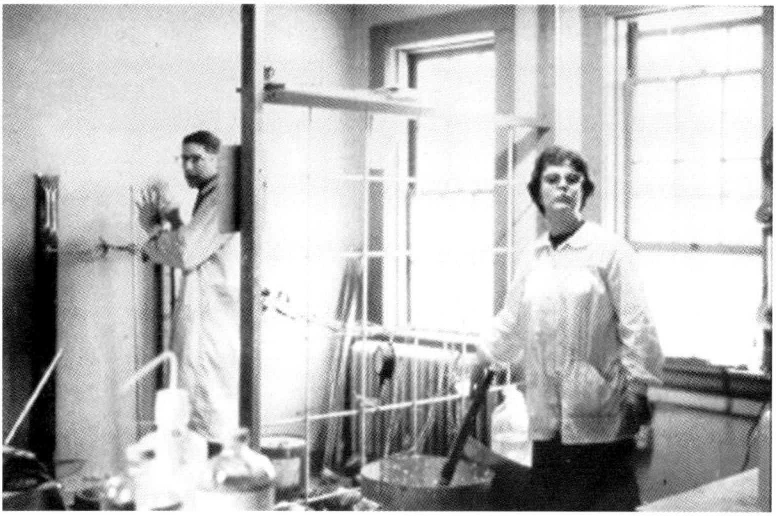

Fig. 28.1. The laboratory of Waters Associates in 1963, established in the female detention area of the former Framingham police station. Right in the front, technician Nicky Anastis is stirring a batch of the PS–DVB mixture used to synthesize the Styragel® column packing. The reaction vessel was fabricated by Jim Waters from a standard, 55-gallon metal drum.

small-volume fractions of the column effluent, but in 1962, Waters personally installed one of his refractometer detectors in Porath's laboratory to monitor the effluent of gel-filtration columns. In contrast to Sephadex, Moore's controlled-porosity rigid PS–DVB copolymer beads could also be used with organic solvents and permitted the separation of sample components with molecular weights ranging from several thousand to several million; and while permitting fraction collection, column effluent was continuously monitored with the differential refractometer detector, producing a true chromatogram.

In 1963 Waters produced five prototypes of the new GPC instrument. These were large, floor-standing units, larger than a kitchen refrigerator. In these prototypes (Fig. 28.2) the column oven (an air thermostat) was made of plywood, with asbestos insulation; this was then changed in the final version to a metal cabinet. Three of these prototypes went to Dow Chemical, and one each to B. F. Goodrich and Mobil Chemical. The final instrument was introduced to the

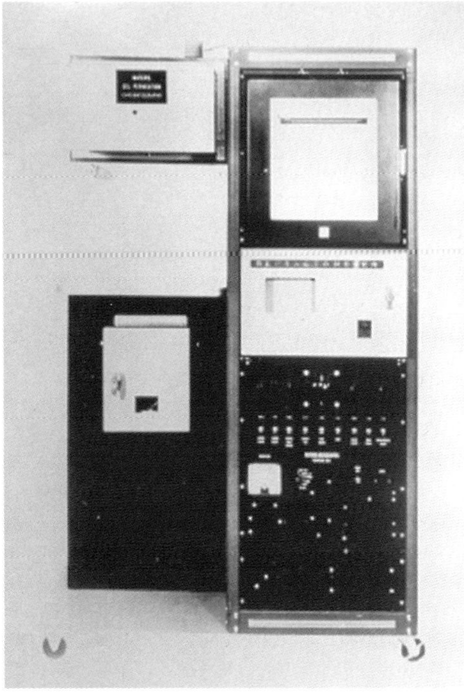

Fig. 28.2. The prototype of the Model GPC-100 gel-permeation chromatograph built in 1963. Top left is the refractometer optics module, while the small metal box in the lower left contains a device continuously measuring the volume flow rate of the pump. The column oven in the back of the left-hand module permits the installation of four 4-ft long columns. All the electronics are in the right-hand module, with a potentiometer recorder at the top.

public at the 1964 Pittsburgh Conference under the name GPC-100; its price was $12,500. By then Moore's fundamental paper was finally published[5] and in this way, chemists interested in the new technique could learn directly from its inventor. (It took well over a year to publish this paper: Moore submitted the manuscript to the editorial office of the *Journal of Polymer Science* in December 1962, but it was published only in the spring of 1964.)

Besides the small-volume differential refractometer key components of the new instrument were the multi-port valves for sample introduction and column switching. The eluent flow was pumped from a reservoir through a heater to degas the solvent and through a filter to a mixing chamber needed to damp out pumping pulsations. A Milton Roy pump was used in the instrument; Waters designed his own pump only later, for high-pressure liquid chromatographs (see below).

The acceptance of the new instrument and of the new technique was great and in 1964 the company sold 40 units. The new technique really filled a gap: using classical methods it took three to four weeks of hard work to establish the molecular-weight distribution of a single sample — now, with the new instrument, one full analysis could be accomplished in about an hour and half.

At the end of 1964 Larry E. Maley, Waters' marketing manager who was very instrumental in demonstrating the new instrument to prospective customers, had the brilliant idea to organize the first GPC symposium. This was held in Cleveland, Ohio, in January 1965. Two plenary lectures were presented: by Fred W. Billmeyer, professor of polymer science at Rensselaer Polytechnic Institute (RPI; Troy, NY) and by John Moore; these were followed by about 20 papers by the users of the instrument and by lengthy discussions (Fig. 28.3). The symposium was so successful that Waters repeated it in 1966, now at

Fig. 28.3. Key participants at the First Waters GPC Seminar, held in Cleveland, Ohio, in January 1965. From the left: Jim Waters, John C. Moore (Dow, Freeport, TX), Prof. Fred V. Billmeyer (R.P.I.) and Larry E. Maley (VP of Sales, Waters Associates). They are holding a model representing a 100,000 MW linear polymer.

the Fontainebleau Hotel in Miami Beach (certainly a more hospitable place than Cleveland in the middle of winter!). From then on these symposia became regular yearly events.

In subsequent years Waters further improved the instrument, introducing a number of improved models and the Ana-Prep that permitted both analytical and semi-preparative work. Without any question, making size-exclusion (gel-permeation) chromatography a routine laboratory tool for the characterization of polymers was his merit.

28.3. Liquid Chromatography

In the early 1960s gas chromatography (GC) was the prevailing chromatographic technique and, as mentioned earlier, liquid chromatography (LC) was still done essentially in the classical way, with very little change from Tswett's original description of the technique. However, slowly researchers well versed in the theory and practice of GC realized that changes were needed to make LC comparable with regard to both speed and resolution. The discussion of Csaba Horváth's early activities in the previous chapter has dealt with this trend. Jim Waters also was thinking along this line: in an internal document from this period he stated that "we believe LC can become a mass market which will extend far beyond the research laboratory into production, quality control, and clinical testing".[7] Waters Associates already sold a limited number of their small-volume refractometers to chromatographers, to monitor the effluent of classical LC columns. Then, in 1965, Waters became involved in the development of a liquid chromatograph for Shell Development Co.

Shell Development Co. in Emeryville, California, had been involved in the development of new instrumentation for their own use and usually licensed smaller companies to produce these. For example, in the early stage of GC development they licensed a short-lived California company, Hallaikainen Instruments, to build gas chromatographs based on their design and Hallaikainen also sold a few to customers other than Shell (see Chapter 19). The situation was similar with LC: by the mid-1960s Shell developed an LC system and in

1965 Jim Waters secured a license to construct instruments based on this design. Shell's system was based on liquid–liquid chromatography (LLC) and Waters built two prototypes, using his refractometer detector: one for Shell and the second for the University of Alberta, in Canada.

Many of the researchers who started to upgrade LC in the 1960s were not sure which approach to take: follow the GC example and use columns coated with a (liquid) stationary phase, or utilize a solid stationary phase (i.e., an adsorbent); in other words carry out liquid–liquid or liquid–solid chromatography (LLC or LSC). Shell's approach was the former. However, Waters soon realized (just as eventually the other pioneers in this early period of modern LC development) that this is not the right approach: the system was not stable and the detector drifted widely. Therefore, he completely redesigned the original Shell system. Meanwhile Karl J. Bombaugh, Waters' director of R&D, learned at the 1966 Chromatography Symposium held in Rome, Italy, about the approach selected by Csaba Horváth (see the previous chapter). As a conclusion Waters switched to the use of solid column packing (an adsorbent), also included a UV detector, and adopted the system for high-pressure operation. The resulting instrument, the ALC-100 analytical liquid chromatograph — the first commercial high-pressure liquid chromatograph — was formally introduced at the 1968 Pittsburgh Conference (Fig. 28.4). It was a benchtop system equipped with a Milton Roy pump, syringe injection, and two detectors: the Waters differential refractometer and a UV detector from Laboratory Data Control (LDC) Co.

Almost simultaneously with the ALC-100, Waters also designed another instrument for large-scale operation, for both GPC and LC: the Chromato-Prep (Fig. 28.5). It was a six-foot tall, floor-standing instrument on casters, equipped with the same two detectors as the ALC-100 and had a multiport valve, with an integral, time-based fraction collector.

As mentioned, Waters Associates first used a pump made by another company, Milton Roy. However, they were aware that the pump is the key component of a liquid chromatograph; therefore, Waters initiated a major project to develop their own system. The

Fig. 28.4. The Model ALC-100 high-pressure analytical liquid chromatograph introduced in 1968. The box on the right top is the compartment for the solvent bottles; below it are the control panels for the refractometer (left) and UV (right) detectors and at the bottom is the pump compartment. The left-hand side of the instrument is the thermostatted column compartment. The operator is injecting a sample with a syringe into the injection port. A separate two-pen potentiometric recorder is used to record the chromatograms.

M6000 pump was announced in December 1972, and formally introduced at the 1973 Pittsburgh Conference (Fig. 28.6). It represented a major innovation, providing pulseless flow at 6000 psi and flow rates between 0.1 and 9.9 mL/min.

A milestone in James Waters' involvement in LC occurred in 1972. At that time Professor Robert B. Woodward of Harvard University (1917–1979), the winner of the 1965 Chemistry Nobel Prize, was involved in the synthesis of Vitamin B-12, but had difficulties in separating two positional isomers of a key intermediary compound along the synthesis. Woodward left for Europe and entrusted Dr. Helmut Hamberger, one of his principal postdoctoral assistants, to try to solve the problem. In turn Dr. Hamberger contacted Waters Associates, asking whether LC could be of help. Waters took a Model ALC-100

Fig. 28.5. The Chromato-Prep GPC–LC instrument for preparative-scale work, introduced in 1968. The column compartment is open, showing four columns installed. Below the space in the middle of the right-side module is a fraction collector, while the two-pen potentiometric recorder is installed in the right top.

liquid chromatograph to Harvard and, in cooperation with Dr. Hamberger, was able to accomplish within a week the separation.[6] Upon the return of Professor Woodward the happy pair could surprise him with 200 mg of the pure compound. A photograph made with Woodward (Fig. 28.7) was used by Waters Associates to promote their excellence in LC among university professors involved in organic synthesis work. (Prof. J. F. K. Huber, then of the University of Amsterdam, The Netherlands, a pioneer in modern LC, was visiting Waters Associates at that time and came along with Jim for the visit to Harvard. In 1974 Dr. Huber became the professor at the University of Vienna, in Austria.)

By the early 1970s Waters Associates was firmly established as the key player in high-performance liquid chromatography (HPLC) and

Fig. 28.6. The Model 6000 solvent-delivery system for liquid chromatographs introduced in 1973.

Fig. 28.7. At Harvard University, in 1972. Left to right: Dr. Helmut Hamberger (Harvard), Prof. J.F.K. Huber (University of Amsterdam), James L. Waters, and Prof. R.B. Woodward (Harvard). An ALC-100 liquid chromatograph can be seen on the right; the pump compartment's door is open.

thus, they rightly trademarked the tagline "The Liquid Chromatography People" for the company. This slogan corresponded to the truth: for years they maintained a dominant role in the field. By the end of the 1970s, when the worldwide LC market was estimated between 100 and 150 million US dollars, Waters had 50 percent of it.[7] In the 25 years since then the use of LC underwent a meteoric rise and it became the most widely used laboratory technique: at present the combined LC and LC–MS market is estimated as about three billion US dollars. Naturally, the relative share of Waters Associates was slowly reduced as the competition "grew up." However, the company is still the world leader in this field.[8]

References

1. H. A. Liebhafsky, *Anal. Chem.* **34**(7), 23A–33A (1962).
2. L. S. Ettre, in *High-Performance Liquid Chromatography — Advances and Perspectives*, ed. Cs. Horváth, Vol. I (Academic Press, New York, 1980), pp. 1–74.
3. J. C. Moore, U. S. Patent 3,326,875 (Filed: 31 January 1963; issued: 20 June 1967).
4. J. Porath and P. Flodin, *Nature* (*London*) **183**, 1657–1659 (1959).
5. J. C. Moore, *J. Polymer Sci.* **A2**, 835–843 (1964).
6. B. J. Murphy, "About Waters" tab in: http://www.waters.com (2003).
7. *Forbes Magazine* **May 1979**, pp. 162–164.
8. *The Laboratory Life Science and Analytical Instrumentation Industry.* Strategic Directions International, Inc., Los Angeles, CA, June 2004; pp. 62–71.

Part Nine

The Most Important Chromatography Meetings

Chapter

29

Two Early Chromatography Symposia*

The rapid growth of modern chromatography was facilitated by the various frequent international meetings which provided a forum where scientists from different geographical areas could meet, present their newest results and discuss them with their peers, learn firsthand the newest achievements of others and apply these immediately in their own work. These meetings also permitted the younger members of our discipline to become personally acquainted with the internationally recognized authorities.

These frequent meetings were made possible by the significant improvements in transportation between continents and by the beginning of the internationalization of economy, commerce, and science in the postwar years. These facilitated the organization of the well-known symposium series from the mid-1950s on. There were, however, two international meetings held in the second part of the 1940s which are by now mostly forgotten, although they served as a kind of

*Based on the articles by L. S. Ettre published in *LCGC* (North America) **17**, 524–531 (1999) and *LCGC Europe* **12**, 768–773 (1999).

springboard for future development: without them the growth of our field probably would have advanced at a slower pace.

29.1. The 1946 Conference on Chromatography

The first of these symposia was the *Conference on Chromatography* organized on November 29–30, 1946, by the New York Academy of Sciences, in New York City; the proceedings of this meeting were published in the first part of 1948.[1] Held just one year after the end of the Second World War, it was mainly restricted to American participants. However, fortunately, two key scientists from Europe had the possibility to participate and summarize their work carried out during the war years about which American scientists knew only little. They were Stig Claesson from the University of Uppsala, in Sweden, a former close associate of Arne Tiselius in whose institute the basic chromatographic techniques were systematized and their common principles elaborated,[2–4] and A. J. P. Martin from England who in 1941, with R. L. M. Synge, invented partition chromatography, and then, in 1944 developed paper chromatography (see Chapters 14 and 15). The major papers of Tiselius and Claesson were originally published in *Arkiv för Kemi, Mineralogi och Geologi*, the periodical of the Swedish Academy of Sciences, which was available only in a few libraries outside Sweden. Therefore, although the language of these papers was English, they were not known to most chemists, and particularly not in the United States. Thus, the publication of Claesson's lecture in the *Annals of the New York Academy of Sciences* has served for many as the primary source of information on their achievements. Claesson's lecture on the displacement and frontal techniques of chromatography and Martin's summary on the various aspects of partition chromatography were very important for the future growth of chromatography in the United States. As a historical document the program of this meeting is presented in Table 29.1.

29.2. The 1949 Faraday Society Symposium

The second international chromatography symposium of this period was organized in 1949, by the Faraday Society. This Society, named

Table 29.1. Program of the Conference on Chromatography Organized by the New York Academy of Sciences, in New York, NY, November 29–30, 1946.

H. G. Cassidy (Yale University, New Haven, CT): Introduction to the Conference on Chromatography.
L. Zechmeister (California Institute of Technology, Pasadena): History, scope and methods of chromatography.
H. C. Thomas (Yale University, New Haven): A Problem in kinetics.
S. Claesson (Uppsala University, Uppsala, Sweden): Frontal analysis and displacement development in chromatography.
W. A. Schroeder (California Institute of Technology, Pasadena): Some experiments in systematic quantitative chromatography.
B. J. Mair (National Bureau of Standards, Washington, DC): Fractionation and analysis of hydrocarbons by adsorption.
L. Zechmeister (California Institute of Technology, Pasadena): Stereochemistry and chromatography.
P. L. Peck (Merck and Co., Rahway, NJ): Chromatography in the Streptomycin problem.
A. J. P. Martin (Boots Pure Drug Co.,Ltd., Nottingham, UK): Partition chromatography.[a]
S. Moore and W. H. Stein (Rockefeller Institute for Medical Research, New York City): Partition chromatography of amino acids on starch.
L. Shedlovsky (Colgate-Palmolive-Peet Co., Jersey City, NJ): Review of fractionation by foam formation.
N. Applezweig (New York University, New York City): Ion-exchange adsorbents as laboratory tools.
W. R. Deitz (National Bureau of Standards, Washington, DC): The surface areas of some solid adsorbents of possible use in chromatography.
H. G. Cassidy (Yale University, New Haven): Concluding remarks.

[a]Included also a detailed discussion of paper chromatography.

after Michael Faraday (1791–1867), one of the greatest scientists of the first part of the 19th century, was founded in 1903: it always had a high reputation and was regarded as the foremost society in physical chemistry. In 1971 it became victim of the amalgamation of a number of British chemical societies, forming the Royal Society of Chemistry; however, its functions continued as the Faraday Division (in physical chemistry) of this new society.

The Faraday Society had its own journal, the *Transactions*, and it also organized each year two symposia (in April and September) the proceedings of which were published in the series *Discussions of the Faraday Society*. In 1949 they felt that the accumulated new results in the field of chromatography would warrant to devote one symposium to this field. Thus, the *General Discussion on Chromatographic Analysis* of the Society was held during September 22–24, 1949, at the University of Reading, just west of London. Over 250 persons participated at the meeting at which a total of 43 papers were presented: of the speakers 24 were from the United Kingdom while 19 (44.2%) papers were presented by foreign scientists from nine countries (six from Sweden, five from the United States, two from The Netherlands, while one of each from Germany, Greece, Italy, Norway, Switzerland, and Australia.) The Symposium was introduced by Arne Tiselius, the head of the Biochemical Institute at Uppsala University, who one year earlier received the Nobel Prize in Chemistry for his "research on electrophoresis and adsorption analysis" (i.e. chromatography); introductory papers were also given by the session chairmen at the start to their session, and each session was then concluded by a noted chromatographer's summarizing remarks. These introductory and summarizing papers were not simply brief comments, but also contained new ideas presented by a leading scientist to his peers. After each group of papers there was a very extensive discussion that was also facilitated by the availability of the preprint of the lectures; some of the comments were communicated *in absentia*. Soon after the meeting the final texts of the papers and the transcript of the discussion sessions were published in the Proceedings of the Symposium.[5]

The participants of the discussions included all the well-known scientists who already had important contributions to chromatography and who continued to have major role in its future evolution. It is, therefore, worthwhile to list some of them here: Arne Tiselius and Stig Claesson, his former associate who by then already occupied another chair at the university; A. J. P. Martin and R. L. M. Synge, the inventors of partition chromatography for which three years later they also received the Nobel Prize in Chemistry; A. Klinkenberg of Koninklijke/Shell in Amsterdam who, a few years later, had a major role in the development of the rate theory of gas chromatography (known generally as the Van Deemter equation, from the first author of the paper[6]), Stanford Moore of the Rockefeller Institute in New

York City who, at that time, just started to investigate, together with W. H. Stein, the separation of amino acids on starch (their first report was actually presented at the 1946 Conference of the New York Academy of Sciences[7]), followed in the next decade by the development of the first automated instrument permitting the full analysis of naturally occurring amino acids (see Chapter 18), a work which, in 1972, was finally honored by the Nobel Prize; C. S. G. Phillips, then a young scientist at Oxford University who at this meeting reported the first time on gas chromatography; E. Glueckauf of the United Kingdom's Atomic Energy Research Establishment, in Harwell, one of the principal contributors to the theory of separation by adsorption and ion exchange;[8–11] Edgar Lederer of the Institut de Biologie Physico-Chimique in Paris who, in 1931, in Heidelberg, in the laboratory of Richard Kuhn, resurrected chromatography (see Chapter 12); and Georg-Maria Schwab of a research institute in Greece who, in 1937, at the University of Munich, first described the use of chromatography for the separation of inorganic ions (see Chapter 13). Glueckauf, Lederer, and Schwab had to leave Germany in the 1930s because of the advent of Nazism and by now became distinguished members of the scientific establishment in their adapted countries. Last, but not least, we should mention the two major American representatives of classical chromatography: László Zechmeister of the California Institute of Technology (another emigrant from Europe) who, in 1936, published the first monograph on chromatography (see Chapter 13), and Harold Cassidy of Yale University, who few years later published a major book on the interrelationship between adsorption and chromatography.[12]

The 1949 Faraday Society Symposium represented a milestone in the evolution of chromatography. As the first major, truly international gathering after the Second World War, it provided a forum to review the status of what we may call classical chromatography and particularly, the achievements of the last decade. In his introductory remarks Tiselius correctly summarized this role of the Symposium by saying that "the progress in the field of chromatography during the last 5–10 years has been so striking both with regard to the method itself and to its scope of application" that one can speak about a second exponential advance in the evolution of chromatography after the first, rapid development period following its rediscovery in 1931 by Kuhn, Winterstein, and Lederer. He then continued by saying that

> The time chosen for this Discussion is therefore particularly suitable (because) there has hardly been time enough to compare experiences ... or to discuss the numerous new applications which now seem possible. This meeting should afford an opportunity for this and other valuable discussions in the field.

The Faraday Society Symposium fulfilled these expectations and presented a true review of the status of chromatography in the middle of the 20th century. This chapter summarizes the most important results discussed there. It is not our purpose to give a thorough review of all the papers: rather, we only want to point out those which have served as the basis of the development in the next decade, completely changing the scope of chromatography.

29.2.1. *Theory*

A large part of the program was devoted to theoretical aspects. The mathematical treatments included in these papers were fairly complicated and this was mainly related to the fact that they have dealt with adsorption chromatography, with the concomitant nonlinear adsorption isotherms. In fact Synge, in his paper summarizing the first part of the Symposium, mildly complained that in some of the treatments physical principles were left behind and replaced by mathematical manipulations; this then further complicated the matter and the understanding of the presented thoughts. It is interesting to note that although the theory of liquid–liquid partition chromatography had been fully described eight years earlier by Martin and Synge, it was barely mentioned: could it be that this was so because there was not anything what could be added to it?

I already mentioned earlier that in Tiselius' laboratory the techniques of frontal analysis[2] and displacement development[3] were elaborated during the war years and clearly differentiated from the elution mode of chromatography. In his paper presented at the present symposium Stig Claesson gave a very clear summary of the theory of the two former techniques. These theories are fully valid even today.

29.2.2. *Partition Chromatography*

A few papers have dealt with the application of partition chromatography on paper, further extending the range of the technique. A. H. Gordon — one of the co-authors of the original publication on paper chromatography now associated with the Karolinska Institute in Stockholm — now also introduced the term reversed-phase chromatography, pointing out that the principles of this version of partition chromatography have already been utilized by others. Tiselius, in his introductory remarks, also mentioned that "the change from non-polar to polar solvent (or the reverse) has been a standard practice in chromatography for many years." Monographs and discussions of the evolution of liquid chromatography generally refer to a 1950-paper by Howard and Martin[13] as the start of reversed-phase chromatography: based on the proceedings of the Faraday symposium, this priority certainly has to be changed.

It is interesting to note the split between the traditional chromatographers and those who realized the advantages of the new technique, partition chromatography. For example, when Zechmeister, the great old man of classical chromatography, described the separation of *cis/trans* isomers based on their adsorption affinity, Martin asked whether such separation could be performed in a partition system. Zechmeister interpreted this question as an inquiry whether partition chromatography would not be more efficient for such a separation, and his answer was that since adsorption chromatography had sufficient high efficiency, he did not bother to try the partition technique.

29.2.3. *Adsorbents*

Most of the papers dealing with adsorption chromatography mentioned the lack of available adsorbents with reproducible characteristics and the required specificity. Extensive work has been carried out on the study of various adsorbents and their controlled modification during preparation. However, as pointed out by Claesson, still no standardized material of constant quality exists and it should be the collective duty of those present at the Conference to press the importance of these questions to the manufacturers.

29.2.4. *Ion-Exchange Chromatography*

A major subject discussed at the Symposium was ion-exchange chromatography and a number of papers have dealt with its theory and application for the separation of both organic and inorganic substances. The new synthetic ion-exchange resins became commercially available in the early 1940s and their first major application was in the separation of rare-earth elements in connection with the Manhattan Project. This work was declassified only in 1947 and the first detailed reports were published in the November 1947 issue of the *Journal of the American Chemical Society* which included 13 papers. These activities were summarized outside the United States the first time at this Symposium in the papers of F. H. Spedding (Iowa State College, Ames, Iowa) and E. R. Tompkins (Oak Ridge National Laboratory, Oak Ridge, Tennessee) which were the highlights of the meeting. In Chapter 17 these two presentations were already mentioned in more detail.

Besides this combined report on the work related to the separation of rare-earths by ion-exchange chromatography, the paper by S. M. Partridge (Cambridge University) is particularly noteworthy: he discussed the use of displacement methods for the preparative-scale separation of biological substances by ion-exchange chromatography. Synge, in his "retrospect on liquid chromatography" published 20 years later[14] called it of "enormous value." The other important new use of ion-exchange chromatography in this field was the separation of amino acids by Moore and Stein, at the Rockefeller Institute in New York City who just started to work in this field, and one would have expected a paper by them at the present Symposium. It is, however, interesting to note that although Moore is listed as a participant, he neither presented a paper nor did he participate in the discussion with a prepared contribution; however, their results were briefly mentioned in the review paper by T. S. G. Jones (Wellcome Research Laboratories, Beckenham, United Kingdom).

29.2.5. Separation by Molecular Size

At the Symposium attention was focused a number of times on the possibility of separation by molecular size. Tiselius, in his introductory lecture, already pointed out this possibility and three of the papers actually described the use of these principles. R. M. Barrer (University of Aberdeen, Scotland) demonstrated that on certain zeolites the effect depends on the shape (the cross-section) of the molecules: e.g., *n*-paraffins are occluded regardless of their chain length while isoparaffins are not because their molecules cannot enter the interstitial channels of the particles. R. Kunin and R. J. Myers of Rohm and Haas (Philadelphia, PA), the manufacturer of the Amberlite-type ion-exchange resins, noted that the extent of cross-linking of the polymer influences its porosity and exchange capacity toward large ions hereby creating an "ionic-sieve" effect, analogous to the molecular-sieve effect of the zeolites. Finally Claesson reported on some preliminary investigations concerning the frontal chromatography of high-molecular-weight polymers on activated carbon which indicated that retardation depends on the molecular-weight distribution of the polymer. He expressed his opinion that chromatography may eventually give information about molecular weight distribution, a prediction fulfilled in the 1950s. As pointed out by Synge in a review article,[15] "the message had gone out" at the meeting, and within a decade led to the development of cross-linked dextran gels of the Sephadex type with controlled porosity by Flodin and Porath,[16] and of the technique of gel-permeation chromatography (GPC) by Moore,[17] permitting the determination of the molecular size distribution of proteins and other high polymers (see Chapter 29). I should also mention that soon after the advent of gas chromatography, the use of synthetic zeolites for the selective retardation of certain compounds from multi-component mixtures based on their molecular shape, has also been described.[18,19]

29.2.6. Gas Chromatography

In 1949 when the Faraday Society Symposium was organized, "chromatography" in the people's mind automatically meant *liquid* chromatography; although gas chromatography had been tried by a few researchers, it did not exist in practice as yet. As discussed in Chapter 19, in 1946–1947 E. Cremer in Innsbruck, Austria, carried out some gas (adsorption) chromatography experiments in the elution mode with her graduate student F. Prior, but the political situation in Europe was not yet stable enough for scientists from Austria to participate at an international symposium, and their results were only published in 1951.[20] In the early 1940s Claesson, in Uppsala, also carried out some preliminary work on gas chromatography in the displacement mode, and its summary was included in his very long report published in 1946,[4] known only to a limited number of people (see Chapter 19).

Parallel to the work of Claesson and Cremer, pioneering work on the use of gas-adsorption chromatography was also carried out by C. S. G. Phillips at Oxford University (see Chapter 19). His first report was presented here, at this Symposium, demonstrating the separation and determination of lower hydrocarbons. It is interesting to note, that in the general discussion that followed, nobody commented on his work, not even participants affiliated with the petroleum industry, although a paper presented at this Symposium by W. M. Smit of the research laboratories of Koninklijke/Shell, in Amsterdam, The Netherlands — probably the most important European research establishment in this field — clearly demonstrated that they were already engaged in studying the possibilities of using (liquid adsorption) chromatography for the analysis of petroleum fractions. We may conclude that at that time, the petroleum chemists were simply not yet ready for the new technique. However, probably they were alerted enough so that three years later, after the publication of the fundamental paper of James and Martin on gas–liquid partition chromatography (GLPC), the petroleum laboratories adapted almost immediately the new technique to solve their analytical problems.

References

1. Conference on Chromatography, *Ann. N.Y. Acad. Sci.* **49**, 141–326 (1948).
2. A. Tiselius, *Ark. Kem. Mineral. Geol.* **14B**(22), 1–5 (1940).
3. A. Tiselius, *Ark. Kem. Mineral. Geol.* **16A**(18), 1–11 (1943).

4. S. Claesson, *Ark. Kem. Mineral. Geol.* **23A**(1), 1–133 (1946).
5. Chromatographic Analysis, *Discussions of the Faraday Society*, **1949**(7), 1–336.
6. J. J. Van Deemter, F. J. Zuiderweg and A. Klinkenberg, *Chem. Eng. Sci.* 5, 271–289 (1956).
7. S. Moore and W. H. Stein, *Ann. N.Y. Acad. Sci.* **49**, 265–278 (1948).
8. E. Glueckauf, *Proc. Roy. Soc. (London) A* **186** 35–37 (1946).
9. E. Glueckauf, *J. Chem. Soc.* **1947**, 1302–1308.
10. J. I. Coates and E. Glueckauf, *J. Chem. Soc.* **1947**, 1308–1314 and 1315–1321.
11. E. Glueckauf, *J. Chem. Soc.* **1947**, 1321–1329.
12. H. G. Cassidy, *Adsorption and Chromatography* (Interscience, New York, 1951).
13. G. A. Howard and A. J. P. Martin, *Biochem. J.* **46**, 532–538 (1950).
14. R. L. M. Synge, in *British Biochemistry Past and Present* (*Biochemical Society Symposium No. 30*), ed. T. W. Goodwin (Academic Press, London, 1970), pp. 175–182.
15. R. L. M. Synge, *J. Chromatogr.* **215**, 1–6 (1981).
16. P. Flodin and J. Porath, *Nature* (London) **183**, 1657–1659 (1959).
17. J. C. Moore, *J. Polymer Sci.* **A2**, 835–843 (1964).
18. N. Brenner and V. J. Coates, *Nature* (London) **181**, 1401–1402 (1958).
19. N. Brenner, E. W. Cieplinski, L. S. Ettre and V. J. Coates, *J. Chromatogr.* **3**, 230–234 (1960).
20. E. Cremer and F. Prior, *Z. Elektrochem.* **55**, 66–70 (1951).

Chapter

30

Early European Symposia Showing the Direction for the Evolution of Gas Chromatography*

As we have already mentioned in the previous chapter, the frequently organized international symposia played an important role in the rapid dissemination of chromatography and still are serving as the places where the newest developments are reported.

Gas chromatography (GC) started its meteoric rise with the four seminal papers of A. J. P. Martin and A. T. James, published in 1952 (see Chapter 14). First used only by a few laboratories, within a few years the interest was so great that after an initial one-day symposium in England, in 1955, two international symposia could be held in 1956 and 1958. Gas chromatography also spread rapidly in the United States and there also, representative symposia played an important part in the dissemination of the technique (see Chapter 31).

*Based on the article by L. S. Ettre, *LCGC Europe* **16**, 346–353 (2003).

Today, more than 50 years later, the papers presented at these symposia provide us a good basis for understanding how GC evolved. Even for present-day chromatographers it is highly beneficial to study the proceedings of these symposia, including not only the presented papers but also the transcripts of the extensive discussions which sometimes were more interesting than the formal papers.

This chapter reports on the three European GC symposia that set the stage for future development. As a brief introduction I shall outline the events that led to these meetings and explain the role of individuals and associations involved in their organization. The early major American symposia will be the subject of the next chapter.

30.1. The Start of GC in England

A. J. P. Martin and A. T. James presented their first preliminary report on the new technique at the 290th meeting of the British Biochemical Society on 20 October 1950.[1] After some additional work, they submitted three fundamental papers on the theory of the technique and its applications for the separation of the lower fatty acids, volatile amines, and pyridine homologues.[2–4] These papers were published in 1952, in *Biochemical Journal* (see Chapter 14).

After successfully separating these biochemically important substances, Martin wanted to investigate the separation of some other closely related volatile compounds, and hydrocarbons seemed to be an obvious choice. However, the British National Institute for Medical Research, Martin's workplace, did not have pure hydrocarbon samples; so, he had to find an institution which could supply them.

During the Second World War an organization was established in England to prepare pure reference hydrocarbons and this organization, the Hydrocarbon Research Group operating within the Institute of Petroleum (IP), continued its function even after the war. At that time the head of its Hydrocarbon Chemistry Panel was Dr. S. F. Birch, a research manager at British Petroleum Co. (BP) at its Sunbury-on-Thames laboratories. Therefore, in the early summer of 1952 Martin wrote him requesting some samples. This sounded as a very unusual request: after all, why would somebody from a medical institution ask for hydrocarbon samples? Therefore, Dr. Birch sent Denis H. Desty, one of his young associates at BP, to visit Martin in order to find out more. When learning about the work of Martin and James, Desty immediately recognized the great potential of GC for petroleum chemists and he became the driving force in spreading its use.

Meanwhile, the First International Congress on Analytical Chemistry was held on 4–9 September 1952 in Oxford, England, where Martin presented a major paper on GC.[5] H. Boer of the Shell Research Laboratories, in Amsterdam, who was present at the meeting called it "an historical lecture"[6] and indeed, this was true. The Congress was attended by nearly 700 people from 26 countries, including representatives of the most important industrial and research laboratories. The participants returned home excited by the potential of the new technique.

The first industrial laboratories active in GC were of the petroleum companies and the divisions of Imperial Chemical Industries (ICI). In fact, chemists at ICI learned about GC through visits to Martin's laboratory while its development was still in progress. Thus, soon after the seminal papers of James and Martin, the first reports on the application of GC were authored by chemists from ICI.[7–10]

The new technique spread so rapidly in the United Kingdom that within three years a symposium on GC could be organized by the Physical Methods and Microchemistry Groups and the Scottish Section of the Society for Analytical Chemistry; it was held on 20 May 1955 in Ardeer, Scotland, and hosted by ICI's Nobel Division. Encouraged by the success of this meeting it was felt that the time is ripe for a larger meeting with international participation, in order to survey the newest developments. The Hydrocarbon Research Group agreed to sponsor this symposium which then took place 30 May–1 June 1956 in London. Following this symposium a new organization, the Gas Chromatography Discussion Group was formed, and two years later it decided to organize again a similar symposium at that time in Amsterdam. This meeting took place 19–23 May 1958, with an even greater interest. From then on the group (renamed in 1972 as the Chromatography Discussion Group and in 1984 as the Chromatography Society, indicating the expanded field and role of the group) continued to organize such biannual symposia, a series that is still alive.

These three symposia, the papers presented there and the extensive discussions had a most significant impact on the evolution of GC, particularly since the information presented at the symposia was generally available soon after the meetings. The abstracts of the papers presented at the Ardeer Symposium were published in *Analytical Chemistry*[11] and a detailed report was also provided in *The Analyst*.[12] In the case of the 1956 and 1958 symposia, the preprints were sent in advance to the participants and a few months after the meetings their proceedings including the final text of the papers and the transcript of the discussions were published in a book form.[13,14] Even today, 50 years later, these books contain valuable information to present-day chromatographers.

30.2. The Ardeer Symposium

About 130 analytical chemists — mostly from Great Britain, with a few from the Koninklijke/Shell Laboratory in Amsterdam, The Netherlands — attended this meeting. Besides five formal papers presented by A. J. P. Martin, E. Chalkey (I.C.I, Billingham), B. Littlewood and C.S.G. Phillips (both Merton College, Oxford University), and N. H. Ray (I.C.I, Winnington), extensive discussions were held and a demonstration of home-built gas chromatographs was also included in the program.

Martin summarized some of the rules influencing column efficiency and surveyed the various possibilities for detecting devices, with special emphasis on those measuring thermal conductivity and his newly developed gas density balance. He also formulated the linear relationships between the logarithm of the retention volumes of members of homologous series and the number of carbon atoms in their molecules. Chalkey continued the discussion of the analysis of hydrocarbon mixtures first presented six months earlier at a meeting of the Institute of Petroleum[10]: he now concentrated on the analysis of C_1–C_4 hydrocarbons and the quantitative aspects of analysis. Littlewood elaborated on the possibility of the analysis of boron and silicon hydrides, while Phillips compared GC in the adsorption and partition modes and discussed the possibilities of thermodynamic calculations using retention volumes. Finally, Ray described an ingenious way of determining non-olefinic impurities present in ethylene, using a Janák-type gas chromatograph.[15]

After Martin's paper, E. F. G. Herington (Chemical Research Laboratory, Teddington) presented relationships for the vapor pressures and other physico-chemical constants versus retention volumes and the carbon number of members of homologous series. He also illustrated the linear relationship of the logarithms of retention volumes of homologues measured on two different stationary phases.

A major part of the program of the Ardeer Symposium was devoted to a discussion session dealing with the construction and characteristics of thermal-conductivity detectors. The discussion was opened by A. I. M. Keulemans (Koninklijke/Shell Laboratories, Amsterdam) who enumerated the requirements for such detectors and described the construction of devices developed at Shell. A number of other participants also described their own home-built systems (made of glass or metal) and their experiences in their use.

30.3. The 1956 London Symposium

This Symposium on Vapor Phase Chromatography (as the technique was then called) was organized by Desty assisted by C. L. A. Harbourn, his associate at BP. Nearly 400 scientists from 13 different countries attended the meeting and 34 original papers were presented. Table 30.1 lists the number of papers presented by participants according to their countries of origin; as seen two-third were by British authors. In addition to the formal papers both Martin and James addressed the symposium: Martin spoke about trends in GC, predicting some future developments, while James surveyed the various detection methods, presenting also some ideas for the future.

Table 30.2 presents the distribution of the papers according to the affiliation of the speakers. The overwhelming majority of the speakers came from the petroleum industry; of these 14 papers, six were presented by chemists from the Shell laboratories in the UK and The

Table 30.1. Number of papers presented at the 1956 and 1958 International GC Symposia according to the speakers' country of origin.

Country	1956 London	1958 Amsterdam
United Kingdom	23	15
The Netherlands	3	4
USA	2	2
Czechoslovakia	2	1
Germany	—	2
Australia	—	1
Belgium	—	1
Canada	1	—
France	1	—
Hungary	1	—
Italy	—	1
New Zealand	1	—
Soviet Union	—	1
Total	34	28

Table 30.2. Distribution of the papers presented at the 1956 and 1958 International GC Symposia according to the speakers' affiliation.

Affiliation	1956 London		1958 Amsterdam	
	No.	%	No.	%
Petroleum and related industries	14	41.2	9	32.2
Other industrial laboratories	10	29.4	7	25.0
Universities	4	11.8	7	25.0
Government laboratories	5	14.7	2	7.1
Scientific instrument industry	1	2.9	3	10.7
Total	34	100.0	28	100.0

Netherlands. Among speakers from industrial laboratories, four came from ICI.

The papers can be divided into five categories: theory (five papers), instrumentation (seven papers), detector technology and operation (seven papers), column technology (three papers), and applications (12 papers). In the theory group the two most important papers were presented by E. F. G. Herington (Chemical Research Laboratory, Teddington, UK) and by A. I. M. Keulemans and A. Kwantes

(Shell Laboratory, Amsterdam). Herrington's lecture represented a detailed discussion of the thermodynamics of GC. In it he further elaborated on the relationships already mentioned one year earlier, at the Ardeer Symposium, but now in a more formal way, with examples. Keulemans and Kwantes presented a thorough discussion of the rate theory and what we call today the Van Deemter equation. At that time the fundamental paper of J. J. van Deemter, F. J. Zuiderweg and A. Klinkenberg was not yet published[16] and thus, this was the first time that chromatographers were exposed to the details of this fundamental relationship. For present-day chromatographers the conclusion of this presentation might be interesting: the speakers postulated that, by optimizing the column parameters and operating conditions, one might be able to achieve a plate height (HETP) of 2 mm: assuming a 20-m long (packed) column, this would give 10,000 theoretical plates. As all our readers know, within two years this limit had been broken.

A further interesting theoretical paper was by J. Janák (Institute for Petroleum Research, Brno, Czechoslovakia) who introduced the concept of the so-called chromatographic spectrum, a scale based on the specific retention volumes. This probably was the first systematic discussion of the differences in separation on polar vs. non-polar stationary phases and different adsorbents (silica gel, active carbon, zeolites). An important aspect pointed out in the discussion was the role of the carrier gas when using adsorbents as the stationary phase. In his system (see Chapter 21) Janák was using carbon dioxide as the carrier gas, and the behavior of these adsorbents was found quite different in such a case. For example, C_2–C_3 saturated/unsaturated hydrocarbons could not be separated on the zeolites when using hydrogen and nitrogen (or helium) as the carrier gas, but their separation was possible using carbon dioxide.

Commercial gas chromatographs just started to be marketed at the time of the symposium, and all the investigations reported at the symposium were carried out on self-constructed instruments which were briefly described in the papers. Five papers specifically dealt with the design and performance of gas chromatographs: J. C. Hawkes (Monsanto Chemicals, Ruabon, UK) discussed an instrument equipped with the gas density balance as the detector and J. Brooks, W. Murray and A. F. Williams (ICI, Nobel Division, Stevenson, UK) described a GC system with a special inlet permitting the selective evaporation of only a part of the sample (a precursor of present-day programmed-temperature (PTV) injectors!). Particularly interesting was the description of a preparative gas chromatograph by B. T. Whitham (Shell Research Center, Thornton, UK) using 27-ft long packed columns of 1/2-in. diameter, with a column packing consisting of 44/52 mesh support particles coated with 26% silicone oil. The maximum operating temperature was 300°C and the liquid sample volumes varied between 0.01 and 3.0 mL. A laboratory gas chromatograph for use up to 300°C, equipped with a thermal-conductivity detector was also described by J. G. Keppler, G. Dijkstra and J. A. Schols (Unilever, Vlardingen, The Netherlands). Finally, H. H. Hausdorff (Perkin-Elmer, Norwalk, Connecticut, USA) described the Model 154, the company's gas chromatograph introduced in the United States in 1955 which at that time was still unknown in Europe. A seemingly harmless remark by the speaker comparing the apparent interest of chromatographers in the United States and Europe initiated a vitriolic answer by one of the Dutch participants; this was the first indication of the aversion by European chromatographers against the description of instruments developed by commercial companies in the form of a presentation at these symposia (while welcoming such reports from a university or an industrial research laboratory).

We should mention here one more paper related to instrumentation: it was presented by C. M. Drew and J. R. McNesby (US Naval Ordnance Test Station, China Lake, California, USA) and it utilized temperature programming. Temperature programming was first described in 1952 by J. Griffiths, D. James and C. S. G. Phillips of Oxford University,[17] but its usefulness was only recognized around the 1958 Amsterdam Symposium (see below).

In 1956 the most frequently used detector was the katharometer (thermal-conductivity detector) and a few papers discussed questions associated with it. At the time of the symposium three additional detectors have already described in the literature: the automatic titrator of James and Martin,[2] the hydrogen flame (thermocouple) detector of R. P. W. Scott (Benzole Producers, Watford, UK) first reported in

1955,[18] and Martin's gas-density balance which at the time of the symposium was not yet described in a formal journal publication, although — based on information from Martin — it had already been used in a few laboratories (a description was finally published soon after the symposium[19]). Now, Scott further elaborated on his detector and M. M. Wirth (British Hydrocarbon Chemicals, Grangemouth, UK) also reported on its use. With respect to the gas-density balance C. W. Munday and G. R. Primavesi (The Distillers Co., Epson, UK) described their experience with it, pointing out some of its shortcomings. At that time these detectors represented a possible alternative to the katharometer; however, within two years the development of the ionization detectors made them obsolete.

Only two papers dealt with column technology. E. R. Adlard (Shell Research Center, Thornton, UK) presented a thorough discussion of the various polyglycols with different molecular weights, used as stationary phases. Even today, these polar phases are used widely and the information presented in this paper is still of interest to us. The other presentation in this field was by Janák who described the characteristics of zeolites (molecular sieves) as adsorbents for the analysis of C_1–C_4 hydrocarbons.

Several papers dealt with the application of GC in a wide variety of fields such as the analysis of fatty acids and fatty alcohols, chlorocarbons and fluorocarbons, industrial solvents, and hydrogen isotopes. These papers demonstrated how widespread GC's use became in only four years.

Finally, I should mention some questions which were extensively discussed during the symposium, but are mostly forgotten today. For example, a number of chromatographers advocated the use of reduced column outlet pressure. Even Martin, in his introductory lecture, indicated the possible improvement in the sensitivity of the katharometer if used at subambient pressure. Another concern was the selection of the carrier gas with this detector. At that time helium was practically unavailable in Europe, so, most chromatographers used nitrogen, which did not permit the full utilization of the detector's sensitivity due to the relative closeness of the thermal conductivities of nitrogen and organic vapors. A further question connected to this detector was its quantitative response: chromatographers were not sure about the best way the response factors could be established and whether calculation as weight or as volume percent was to be preferred. A peculiar question hotly debated was column configuration. Some prominent chromatographers strongly felt that column efficiency is reduced if the (packed) column is coiled and therefore, it should be made of straight tubes; on the other hand, other equally prominent chromatographers had no objection against coiled columns. There was also an uncertainty about the way a column should be packed with the coated support and adsorbent particles. In this respect we must not forget that at that time, everybody made his or her own columns, including the preparation of the support, sieving, and coating; companies specializing in column manufacturing did not exist for a long time. (In this respect it is worthwhile again to mention the difference between Europe and the United States: by 1957 the major American instrument manufacturers already supplied standardized (packed) columns to their customers, a service unknown in Europe.) Although these questions have since been settled, it is interesting to read these discussions, presenting a good base to understand the state of GC in the mid-1950s.

At the beginning of the symposium a committee was established that presented its recommendations for proper chromatographic nomenclature and for the universally applicable general relationships. The use of *gas chromatography* instead of *vapor phase chromatography* was approved unanimously and a list of standard phases and test substances to be used as standards was established. It is interesting to note that this was the first time that the compressibility correction factor was given as a separate term, however, it was called a *reducing factor:* its present name was introduced only later. A shortcoming of this committee's recommendation was the confusion of the terms *efficiency* and *resolution*. This problem was corrected two years later at the Amsterdam Symposium.

30.4. The 1958 Amsterdam Symposium

This symposium was organized by the Gas Chromatography Discussion Group (formed after the 1956 London Symposium) together with the Royal Dutch Chemical Society and held on May 19–23, 1958 in Amsterdam, in the Royal Tropical Institute. The secretary of the joint organizing committee was G. Dijkstra (Unilever, Vlaardingen, The Netherlands), and Desty was in charge of editing the submitted papers, and preparing the preprints and the final proceedings.[14] Nearly 500 participants from 18 different countries attended the symposium, the most memorable in the Group's long series of biennial meetings. For me, this symposium also evokes personal memories because it was the first I attended.

The organization of the meeting was slightly different than that of the meeting in London, two years earlier. Because now the papers were preprinted well before the meeting and mailed to the participants, the authors only presented a brief summary and some additional information not included in the printed text. This new material was then printed in the final proceedings as "authors additional comments," and in many cases, represented important additions to the original text. Each paper was followed by a very intensive discussion (transcribed in the Proceedings) and some of the remarks can almost be considered as mini-lectures. The intensity of the discussions is best demonstrated by the fact that a total of 77 persons participated in them; that is, more than 15% of the participants had an active role in the meeting.

Table 30.2 presents the distribution of the papers according to the affiliation of the speakers. As can be seen the earlier dominance of the petroleum industry was reduced although still remained substantial. It is interesting to observe the significant increase of papers from universities: their number, when added to those from government laboratories, was now equal to the number of papers from the petroleum industry. Again, a Nomenclature Committee was formed during the symposium; it reviewed the draft of two papers on the way GC data should be presented and on the definition of some terms. These papers have already been presented in the United State at various meetings, but the authors proposed that this symposium also endorses the recommendations outlined in them. After some minor modifications these papers were published soon after the symposium.[20,21] The committee also clarified the difference between efficiency (plate number) and peak resolution, and introduced for the latter the formula

$$R_s = \frac{2(t_{R2} - t_{R1})}{w_{b1} + w_{b2}},$$

where t_R is the retention time and w_b is the peak width at base.

A total of 28 papers were presented at the symposium; from these 15 (53.6%) were by British, and four (14.3%) by Dutch scientists. In addition to the formal papers brief addresses were presented by J. J. van Deemter (Shell Laboratory, Amsterdam), A. J. P. Martin, and P. H. Emmett (Johns Hopkins University, Baltimore, MD, USA). Below the most important presentations are discussed briefly.

Without any question the two most important papers presented at this symposium were those by I. G. McWilliam and R. A. Dewar (ICI of Australia and New Zealand, Melbourne) on the flame-ionization detector (FID), and by M. J. E. Golay (Perkin-Elmer, Norwalk, Connecticut, USA), on the theory of open-tubular (capillary) columns. These papers have changed the way GC has been carried out ever since. In October 1957 McWilliam already presented a brief discussion of the FID at an informal meeting of the GC Discussion Group, in Cambridge, UK, and also published a short paper on it in *Nature* (see Chapter 22); now he presented a more detailed report on the detector and its variables. Because of the earlier disclosures, a few laboratories already experimented with the FID and thus, after McWilliam's presentation, H. Boer (Koninklijke/Shell Laboratory, Amsterdam) could report on their results. The ensuing discussion between them, and McWilliam's answers to Golay's questions (who immediately recognized that the FID was the ideal detector for his columns) further qualified a number of practical questions.

Golay had already presented a crude theory of the open-tubular (capillary) columns one year earlier at the 1957 Lansing Symposium of the Instrument Society of America (see Chapter 24). However, that paper cannot be compared to this presentation, which now gave a most complete theory, more elaborate than the Van Deemter equation was for packed columns. (The original form of Van Deemter's theory and equation described at the 1956 Symposium by Keulemans and Kwantes (see above), and in their detailed journal publication[16] neglected the resistance to mass transfer in the gaseous phase). In the form of the text published in the preprints, in which 92 equations were crammed into 16 printed pages, Golay's paper probably would have created little attention: as noted by Desty,[22] it "was totally beyond almost all the participants of the symposium." However, a sensation was created by Golay's additional comments in which he showed two chromatograms obtained on the new capillary columns: the separation of all the C_6 hydrocarbons in less than 9 min, and the resolution of *m*- and *p*-xylenes. A 150 ft (45.7 m) long column, with an i.d. of 0.25 mm and coated with dodecyl phthalate was used in both cases, and over 50,000 theoretical plates were obtained using a flow rate of 0.5 mL/min. These results were even more remarkable because they were obtained using a micro-thermal-conductivity detector specially constructed by Golay for this purpose. It was immediately clear at the symposium that McWilliam's FID would be the ideal detector for such columns. The general feeling was best described by Desty, reminiscing about his own reaction[23]:

> We dashed back from Amsterdam to setup a capillary column apparatus within a few days ...(consisting of) the very versatile hydrogen flame (ionization) detector described by McWilliam ... (and using) thick-walled lengths of 50 metres of stainless steel tubing.

It should be mentioned that G. Dijkstra and J. de Goey (Unilever, Vlaardingen) also presented a paper on coated capillary columns at the symposium. It mentioned the potentials of such columns, however, without any theoretical treatment and showing very poor results: the maximum efficiency obtained for a 120 m long, 0.3 mm i.d. column was only 2500 theoretical plates (as compared with Golay's over 50,000 plates obtained on a 45-m long column). It was pointed out in the discussion that these extremely poor results were obviously due to too large volumes of both the injector and the detector, too large sample sizes, and I might add, a much too thick stationary phase film. (Based on the information presented in their report, their columns would be equivalent to a present-day "megabore" column of 0.53 mm i.d., having a film thickness of 9 μm.)

Dijkstra's work had been carried out independently of Golay's and he did not continue to work in this field. However, one of his developments survived: the way the capillary tubes were coated. He was the first to use the dynamic coating technique, wetting the inner wall of the tube with the solution of the phase and then evaporating the residual solvent by maintaining a nitrogen stream through the tube for some time.

Some theoretical papers further consolidated our knowledge on the Van Deemter equation. J. Bohemen and J. H. Purnell (Cambridge University, UK) and A. B. Littlewood (University of Durham, Newcastle, UK) gave detailed data on measurements related to the individual terms of this relationship obtained under different conditions, drawing conclusions about how (packed) column efficiencies could be improved. The very detailed discussion following these papers dealt with every aspect of column preparation and use. Other theoretical papers dealt with the relationship between the adsorption isotherms and chromatographic separation on adsorption columns and various other aspects of gas adsorption chromatography, as well as with the use of GC for the determination of activity coefficients.

Two papers dealt specifically with the design of gas chromatographs. The first, an automated preparative gas chromatograph, was developed by E. P. Atkinson and G. A. P. Tuey (May & Baker, Dagenham, UK). It utilized two 175-cm long, 2-cm i.d. columns in series, with a column packing consisting of 60/100 mesh kieselguhr support coated with 20–30% stationary phase. The important feature of the instrument was that it permitted the automated repetitive injection of the same sample and the joint collection of the same fractions from the repetitive injections. The second instrument, described by J. Hooimeijer, A. Kwantes and F. van de Craats (Koninklijke/Shell Laboratory, Amsterdam), was a process GC for automated control of pilot plant streams. It should be noted that by this time process gas chromatographs already were commercially available in the United States and some of these have been described at the 1957 Lansing Symposium of the Instrument Society of America (see Chapter 31).

It is interesting to note that although by 1958 a number of gas chromatographs had became commercially available both in Europe and the United States, none of these were described at the symposium in a contributed paper. During the symposium an instrument exhibit was also held and the commercially available instruments were displayed there. The sensation of the exhibition was the Argon Gas Chromatograph of W.G. Pye & Co. (Cambridge, UK) that utilized the argon-ionization detector developed by J. E. Lovelock (National Institute for Medical Research, London) (see Chapter 23). In fact, the interest in this system was so great that Lovelock was asked to present an unscheduled, brief description during one of the sessions.

The presentation of L. Guild, S. Bingham and F. Aul (Burrell Corp., Pittsburgh, Pennsylvania, USA) dealt with various instrumental problems encountered during the operation of a gas chromatograph, such as an unstable baseline due to impurities in the carrier gas, fluctuation of its flow rate, and stationary phase bleeding when programming the temperature of the column. Particularly, the question of impurities present in helium was discussed in considerable detail. Everybody was smiling when E. Glueckauf of the British Atomic Energy Research Establishment mentioned that helium can easily be purified by conducting it over uranium turnings at 700°C, and as an answer, the session chairman (A. J. P. Martin) spontaneously remarked that unfortunately, chromatographers generally do not have uranium turnings in their laboratories.

Temperature programming was intensively discussed; besides the paper of Guild *et al.*, prepared contributions were presented by S. Dal Nogare (DuPont, Wilmington, Delaware, USA), C. L. A. Harbourn (BP, Sunbury-on-Thames,), and K. E. Murray (Commonwealth Industrial Research Organization, Melbourne, Australia). In addition, in their formal presentation G. F. Harrison, P. Knight, R. P. Kelly, and M. T. Heath (Associated Ethyl Co., Ellesmere Port, UK) also described a device to assure the controlled programming of column temperature. They also reported on a system consisting of two columns in series, where the second column was at a lower temperature to provide better separation of the lower-boiling part of a wide-boiling-range sample.

The FID was not the only detector described at the meeting. Improvements to Scott's hydrogen flame (thermocouple) detector were presented by G. R. Primavesi, G. F. Oldham, and

R. J. Thompson (The Distillers Co., Epson, UK), while W. Struwe (Margarine-Union, Hamburg-Bahrenfeld, Germany) described a simplified thermal-conductivity detector in which the platinum resistance wire was replaced by a standard, small light bulb (having its glass envelope removed). Van de Craats (Koninklijke/Shell Laboratory, Amsterdam) examined the quantitative aspects of the thermal-conductivity detector, its linearity, and the ways of calibration, and D. W. Grant (Coal Tar Research Association, Leeds, UK) proposed a new type of detector in which the column effluent was mixed with coal gas, the mixture combusted, and the emissivity of the flame measured.

With respect to column technology, R. P. W. Scott's paper (Benzole Producers, Watford, UK) described highly efficient packed columns. The long (up to 50 ft) columns used with 200 psi inlet pressure could provide efficiencies up to 30,000 theoretical plates. The other report dealing with column performance was by D. H. Desty, F. M. Godfrey, and CV. L. A. Harbourn, who presented detailed data on two materials (Celite 545 and firebrick 22) used as stationary phase supports.

Among the papers describing the applications of GC, the presentation by L. Bovijn, J. Pirotte, and A. Berger (Belgian Electrical Co., Schelle) was the most interesting: using headspace analysis they determined traces of hydrogen in the water present in high-pressure boilers. Among the other papers, A. Liberti and G. P. Cartoni (University of Messina, Italy) surveyed the potential of GC in the analysis of essential oils, E. Bayer (Technical University, Karlsruhe, Germany) demonstrated the analysis of amino acid derivatives, and E. R. Adlard and B. T. Whitham (Shell Research Center, Thornton, UK) illustrated the wide range application of GC for the analysis of high-boiling samples.

At the end of the symposium D. Ambrose (Chemical Research Laboratory, Teddington, UK) and J. H. Purnell (Cambridge University, UK) presented retention data for a large number of compounds. They tabulated the values of the constants A, B, and C in the Antoine-type equation

$$\log V_g = A + \frac{B}{t + C}$$

where V_g is the specific retention volume and t is the column temperature, in °C, for given temperature ranges on three stationary phases.

The Amsterdam Symposium started the tradition of having a reception for the participants and their spouses. Such receptions became a standard feature of future symposia; however, none of them could ever surpass Amsterdam. The reception was hosted by the Dutch government and the City of Amsterdam, and was held in the Rijksmuseum, one of the most famous picture galleries in the world. To stroll among the great paintings of Rembrandt, waiting that maybe the members of the militia will step out from his *Night Watch* and join the party, was an experience I will never forget ...

References

1. A. T. James and A. J. P. Martin, *Biochem. J.* **48**(1), vii (1951) (abstract).
2. A. T. James and A. J. P. Martin, *Biochem. J.* **50**, 679–690 (1952).
3. A. T. James, A. J. P. Martin and G. M. Smith, *Biochem. J.* **52**, 238–242 (1952).
4. A. T. James, *Biochem. J.* **52**, 242–247 (1952).
5. A. T. James and A. J. P. Martin, *Analyst* (*London*) **77**, 915–932 (1952).
6. H. Boer, in *75 Years of Chromatography — A Historical Dialogue*, eds. L. S. Ettre and A. Zlatkis (Elsevier, Amsterdam, 1979), pp. 11–19.
7. F. R. Cropper and A. Heywood, *Nature* (*London*) **172**, 1101–1102 (1953).
8. N. H. Ray, *J. Appl. Chem.* **4**, 21–25 (1954).
9. N. H. Ray, *J. Appl. Chem.* **4**, 82–85 (1954).
10. B. W. Bradford, D. Harvey and D. E. Chalkey, Lecture at the Institute of Petroleum, London, on 10 November 1954; *J. Inst. Petrol.* **41**, 80–91 (1955).
11. *Anal. Chem.* **27**, 1667 (1955).
12. Anon., *Analyst* (*London*) **81**, 52–58 (1956).
13. D. H. Desty and C. L. A. Harbourn, eds., *Vapour Phase Chromatography* (*1956 London Symposium*) (Butterworths, London, 1957), xvi + 436 pp.
14. D. H. Desty, ed., *Gas Chromatography 1958* (*Amsterdam Symposium*). (Butterworths, London, 1958) xiv + 383 pp.
15. N. H. Ray, *Analyst* (*London*) **80**, 853–860 (1955).
16. J. J. van Deemter, F. J. Zuiderweg and A. Klinkenberg, *Chem. Eng. Sci.* 5 271–289 (1956).

17. J. Griffiths, D. James and C. S. G. Phillips, *Analyst (London)* 77, 897–904 (1952).
18. R. P. W. Scott, *Nature (London)* **176**, 793 (1955).
19. A. J. P. Martin and A.T. James, *Biochem. J.* **63**, 138–142 (1956).
20. D. Ambrose, A. I. M. Keulemans and J. H. Purnell, 132nd National A. C. S. Meeting, New York, NY, September 1957; *Anal Chem.* **30**, 1582–1586 (1958).
21. H. W. Johnson and F. H. Stross, 9th Pittsburgh Conference on Analytical Chemistry, Pittsburgh, PA, March 1958; *Anal. Chem.* **30**, 1586–1589 (1958).
22. D. H. Desty, in *75 Years of Chromatography — A Historical Dialogue*, eds. L. S. Ettre and A. Zlatkis (Elsevier, Amsterdam, 1979), pp. 31–42.
23. D. H. Desty, *Chromatographia* **8**, 452–455 (1975).

Chapter

31

Early GC Symposia in the United States*

Gas chromatography started in the United States very soon after the fundamental publications of James and Martin. We can assume that the large research laboratory of Shell Development Co., in Emeryville, CA, had already been involved in such investigations parallel to their sister laboratories in England and Amsterdam; however, the first papers from this laboratory were published only in 1956.[1–4] The other laboratory where such investigations started early was at Tennessee Eastman Co., in Kingsport, Tennessee,[5] and H. W. Patton from this lab presented the first paper on GC entitled Separation and Analysis of Gases and Volatile Liquids by Gas Chromatography at the 126th National Meeting of the American Chemical Society (ACS), held in September 1954, in New York City.[6] According to my best knowledge this was the first paper on GC presented in the United States at an open meeting. It was soon followed by a paper of D. H. Lichtenfels (Gulf Oil, Pittsburgh, PA) on Gas Liquid Partition Chromatography presented in

*Based on the article by L. S. Ettre published in *LCGC* (*North America*) **21**, 144–149, 167 (2003).

March 1955, at the Pittsburgh Conference.[7] A third paper on Vapor-Phase Chromatography was presented by R. H. Munch (Monsanto Chemical Co., St Louis, Missouri) in the spring of 1955 during the 127th National ACS Meeting in Cincinnati, OH. Based on this presentation, *Chemical & Engineering News*, the weekly magazine of ACS, published a report on GC[8] which emphasized that "the separation of gases and volatile liquids by vapor-phase chromatography may add one more powerful tool to the analytical chemist's kit" and expressed the editors' opinion that the new technique "is worth a look."

If we compare the early evolution of GC in the United States with that in Europe, a few major differences can be pointed out. As we have seen in the previous chapter, in Europe early development was concentrated primarily in petroleum and petrochemical companies, while there was a broader base in the United States; this can be seen by comparing the statistics of the early European symposia (see Table 30.2 in Chapter 30) with the data given here in Table 31.1. Another major difference was the availability of commercially manufactured thermal conductivity detectors, and the willingness of even laboratories of large chemical and petrochemical companies to use components available on the market. For example, the book on GC by A. I. M Keulemans published in 1957 in Europe[9] still had a major chapter describing how to construct a thermal conductivity detector in one's own lab. At the same time, H. W. Patton, in the first paper on GC presented in the fall of 1954, already used a modified, commercially available thermal conductivity cell, and its manufacturer almost immediately adapted the suggested modifications; by the time of the publication of Patton's paper a few months later such modified cells were already commercially available. Similarly, in his first self-constructed GC, R. H. Munch also utilized a standard potentiometric recorder and a commercially available thermal conductivity cell.

The availability of helium in the United States may also be the reason for a peculiar difference one can observe by comparing the papers and discussions of the European vs American meetings. In the discussions during the European symposia uncertainties about the proper way to establish quantitative results when using a thermal conductivity

Table 31.1. Distribution of papers presented at the early US symposia on gas chromatography.[a]

Presenter's affiliation	ACS April 1956 Dallas, TX		ISA August 1957 Lansing, MI		ISA June 1959 Lansing, MI		NYAS April 1958 New York, NY	
	No. of papers	%	No. of papers	%	No. of papers	%	No. of papers	%
Petroleum industry and related companies	4	20.0	2	7.4	5	14.7	2	10.0
Other industrial laboratories	8	40.0	14	51.9	8	23.5	4	20.0
Universities	2	10.0	2	7.4	2	5.9	6	30.0
Government laboratories	2	10.0	3	11.1	8	23.5	4	20.0
Scientific instrument industry	4	20.0	6	22.2	11	21.4	4	20.0
Total	20	100.0	27	100.0	34	100.0	20	100.0

[a] **ACS** = American Chemical Society; **ISA** = Instrument Society of America; **NYAS** = New York Academy of Sciences.

detector are often mentioned, however, such discussions were practically absent from the American meetings. This may be due to the fact that with the general availability of helium in America, there was no problem with the closeness of the thermal conductivities of the carrier gas vs sample vapor and there were no erratic results as with the forced use of nitrogen. Also, the way to calculate quantitative results was established in the United States in an early stage of development: the regularity of the response of the thermal conductivity detector was described in the spring of 1957 by Rosie and Grob[10] and their results rapidly became common knowledge. (The acerbic comments at the 1956 London Symposium after the presentation of Hausdorff mentioned in Chapter 30 were actually due to an innocent remark of the speaker pointing out this difference.)

The fourth major difference between early GC development in the United States vs Europe was the major and active role of the then nascent American scientific instrument industry. As discussed in Chapter 19, the first commercial gas chromatographs were introduced in the United States in the spring of 1955; by the end of this year three companies marketed their instruments and these were followed by four more in the first part of 1956. Because of the early availability of commercial gas chromatographs, the technique was able to spread rapidly even to smaller laboratories which did not have the capability of constructing their own instruments. In fact, even laboratories of large chemical and petrochemical companies soon switched from self-built instruments to the standardized gas chromatographs available commercially (see e.g., Fig. 21.4 in Chapter 21), and even those who first constructed their own instruments utilized some commercially available components. It is also interesting to note that in the symposia held in the United States a number of presentations described the development of commercially available instruments: in the United States this was considered a bona fide technical achievement, justifying its discussion at a scientific meeting. Particular characteristics of the ISA symposia and of the meeting of the New York Academy of Sciences discussed below were the papers dealing with the development and use of automated process gas chromatographs, which by 1957, became commercially available.

The young American instrument companies not only supplied commercially available instruments: their associates had a significant role in the advancement of the technique. They carried out basic studies, contributed to the theory of GC, held seminars and instructional lectures, and even published brochures informing the user about the basics of the technique (see e.g., Refs. 11 and 12). As we shall see below, they also actively participated in the organization of the first nationwide symposia dealing with the various aspects of GC.

31.1. The Early American Symposia

In the early development of GC in the United States four symposia had major roles: the symposium held in April 1956 during the 129th National Meeting of the American Chemical Society (ACS), in Dallas, TX, with 20 papers on its program; the two symposia organized in 1957 and 1959 by the Instrument Society of America (ISA) in East Lansing, MI, with 27 and 34 papers, respectively; and finally the representative conference organized by the New York Academy of Sciences, in April 1958, with 20 presentations (see Table 31.1). In this chapter we shall summarize the subjects discussed at these meetings.

Similar to the European symposia the papers presented at these meetings and also the essence of the discussion sessions became available to a wide audience. Over 50% of the papers presented at the 1956 Dallas meeting were published within a year in *Analytical Chemistry*, and the proceedings of the ISA symposia were published in book form.[13,14] Similarly, the papers presented at the Conference of the New York Academy of Sciences were published in a separate issue of the *Annals* of the Academy.[15]

The two proceedings of the ISA symposia also contained some additional materials, useful to chromatographers. Both included a complete bibliography of publications on GC, including papers presented at major meetings. These two bibliographies,[16,17] prepared by S. H. Langer and C. Zahn (US Bureau of Mines, Pittsburgh, PA) and L. S. Ettre (Perkin-Elmer Co., Norwalk, CT), respectively, listed a total of 1975 papers. In addition, the 1959 ISA Symposium proceedings included a compilation of retention data of 69 compounds belonging

Table 31.2. Distribution of the papers presented at the 1958 Conference of the New York Academy of Sciences according to the speakers' country of origin.

Country	No. of papers	%
United States	13	65.0
United Kingdom	4	20.0
Czechoslovakia	2	10.0
The Netherlands	1	5.0
Total	20	100.0

to 18 homologous series, at two temperatures, obtained on standard columns coated with nine stationary phases.[18] This compilation, prepared by P. R. Scholly and N. Brenner of Perkin-Elmer, also presented logarithmic plots of retention times against the boiling points of the individual substances.

Overseas participation at these early American symposia was limited because in the 1950s transatlantic traffic was not as high as it would become later, in the jet age, although the organizers always invited some prominent European chromatographers, in order to provide the participants the opportunity to meet the pioneers and learn firsthand their results. Foreign participation became particularly significant at the Conference organized by the New York Academy of Sciences where 35% of the speakers came from overseas (see Table 31.2).

31.2. The 1956 Dallas ACS Symposium

This symposium was organized by the Division of Analytical Chemistry of the ACS. The interest was great: as commented in *Analytical Chemistry*, "no beautiful movie actress could have drawn a more appreciative and attentive audience"; and the symposium was "a standing-room-only event," with more than 600 participants.[19] The presentations were mainly practical, dealing with investigations of the variables of a GC system, some instrumental aspects and a number of applications.

H. W. Johnson Jr (Shell Development Co., Emeryville, CA), H. W. Patton (Tennessee Eastman), and L. V. Guild (Burrell Corp., Pittsburgh, PA) studied the various liquids and adsorbents useful as stationary phases, while A. K. Wiebe (Hercules Powder Co., Wilmington, DE), and K. V. Wise and G. D. Oliver (Monsanto, St Louis, MO) dealt with conditions such as flow rate and the selection of the carrier gas and temperature, and how these influence retention and resolution. One of the recommendations of Wise and Oliver was strange: the use of dimethyl ether vapor (BP: −24.82°C) as the carrier gas for the analysis of impurities in vinyl chloride monomer.

With respect to instrumental development, G. K. Ashbury, A. J. Davies, and W. J. Drinkwater (Shell Research Center, Thornton, England) described a laboratory GC developed at Shell, in England, while B. W. Taylor (Fisher Scientific, Pittsburgh, PA) presented information on their commercial gas chromatograph. R. H. Munch (Monsanto, St Louis, MO) dealt with various possible detection principles and C. H. Deal, J. W. Otvos, W. N. Smith, and P. S. Zucco (Shell Development, Emeryville, CA) described a new radiological detector.

In 1956, gas chromatography was generally used at temperatures not exceeding 125–150°C. There were two main reasons for this: the unavailability of partition-type stationary phases with low bleed and instrument design problems in providing a constant temperature air thermostat for higher temperatures. Therefore, the paper of E. F. Williams (American Cyanamid Co., Stamford, CT), discussing the extension of the temperature range was of particular interest. R. S. Gohlke and F. M. McLafferty (Dow Chemical Co., Framingham, MA) also extended the temperature range for the analysis of complex samples with boiling points up to 300°C; they also described the possibility of identifying the collected fractions by infrared and mass spectrometry.

One purely theoretical paper was presented at the Symposium: the famous paper of M. J. E. Golay (Perkin-Elmer Corp., Norwalk, CT) on Vapor-Phase Chromatography and the Telegrapher's Equation. It presented a simplified picture of the separation process in the GC column, discussing it based on mathematical considerations with the help of an equation used in electrical engineering. Subsequent attempts to fit this theoretical treatment to experimental results proved difficult for conventional packed columns. With typical insight, Golay then substituted theoretically open capillary tubes for each passage in the packed column; this sequence of thinking led about one year later, to the idea of open-tubular columns eventually replacing the packed columns (see Chapter 24).

The papers on the application of GC covered a wide range of samples. H. J. Dawson (Standard Oil of Indiana) discussed the analysis of C_3–C_5 paraffins in naphthas; the flavor of strawberries and other foodstuff was studied by K. P. Dimick, F. Stitt, and J. Corse (Western Regional Research Laboratory, Albany, CA), and by K. G. Sloman and E. Borker (General Foods), while the analysis of tobacco smoke was explored by B. B. Seligman and associates (Phillip Morris). Pyrolysis and thermal decomposition of various compounds were studied using GC by C. M. Drew and J. R. McNesby (US Naval Ordnance Station, China Lake, CA) and C. B. Colburn (Rohm & Haas, Philadelphia, PA).

At the end of the symposium a panel was assembled consisting of R. H. Munch (Monsanto), H. W. Johnson Jr (Shell Development), H. W. Patton (Tennessee Eastman), D. H. Lichtenfels (Gulf Oil, Pittsburgh) and C. B. Willingham (Mellon Institute, Pittsburgh) who then tried to answer the diverse questions of the audience hoping to learn more about gas chromatography.

31.3. The 1957 ISA Symposium

In the 1950s the Instrument Society of America (ISA) was a very active organization. It had a special Analysis Instrumentation Division and within it a special committee was formed dealing with GC. After the success of the 1956 London Symposium the ISA decided to take over the responsibility of organizing a similar symposium series in the United States. Because the European symposia were scheduled for even-numbered years, the ISA symposia were planned for the odd-numbered years. These meetings were held on the campus of Michigan State University, in East Lansing. Starting in 1957, four such symposia were organized, in 1957, 1959, 1961, and 1963; we are interested here in the first two, held in August 1957 and June 1959, with 27 and 34 papers on their programs, respectively.

The first symposium was held during 28–30 August, 1957, at the Kellogg Center, on the Michigan State University campus. The organizers were V. J. Coates (Perkin-Elmer) serving as the general chairman, H. J. Noebels (Beckman Instruments, Fullerton, CA), and I. S. Fagerson (University of Massachusetts, Amherst). Adjacent to the lecture and conference rooms this Center also included a hotel where the participants were staying. This arrangement provided the opportunity for unlimited informal discussions organized outside the formal program. The organizers invited a number of European pioneers such as A. J. P. Martin, the inventor of partition chromatography, C. S. G. Phillips (Oxford University, United Kingdom), D. H. Desty (British Petroleum, Sunbury on Thames, United Kingdom), and A. I. M. Keulemans (Shell Research Laboratories, Amsterdam, The Netherlands) who actively participated in the discussions, while Martin and Phillips also presented lectures. In his address Martin spoke about the Past, Present and Future of Gas Chromatography, recounting the events which lead to its development and pointing out some intriguing possibilities, from recirculating columns to detectors of biological origin. Phillips surveyed some rules governing separation, and then dealt in more detail with a special field: the use of fused metal salts, coated on the usual support particles, as stationary phases.

The first paper presented on the program of the symposium was by M. J. E. Golay (Perkin-Elmer) who spoke about the Theory and Practice of Gas–Liquid Chromatography with Coated Capillaries. This was his first disclosure about the open-tubular column concept: use of capillary tubes coated with the stationary phase on their inside wall instead of the packed columns used universally at the time. Golay presented a rough theory of these columns and showed one of the first (still very crude) chromatograms he obtained. Martin immediately recognized the importance of this work, calling it the highlight of the symposium. Other theoretical discussions were presented by J. F. Young (Douglas Aircraft Co., Santa Monica, CA), E. A. Hinkle and S. E. Johnsen (Monsanto Chemical Co., Texas City, TX), S. S. Ober (Abbott Laboratories, North Chicago, IL), C. B. Cowan and P. H. Stirling (Canadian Industries Ltd., McMasterville, Québec), and L. F. Hatch (University of Texas, Austin, TX).

In 1957 the number of available support materials were limited and they were not inert. For this reason various chemical pretreatment methods were available and used widely: let us not forget that at that time, many people prepared their own packed columns. (I remember, when I started in GC in 1957, in Europe, I had two Soxleth apparatus continuously working in my laboratory, extracting the iron impurities from the commercially available diatomaceous material.) Also, the particle size of the commercial products was too broad and for better column performance, one had to sieve it prior to coating. Therefore, the paper of Theron Johns (Beckman) was very useful: he discussed in detail the various support treatment methods and the optimization of particle size.

In 1957 the now ubiquitous microsyringes were not yet available and reproducible injection of small liquid sample volumes represented a problem. At this symposium D. W. Clarke (Beckman) described a device — a kind of syringe — for this purpose. It was claimed to enable the precise introduction of liquid volumes between 10 and 50 μL, but it was a complex, bulky system with only limited applications. Similar devices (called "microdippers") were also developed by others such as Fisher Scientific (Pittsburgh, PA) and Consolidated Electrodynamic Corporation (CEC, Pasadena, CA), and the papers presented by B. W. Taylor and R. A. Meyer describing gas chromatographs developed by these companies (see below) also mentioned the development of such devices. However, all these systems were very unreliable and thus, everybody could at least breathe freely when in less than one year Clark Hamilton (Hamilton Co., then in Whittier, CA) developed his precision microsyringes, first with 10 and 50 μL, and soon after with 1 and 5 μL volumes.

By 1957 chromatographers tried to determine smaller and smaller concentrations, and the then universally applied thermal conductivity detectors (TCD) often did not have sufficient sensitivity for such applications. To solve this problem S. D. Norem (Perkin-Elmer) proposed in his paper to combust the fractions in the column effluent and to detect with the TCD the CO_2 so formed. Such a system could act as a "molecular amplifier;" for example one mole of a hydrocarbon with a composition of C_nH_{2n+2} would be converted to n moles of CO_2, thus increasing detector sensitivity. A commercial device based on these principles soon became available, but the advent of the ionization detectors, with their superior sensitivity, within one year made it obsolete.

A number of papers dealt with the design and performance of laboratory gas chromatographs. (All these were operated isothermally; column temperature programming did not exist as yet at that time.) Scientists from E. I. DuPont's Experimental Station (Wilmington, DE) described three instruments: a high-temperature unit capable of operation up to 500°C (H. R. Felton), a preparative unit with 6 ft × 1.25 in. i.d. columns (J. J. Kirkland), and an electromechanical integrator (S. Dal Nogare, C. E. Bennett, and J. C. Harden). G. W. Taylor and A. S. Dunlop (Polymer Ltd., Sarnia, Ontario, Canada) reported on a low-temperature gas chromatograph operated at 0°C, specially developed for the analysis of low-boiling light hydrocarbon samples: the essential problem with such samples was to obtain a representative aliquot from a sample usually under pressure. Finally, R. A. Meyer described the gas chromatographs marketed by CEC, while B. W. Taylor (Fisher Scientific, Pittsburgh, PA) dealt with the gas chromatograph developed by Fisher Scientific in cooperation with Gulf Oil's research laboratories.

The idea of using GC for automated process control is almost as old as the technique itself and a number of companies had been working on its realization. At the 1957 Lansing Symposium reports on two such instruments were presented. T. L. Zinn, W. J. Baker, H. L. Norlin, and R. F. Wall (Monsanto Chemical Co. Texas City) summarized their specifications for an instrument which was then under development; a detailed report was presented next spring, at the conference organized by the New York Academy of Sciences (see below). The second instrument was a joint development of Phillips Petroleum Co. (Bartlesville, OK) and Perkin-Elmer, and it was described in two papers, presented by B. O. Ayers of Phillips, and C. C. Helms and H. N. Claudy of Perkin-Elmer; the instrument already had been field-tested and became commercially available soon after the symposium.

Finally, I should mention three papers dealing with the applications of GC in a wide range of fields: the analysis of C_2–C_5 hydrocarbons was discussed by W. A. Dietz (Esso, Linden, NJ), determination of phthalates and other ester-type plasticizers was the subject of the paper of J. S. Lewis and H. W. Patton, (Tennessee-Eastman), and M. J. Root (G. Barr & Co., Chicago, IL) dealt with the GC investigation of halocarbons.

31.4. The 1959 ISA Symposium

This symposium was held during 10–13 June, 1959, again at the Michigan State University campus. Now the organizers were H. J. Noebels (Beckman), serving as the general chairman, N. Brenner (Perkin-Elmer), and R. F. Wall (Monsanto Chemical Co., Texas City). A total of 34 papers was presented, of which 14 (41.2%) dealt with the design considerations of instruments or accessories.

When the use of GC was extended to the analysis of more complex samples with wide boiling ranges, it was recognized that one single-temperature column could not provide optimum separation for all sample components. The approach followed first was to use two or more columns in series, each at a different temperature; thus, each column was optimized for a different part of the sample. By 1957 such systems already had been described in the literature and even a three-stage commercial instrument (Perkin-Elmer's Model 188) became available.[20] At the present symposium three papers dealt with such multi-column systems; these were presented by K. J. Hughes, R. W. Hurn, and F. G. Edwards (Petroleum Experiment Station, Bartlesville, OK), M. G. Bloch (Socony Mobil Oil, Paulsboro, NJ), and J. A. Perry (Sinclair Research Laboratories, Harvey, IL). For a short time there was a debate about the relative merits of such systems vs temperature programming of a single column. In this respect it is worthwhile to cite the comments by C. L. A. Harbourn (British Petroleum, Sunbury-on-Thames, England) at the 1958 Amsterdam Symposium, that "the practical value of programmed heating is far less than advocated" and that multistage GC "offers many advantages" as compared with temperature programming.[21] Of course we know today, that this prediction was wrong.

At the present symposium A. J. Martin, C. E. Bennett, and F. W. Martinez Jr, former DuPont associates who just formed F&M Corporation a few months earlier to produce and market temperature-programmed instruments, presented a comprehensive discussion on various aspects of temperature programming. In retrospect we can state that within a short time, temperature programming "won," and by 1962 practically every new commercial gas chromatograph provided this capability.

Besides the Phillips Petroleum — Perkin-Elmer process gas chromatograph (reported at the 1957 ISA Symposium), Beckman Instruments also introduced such an instrument and G. Turner discussed the principles of this system. He mentioned one particular application: control of the concentrations of H_2S, O_2, SO_2, and water in the streams of a sulfur recovery plant. Another major paper by associates of Beckman was presented by R. Villalobos, R. O. Brace, and Th. Johns, dealing with the theory and practice of column backflushing.

Three papers dealt with instrumentation for semi-preparative and preparative GC. G. Kronmueller (Department of National Health and Welfare, Ottawa, Canada) developed an automatic fraction collection system used as an accessory to commercial gas chromatographs; the collected fractions were used for confirmatory tests. Special traps were also used by M. C. Simmonds and T. R. Kelley (Shell Oil, Houston, TX) in the analysis of complex petroleum alkylate samples, containing over 30 C_3–C_9 paraffins. Nine fractions — each containing a number of compounds — were collected and analyzed by mass spectrometry to identify the individual sample components. The third paper in this field was presented by Theron Johns, M. R. Burnell, and D. W. Carle (Beckman) describing the company's Megachrom preparative gas chromatograph. This instrument utilized eight 6-ft long, 5/8-in. i.d. columns in parallel configuration; it was equipped with an efficient fraction collection system and permitted the recirculation of the carrier gas.

The new ionization detectors (argon ionization and flame ionization) introduced in Europe in 1958 started to be incorporated into American-made commercial gas chromatographs in the first part of 1959. However, some research was still carried out on other types of sensitive detectors and three papers at this symposium presented such reports. The first detector now discussed in detail by E. A. Hinkle,

H. C. Tucker, R. F. Wall, and J. F. Combs (Monsanto) and by L. V. Guild, M. J. Lloyd and F. Aul (Burrell Co., Pittsburgh) was originally described in 1957 by S. A. Ryce and W. A. Bryce[22,23]: it was a high-vacuum ionization gauge, permitting a 10^3-fold improvement in sensitivity to the thermal conductivity detector. The other new device was a radiofrequency discharge detector, developed by A. Karmen and R. L. Bowman (National Heart Institute, Bethesda, MD). In the end, however, the simplicity and ease of operation of the flame-ionization detector prevailed, and the other types of detectors remained in the research stage. One paper, by N. H. Ray (Imperial Chemical Industries, England), discussed the operation of the thermal conductivity detector, and how changes in temperature, pressure and the choice of carrier gas affect its sensitivity. Finally I should mention the paper of I. G. Young (Leeds & Northrup, Philadelphia, PA) who attempted to standardize the various terms used to express detector sensitivity.

A few papers dealt with new instruments and instrumental systems. D. Alexander and R. F. March described the new gas chromatograph of Consolidated Electrodynamics Co. (Pasadena), operating up to 500°C. O. K. Dooley (Gulf Oil R&D Co., Pittsburgh, PA) and W. L. Perrine (Disc Instruments, Santa Ana, CA) described new mechanical integrators connected to the standard potentiometric recorders. J. J. Davis (Perkin-Elmer) drew attention to the special requirements of capillary columns and the new ionization detectors with respect to the electronics and recording systems of the gas chromatographs.

A couple of papers discussed questions related to the proper selection of column parameters and the operation of gas chromatographs. R. M. Bethea and T. D. Wheelock (Iowa State University, Ames) studied the effect of flow rate, while W. J. Baker, E. H. Lee, and R. F. Wall (Monsanto Chemical, Texas City) investigated the influence of the solid support on column efficiency. M. J. E. Golay (Perkin-Elmer) examined the theory of large GC columns used in preparative instruments, including how the increase of column diameter influences its efficiency.

The keynote lecture at the symposium was presented by A. T. James (National Institute for Medical Research, London, England). The title of his lecture was The Development of an Idea; in it James

Table 31.3. Application papers presented at the 1959 I.S.A. Symposium.

Sample type	Presenters	Affiliation
Rare gases	M. Krejčí, K. Tesařík, and J. Janák	Laboratory for Gas Analysis, Brno, Czechoslovakia
o/p-Hydrogen	E. Erb	Linde Co., Tonawanda, NY
Dissolved O_2 in petroleum	J. P. Paglis	Standard Oil, Whiting, IN
Water in hydrazine	E. F. C. Cain and M. R. Stevens	Rocketdyne, Canoga, CA
C_3–C_4 Olefins	A. Zlatkis and H. R. Kaufman	University of Houston, TX
Alkylbenzenes and cresols	A. R. Paterson	Allied Chemical Co., Morristown, NJ
Trace impurities in "pure" hydrocarbons	W. C. Jones Jr	Humble Oil, Baytown, TX
Halocarbons	J. L. Monkman and L. Dubois	Dept. of National Health and Welfare, Ottawa, Canada
Isomeric derivatives of cyclobutanes	B. C. Anderson	DuPont, Wilmington, DE
Oxygenated combustion products of hydrocarbons	K. J. Hughes, R. W. Hurn, and F. G. Edwards	Petroleum Experiment Station, Bartlesville, OK
C_1–C_7 Hydrocarbons	M. G. Bloch	Socony Mobil Oil, Paulsboro, NJ
Strongly basic nitrogen compounds	A. W. Decora and G. V. Dinneen	Petroleum Research Center, Laramie, WY

described how in 1950–1952 he, with A. J. P. Martin, developed gas–liquid partition chromatography, and then he discussed the trends in the evolution of GC since its inception.

Finally I should mention the many papers dealing with a wide range of applications of GC. These are summarized in Table 31.3.

31.5. The 1958 Conference of the New York Academy of Sciences

In addition to the two biannual ISA Symposia we should mention one more important GC meeting held in the United States in the second part of the 1950s. It was sponsored by The New York Academy of Sciences and held during 10–11 April 1958 in New York City; E. F. Williams of American Cyanamid Co. (Stamford, CT) has served as the conference chairman. In Chapter 29 we have already mentioned the representative conference on (liquid) chromatography organized by the Academy in 1946, just one year after the end of the Second World War. The aim of the present conference was the same as of the 1946 meeting, to bring together key scientists from both the United States and Europe, and in this they succeeded: from the 19 papers presented at the symposium, seven (35%) were by European authors.

The particular aim of the conference was to combine papers on theory, applications, and instrumentation, with emphasis on process control; in fact, four of the papers dealt with process GC. As a continuation of the reports presented at the 1957 ISA meeting, now H. N. Claudy, C. C. Helms, P. R. Scholly, and D. R. Bresky (Perkin-Elmer, Norwalk, CT) provided further details on the instruments introduced for automated stream analysis, and R. F. Wall, W. J. Baker, T. L. Zinn, and J. F. Combs (Monsanto, Texas City) also augmented their report presented 7 months earlier at the ISA Symposium. C. W. Skarstrom (Esso Research & Engineering, Linden, NJ) dealt with the plant-type GC analyzers used at the Esso refinery, and F. C. Snowden and R. D. Ranes (Leeds & Northrup, Philadelphia, PA) discussed the criteria of thermal conductivity detectors used for process control. In addition, R. W. Hurn, J. O. Chase, and K. J. Hughes (Petroleum Experiment Station, Bartlesville, OK) described multistage GC analyzers used for exhaust gas analysis.

Three papers dealt with GC detectors. W. Wiseman (Gas Chromatography Ltd., London, United Kingdom) spoke about some problems encountered with the katharometer; J. E. Lovelock (National Institute for Medical Research, London, United Kingdom) described the new argon-ionization detector; and A. Karmen and R. L. Bowman (National Heart Institute, Bethesda, MD) outlined the design-performance of a radiofrequency glow detector.

In the second part of the 1950s the introduction of improved stationary phases for the analysis of fatty acids (in the form of their methyl esters) represented a major development, and at this Symposium three papers dealt with this subject, presented by C. H. Orr and J. E. Callen (Procter & Gamble, Cincinnati, OH), S. R. Lipsky and R. A. Landowne (Yale University Medical School, New Haven, CT), and R. K. Beerthuis, G. Dijkstra, J. G. Keppler and J. H. Recourt

(Unilever, Vlaardingen, The Netherlands); in addition the determination of plasma cholesterol fatty acids was the subject of the presentation of G. D. Michaels, P. Wheeler, G. Fukayama, and L. W. Kinsell (Highland Alamada County Hospital, Oakland, CA).

Among the other papers presented at this Symposium we should mention the discussion of questions associated with highly efficient (packed) columns and the relationship between retention, column efficiency, and resolution by J. H. Purnell (Cambridge University, United Kingdom), and the two presentations by J. Janák (Laboratory for Gas Analysis, Brno, Czechoslovakia): one on the use of zeolites as the stationary phase, and the other on retention relationships in gas adsorption chromatography.

Following the tradition set by the 1946 Conference of the Academy, the papers presented at the present Symposium were also published as a special issue of the Academy's *Annals*,[15] but at this time, the editors were faster: the special issue was already published in the fall of 1958. Thus, chromatographers interested in the activities reported at the Conference could read the presentations while the subjects were still "hot."

References

1. P. E. Porter, C. H. Deal and F. H. Stross, *J. Am. Chem. Soc.* 78, 2999–3006 (1956).
2. G. J. Pierotti, C. H. Deal, E. L. Derr and P. E. Porter, *J. Am. Chem. Soc.* 78, 2989–2998 (1956).
3. M. Dimbat, P. E. Porter and F. H. Stross, *Anal. Chem.* 28, 290–297 (1956).
4. E. M. Fredericks and F. R. Brooks, *Anal. Chem.* 28, 297–303 (1956).
5. H. W. Patton, in *75 Years of Chromatography — A Historical Dialogue*, eds. L. S. Ettre and A. Zlatkis (Elsevier, Amsterdam, 1979), pp. 309–313.
6. H. W., Patton, J. L. Lewis and W. I. Kaye, 126th National A.C.S. Meeting, New York, NY, September 1954; *Anal. Chem.* 27, 170–174 (1955).
7. D. H. Lichtenfels, S. A. Fleck and F. H. Burow, 6th Pittsburgh Conference on Analytical Chemistry, Pittsburgh, PA, March 1955; *Anal. Chem.* 27, 1510–1513 (1955).
8. *Chem. Eng. News* 33(15), 1510–1511 (April 11, 1955).

9. A. I. M. Keulemans, *Gas Chromatography* (Chapman & Hall, London, 1957).
10. D. M. Rosie and R. L. Grob, *Anal. Chem.* **29**, 1263–1264 (1957).
11. H. H. Hausdorff, *Vapor Fractometry (Gas Chromatography) — A Powerful New Tool in Chemical Analysis* (The Perkin-Elmer Corp., Norwalk, CT, 1955).
12. Th. Johns, *Beckman Gas Chromatography Application Manual* (Beckman Instruments, Fullerton, CA, 1956).
13. V. J. Coates, H. J. Noebels and I. S. Fagerson, eds., *Gas Chromatography (1957 I.S.A. Symposium)* (Academic Press, New York, 1958).
14. H. J. Noebels, N. Brenner and R. F. Wall, eds., *Gas Chromatography (1959 I.S.A. Symposium)* (Academic Press, New York, 1961).
15. *Ann. N.Y. Acad. Sci.* **72** (Art. 13), 359–785 (1959).
16. S. H. Langer and C. Zahn, in Ref. 13, pp. 287–313.
17. L. S. Ettre, in Ref. 14, pp. 375–455.
18. P. R. Scholly and N. Brenner, in Ref. 14, pp. 263–309
19. *Anal. Chem.* **28**(5), 13A–14A (May 1956).
20. *Anal. Chem.* **29**(9), 57A (September 1957).
21. C. L. A. Harbourn, in *Gas Chromatography 1958 (Amsterdam Symposium)*, eds. D. H. Desty (Butterworths, London, 1958), pp. 244–245.
22. S. A. Ryce and W. A. Bryce, *Can. J. Chem.* **35**, 1293–1297 (1957).
23. S. A. Ryce and W. A. Bryce, *Nature (London)* **179**, 541 (1957).

Chapter

32

Two Symposia, When HPLC was Young*

In Chapters 30 and 31 we have seen how the organization of frequent symposia played an important role in the rapid dissemination of gas chromatography. The first two symposium series — in Europe and in the United States — were soon followed by other regional and international meetings, serving as the places where key chromatographers met, presented their newest results and learned about the achievements of their peers. An important regular meeting series was the International Symposia on Advances in (Gas) Chromatography, organized by Professor Al Zlatkis of the University of Houston: the first such symposium was held in 1963 and the series continued regularly until 1988.[1]

Modern liquid chromatography (LC) started in the beginning of the 1960s when a few key scientists began to explore the possibility of overcoming the inherent slowness of classical column liquid chromatography. First they considered theoretical questions, followed by

*Based on the articles of L. S. Ettre and V. Meyer, published in *LCGC* (*North America*) **18**, 704–714 (2000) and *LCGC Europe* **14**, 314–320 (2001).

the development of columns in which the slowness of diffusion in the liquid medium could be overcome by using uniform small particles or other means to reduce the length of the diffusion path and by high pressures in the system. It was also realized that this new variant of LC would need sophisticated instrumental systems, a far cry from the classical, gravity-flow columns used for many decades (see Chapters 26–28). A few preliminary reports were presented in the 1960s at various symposia, and the first comprehensive discussion of the potentialities of this new variant of LC took place at one of the discussion sessions during the International Symposium on Gas Chromatography held in September 1966, in Rome, Italy, organized by the GC Discussion Group as one of the biannual meetings in the series which started in 1956, in London.[2] Finally, a representative collection of reports on the new developments was presented in 1969, at the Fifth International Symposium on Advances in Chromatography, held in Las Vegas, Nevada, which can be considered as the start of modern, high-performance liquid chromatography (HPLC). The technique advanced very rapidly and soon started to stand on equal foot to GC. After a few more years it was felt that instead of representing just a session of a meeting still dominated by GC, the new advances in HPLC deserve their own symposium. As a conclusion an International Symposium on Column Liquid Chromatography was organized in May 1973 in Interlaken, Switzerland. This symposium was so successful that it represented the start of the International Symposia on High-Performance Liquid Chromatography, later extended to encompass all liquid-phase separation techniques. This symposium series is still alive, and the 32nd symposium is scheduled for May 2008, in Baltimore, MD, USA.

In this chapter, we want to round up our reporting on the pioneering symposia setting the direction of chromatography developments for decades. We shall outline the events which led to these two symposia and summarize the status of LC at the beginning of its meteoric development.

32.1. The 1969 Las Vegas Symposium

Above we already mentioned the international symposia series started in 1963 in Houston by Professor Al Zlatkis of the University of

Houston, originally dealing with GC only. Subsequent symposia in this series were held in 1964 and 1965 in Houston, Texas, and in 1967 in New York City. At that time the growing interest in modern LC induced Professor Zlatkis to enlarge the scope of the meeting and also include a few reports on this new variant of chromatography. In the next couple of years the interest in modern LC grew rapidly, and therefore the organizers of the fifth symposium of this series scheduled for 20 23 January 1969, in Las Vegas, Nevada, decided to devote a large part of the meeting to LC. As a conclusion almost one-third of the papers presented at this symposium (16 out of a total of 52) dealt with new developments in LC. Lloyd Snyder, one of the pioneers of modern LC, characterized the Las Vegas Symposium as the place where[3]

> for the first time, a critical mass of HPLC practitioners was assembled in one location. The resulting exchange of information and enthusiasm made this a truly exciting event.

Indeed, the Las Vegas Symposium represented a milestone in the evolution of modern LC.

At the early stage of modern LC, essentially two main problems concerned those who investigated the possibilities to improve LC: the proper column packing and the selection of the proper device which can be used to detect the solutes present in the column effluent resulting in a true chromatogram. The bulk of the papers presented in Las Vegas concentrated on these questions: from the 16 papers eight (50%) concerned the columns and their operation and five (31%) reported on investigations of the various detectors. Let us investigate the questions discussed in these papers.

At the end of the 1960s, most LC columns contained an adsorbent-type packing. Since particle technology was not advanced enough to provide fully porous particles with uniform sieve diameter, the so-called pellicular particles were developed, consisting of a thin — a few μm — coating of a porous sorbent layer on an impermeable siliceous core of 30–40 μm diameter, providing in this way the short diffusion path necessary for rapid analysis (see Chapter 27). Pellicular type column packing was used in the report of Cs. Horváth and S. R. Lipsky (Yale University, New Haven, Connecticut), with both alumina and ion-exchange resin outer layer, by I. Halász and P. Walking

(University of Frankfurt, Germany) who utilized particles with a thin alumina coating, and in the paper of J. J. Kirkland (E.I. duPont de Nemours Experimental Station, Wilmington, Delaware), describing columns containing a support-type particles coated with a liquid stationary phase (i.e., true liquid–liquid partition chromatography, LLC), similar to the packed columns used in GC. In 1969 such columns were still used frequently in LC. The best illustration of the situation was given by L. R. Snyder[3] who compared the relative usage of the different HPLC methods in papers published in 1971 and 1994. In 1971 (i.e., two years after the Las Vegas Symposium) 43% of the papers still used liquid–liquid chromatography, while by 1994 LLC was not used at all. The demise of LLC was simply due to the fact that such columns were not stable enough: they were eventually replaced by columns containing the so-called bonded phases.

It should be noted that the paper of Halász and Walking (discussing in general the various column types and their underlying theoretical principles) published among the symposium papers was not presented at the symposium. Instead of it, the actual oral presentation of Halász dealt with a new bonded-phase column packing developed by his group. However, apart of a short publication in a German journal[4] the text of his presentation was published only five years later.[5] These packings were marketed for a limited time under the trade name Durapak® by Waters Associates for use in both GC and LC. However, it was soon realized that their use in LC was limited because of the nature of the chemical bonding between the surface of the siliceous support particles and the actual stationary phase molecules (called "brushes" or "bristles" by Halász): the ester-type bond used here had poor hydrolytic stability. As stated by Halász himself one year later in a lecture, "the use of water as the mobile phase results in the rapid decomposition of this stationary phase: the 'bristles' are shaved".[6] The stable bonded-phase column packings widely used today were developed after the Las Vegas meeting: such materials were described in 1969–1970 by Aue and Hastings[7] and particularly by Kirkland and DeStefano,[8] and marketed from about 1970 on by DuPont under the trade name Permaphase® (see Chapter 26).

The second largest group of papers presented in Las Vegas dealt with detectors. Sensitive, variable-wavelength photometric detectors in general use today were not yet available and chromatographers were still in search of the proper detector suited for the modern LC systems. At that time two new detectors (since then long forgotten) drew much attention and papers dealing with them were also presented in Las Vegas. The first was the micro-adsorption detector originally developed two years earlier by Hupe and Bayer[9] and at the symposium papers by T. W. Smuts, F. A. Van Niekerk and V. Pretorius (University of Pretoria, South Africa) and M. N. Munk and D. N. Raval (Varian Aerograph, Walnut Creek, California) discussed these detectors providing the necessary low detection limit but, as soon realized, were too sensitive to flow fluctuations. The other LC detector tried to utilize the high sensitivity of the ionization detectors, by then well established in GC. Because of the presence of the liquid mobile phase, the column effluent could not be directly conducted into the detector. Since, however, in LC the mobile phase has a higher volatility (lower boiling point) than the analyzed solutes, it could be evaporated prior to the detector. The first such system was described by Haahti, in 1963.[10,11] In it the column effluent was wetting a circulating gold wire which went through a heated zone where the mobile phase was evaporated, and then entered a flame-ionization detector where the solutes were combusted, giving rise to the usual ionization current. The system of James *et al.* described in 1965[12] was based on the same principles; however, because it utilized an argon-ionization detector, the solutes on the wire (now made of iron) had to be combusted prior to the detector and the gaseous combustion products conducted into it. A few years later researchers at Shell Development Co. (Emeryville, California) also constructed a detector similar to the simpler Haahti design, with an FID, but using a moving belt instead of a wire.[13] The combination of this detector with a gel-permeation chromatograph was now reported in Las Vegas by H. Coil, H. W. Johnson, A. G. Polgar, E. E. Seibert and F. H. Stross of Shell Development Co. Ultimately, however, these detectors were found too cumbersome and were not accepted in the general practice.

Two papers presented at Las Vegas dealt with gel-permeation chromatography (GPC), a form of size-exclusion chromatography. In the mid-1960s special GPC instruments had already been introduced for the determination of the molecular-weight distribution of polymers (see Chapter 28). Besides the paper by associates of Shell Development Co. (mentioned above) the other report on GPC was presented by K. J. Bombough, W. A. Dark and R. F. Levangie (Waters Associates, Framingham, Massachusetts, the manufacturer of GPC instruments), giving an overview of the technique and its use, and describing the Waters system which was equipped with a differential refractometer detector.

Last, but not least, we should mention the paper by J. C. Giddings, M. N. Myers and J. W. King (University of Utah, Salt Lake City, Utah), dealing with supercritical-fluid chromatography. Giddings had been interested for years in chromatography at very high pressures; as indicated by him in a paper presented five years earlier,[14]

> ... at 1000 atmospheres ... gas molecules crowd together with a liquid-like density. At such densities ... in effect, non-volatile components are made volatile.

Now, in the paper presented at the Las Vegas meeting, he provided a theoretical treatment for dense gas conditions, based on the Hildebrand solubility parameter and derived the concept of threshold pressure, the pressure at which under given conditions a non-volatile solute start to volatilize. The solutes investigated were large biomolecules, as well as some of the high-molecular-weight substances used normally as liquid phases (e.g., silicone oils, Carbowaxes), and both carbon dioxide and ammonia were used as the mobile phase. With CO_2 as the mobile phase pressures up to 1350 atm were utilized at 40°C, while with NH_3 the experiments were carried out at 200 atm and 140°C (critical values are 31°C and 72.9 atm for CO_2, and 132.5°C and 112.5 atm for NH_3).

When discussing the papers presented at Las Vegas, the absence of discussions on two techniques in universal use today may be conspicuous to the reader: these are reversed-phase chromatography (RPC) and changing the selectivity through changing the mobile phase during a run (gradient elution). The scarcity of the use of RPC can be explained by the absence of the stable bonded phases so common today and the lack of understanding of the underlying principles. In their paper presented at the Las Vegas Symposium S. T. Sie and N. van den Hoed (Koninklijke/Shell Laboratory, Amsterdam, The Netherlands) reported on an RPC column by coating Sil-O-Sel particles with *n*-dodecane, but this was an isolated example, without followers. The situation was well illustrated by Snyder in the already quoted survey,[3] comparing the relative usage of the different HPLC methods in papers published in 1971 and 1994: in 1971 only 9% of the papers used reversed-phase liquid chromatography, but by 1994, their use was reported in over three-quarters of the papers, with a concomitant reduction of the use of liquid–solid chromatography (i.e., using an adsorbent as the stationary phase) from 36% to 10%. It is typical for the situation around 1970 that a nomenclature recommendation of the I. UPAC dated February 1972[15] characterized RPC as "a technique of only historical interest." The renaissance of RPC started only after the introduction of the stable bonded phases (already mentioned earlier) and the seminal work of Horváth on both the theory and use of RPC.[16]

With respect to gradient elution or more correctly, the lack of its general use, the mentality of early liquid chromatographers was influenced by their GC experience where selectivity is adjusted by changing the stationary phase. Also, in LLC, it is not easy to change the mobile phase with a given stationary phase and still maintaining phase inmiscibility. One may add to this the technical problems: early LC pumps were fairly cumbersome even when using in a single mode and the difficulties exponentially increased when the pumps would have to be combined to provide two flows in changing proportions to be mixed prior to the column. The use of gradient elution only became common later in the 1970s, after the availability of the new column packings, new improved pump systems, and a better education of the chromatographers.

The papers of the Las Vegas Symposium were published in a number of issues of the *Journal of Chromatographic Science* (Vol. 7, 1969) and later were also collected in a limited-edition bound volume published by Preston Technical Abstracts Co. (Evanston, IL). In this way, the information presented at the Symposium became available to a wide audience.

32.2. The 1973 Interlaken Symposium

At the beginning of 1969 when the Las Vegas Symposium was held, not too many scientists were involved in modern LC and the investigations were mostly of exploratory nature. This is also evident from the published chromatograms: the samples were usually artificial mixtures, used to demonstrate a point and not to solve an analytical problem. The situation changed rapidly, and by 1972 three leading European scientists felt that it is time to convene a meeting dealing only with the new LC: they were Georges Guiochon (Ecole Polytechnique, Paris), Joseph F. K Huber (Free University of Amsterdam), and Willy Simon (the University of Zurich). The symposium took place on 2–4 May 1973, in Interlaken, the beautiful Swiss mountain resort between the two lakes, the Thunersee and the Brienzersee, at the feet of the three giants of the Swiss Alps: the Jungfrau (13,642 ft), Mönch (13,448 ft), and Eiger (13,025 ft). The interest exceeded all expectations: 450 participants from 22 countries were registered and the local hosts had some trouble to find the flags of all these countries. At the last moment the Russian scientists scheduled to present papers (professors Kiselev and Eltekov) were not permitted to leave their country (the Cold War was still on); however, as a report on the meeting stated, "the red flag with the hammer and sickle waved in the Aula — a quite unaccustomed view." We should also mention that in order to provide an authentic local touch, a cowbell has served to warn the speakers who overstepped their allocated time.[17]

There was a debate prior to the meeting whether the papers can also be presented in languages other than English, and those favored multilingual presentation argued that not everybody has a good command of English. However, the controversy was solved by Willy Simon, the principal organizer of the symposium, who in his opening words declared that "the official language of the meeting is broken English".[18]

The original program of the symposium listed 60 papers; from these 55 were actually presented and 45 were included in the proceedings published later in the year as Vol. 84 of the *Journal of Chromatography*. A selective summary of 27 papers was also published in the July 1973 issue of *Chimia*, the journal of the Swiss Chemists' Association[17] and we also have a retrospective discussion on the importance of the meeting, published in 1991 on the occasion of holding in Basel, Switzerland, the 15th symposium of this series which started with the Interlaken meeting.[19]

Because of the large number of papers their comprehensive discussion would be impossible; we shall only mention some of the most important presentations and outline the trends indicated by the papers.

As mentioned the Interlaken Symposium differed from the Las Vegas meeting in that by now modern LC has found application in a number of laboratories. Thus, a number of papers described real applications, and even the reports dealing with technical questions or column development utilized more or less "real" samples the analysis of which had practical importance.

With regard to column packings, the preparation and characteristics of the brand new, fully porous silica spheres with a narrow size range and particle diameters below 10 μm were described by K. Unger and J. Schick-Kalb (Technical University, Darmstadt, Germany) and by J. J. Kirkland (E.I. duPont de Nemours, Experimental Station, Wilmington, Delaware), while R. E. Majors and F. R. MacDonald (Varian Instrument Division, Walnut Creek, California) have shown the superiority of these packings over the porous-layer (pellicular) beads. A question mentioned by a number of lecturers was the problem of how to pack the new small particles into columns, with reproducible qualities, and Kirkland described a high-pressure slurry packing procedure. While we still had reports on the less useful bonded phases with ≡Si–N=bonding by O. E. Brust, I. Sebastian and I. Halász (University of Saarbrücken, Germany), it became apparent that the stable ≡Si–O–Si–C≡ type bonded phases are much more superior, as discussed by M. Novotny and S. L. Bektesh (Indiana University, Bloomington, Indiana) and J. H. Knox and G. Vasvari (University of Edinburgh, Scotland). Such packing was also used in preparative columns of 23 mm i.d. and 50 cm length, as reported by a multiauthored paper from the laboratories of E. I. duPont de Nemours in Wilmington, and Hitchin, U. K. (D. R. Baker *et al.*). This trend was also recognized in the report on the symposium published in *Chimia*[17] which stated that "according to the now widely held opinion, the future belongs to the chemically bonded, well-defined phases"; however, this report also complained that these new phases command a high price.

New stationary phases were also introduced for size-exclusion (gel permeation, gel filtration) chromatography. The availability of rigid, porous glass column packings extended the field of application of GPC, permitting not only the rapid determination of molecular-weight distribution of polymers but also the analysis of a wide variety of samples, as demonstrated by M. J. Telepchak (Perkin-Elmer Corp., Norwalk, Connecticut). Similarly, the introduction of hydroxy-alicyclic derivatives of Sephadex resulted in improved separation of epimeric 3-hydroxysteroids, as reported by R. A. Anderson, C. J. W. Brooks and B. A. Knights (University of Glasgow, Scotland).

Special stationary phases containing complex-forming organic metal salts were described by F. Mikeš, V. Schurig and E. Gil-Av (Weizman Institute, Rehovot, Israel): these offered good resolution of *cis/trans* isomers. A particularly interesting separation was the full resolution of stearic, oleic, and elaidic acid methyl esters in about 6 min.

Although by now stable bonded phases started to be used, a kind of aversion still existed against gradient elution. This is illustrated by the fact that only five papers mentioned the use of gradient programming.

It is interesting to note that one paper picked up the presentation of Giddings at Las Vegas on carrying out chromatography under supercritical conditions and his theoretical treatment based on Hildebrand's solubility parameter. This was the presentation of D. Bartmann and G. M. Schneider (University of Bochum, Germany) considering isobaric and pressure-programmed separations with supercritical CO_2. The paper also investigated the validity of a modified Van Deemter equation incorporating pressure-induced terms.

With respect to instrumentation it should be noted that most groups presenting their results in Interlaken still used LC systems assembled by themselves from a pump and a detector, acquired separately. This was illustrated by the 11 papers discussing various applications of LC: nine were using self-assembled systems from individually acquired parts and only two were utilizing liquid chromatographs manufactured by an instrument company, although at that time such instruments were already marketed by a number of companies. The paper by Schreiber[20] published in 1971 was typical of the general situation, working with "the poor man's liquid chromatograph." In most of the applications UV detectors were used, some of them already providing variable-wavelength capability, while a rapid-scanning detector was described by A. Bylina (Institute of Physical Chemistry, Warsaw, Poland). It is interesting to note that only one paper presented here dealt with commercially available instrumentation: it was discussing the combined development carried out in Japan by H. Hatano (Kyoto University) and a group from Japan Spectroscopic Co., Tokyo.

As far as we know the paper of H. R. Schulten and H. D. Beckey (University of Bonn, Germany) represented the first presentation combining LC with mass spectrometry. However, this was not done on line: first the LC fractions were collected, dried and only then introduced into the mass spectrometer. Direct coupling of the two systems was still far away.

We should finish this summary of the papers presented at the Interlaken Symposium, by mentioning three papers reporting on work in which the analysis time was significantly *increased* relative to state-of-art analysis of that time, in order to improve resolution. Two of these papers used GPC, and the third used ion-exchange chromatography. Instead of physically increasing column length to improve resolution, S. Nakamura, S. Ishiguro, T. Yamada and S. Morizumi (Showa Denko, Tokyo, Japan) devised a system in which the sample stream could be recycled a number of times through a standard, 1.2-m long column containing an improved cross-linked polystyrene gel packing. Impressive separations were demonstrated, not only of polymers but also of relatively low molecular-weight substances; however, one determination took 2–3 h. Even longer analysis times were needed in the very unusual system described by W. Heitz (University of Mainz, Germany) who utilized very long (10–20 m!) Teflon tubes of 2 mm i.d., packed with 40-μm particles of a poly(vinyl acetate) gel. The columns were coiled on a cage of 10–20 cm diameter, similar to the configuration of the capillary columns used today in GC. Naturally, the price to be paid was the time of analysis: the presentation showed the chromatogram of a polystyrene mixture with n=2–36 (n = degree of polymerization) in which each oligomer was separated; however, the total analysis time was 20 days! In the third paper C. D. Scott and N. E. Lee (Oak Ridge National Laboratory, Oak Ridge, Tennessee) carried out ion-exchange chromatography with pellicular resins. Because such material has a low sample capacity, sequential columns were used, with a short pre-column containing the usual microreticular resin in which preliminary separation of the sample components took place. The system permitted the determination of over 50 compounds in a urine sample, however, again, the cost was an excessive analysis time, up to 80 h.

The Interlaken Symposium already indicated a matured technique standing on its on feet. From then on, HPLC grew very rapidly, eventually surpassing gas chromatography.

References

1. L. S. Ettre, *J. Chromatogr.* **468**, 11–34 (1989).
2. I. V. Mortimer, Reporter, in *Gas Chromatography 1966* (*Rome Symposium*), ed. A. B. Littlewood (Institute of Petroleum, London, 1967), pp. 414–418.
3. L. R. Snyder, *J. Chem. Ed.* **74**, 37–44 (1997).
4. I. Halász and I. Sebastian, *Angew. Chem. Int. Ed.* **8**, 453–454 (1969).
5. I. Halász and I. Sebastian, *J. Chromatogr. Sci.* **12**, 161–172 (1974).
6. I. Halász, in *Modern Practice of Liquid Chromatography*, ed. J. J. Kirkland (Wiley-Interscience, New York, 1971), p. 336.
7. W. A. Aue and C. R. Hastings, *J. Chromatogr.* **42**, 319–335 (1969).
8. J. J. Kirkland and J. J. DeStefano, *J. Chromatogr. Sci.* **8**, 309–314 (1970).
9. K. P. Hupe and E. Bayer, *J. Gas Chromatogr.* **5**, 197–201 (1967).
10. E. Haahti and T. Nikkari, *Acta Chim. Scand.* **17**, 2565–2568 (1963).
11. E. Haahti, T. Nikkari and J. Karkkainen, in *Gas Chromatography 1964* (*Brighton Symposium*), ed. A. Goldup (Institute of Petroleum, London, 1965), pp. 190–196.

12. A. J. James, R. Ravenhill and R. P. W. Scott, in *Gas Chromatography 1964* (*Brighton Symposium*), ed. A. Goldup (Institute of Petroleum, London, 1965), pp. 197–206.
13. H. W. Johnson Jr., E. E. Seibert and F. H. Stross, *Anal. Chem.* **40**, 403–408 (1968).
14. J. C. Giddings, in *Gas Chromatography 1964* (*Brighton Symposium*) ed. A. Goldup (Institute of Petroleum, London, 1965), pp. 3–24.
15. *Recommendations on Nomenclature for Chromatography*. Information Bulletin, Appendices to Tentative Nomenclatures, Symbols, Units and Standards, No.15, I.U.P.A.C. Secretariat, Oxford, February 1972.
16. Cs. Horváth, M. Melander and I. Molnár, *J. Chromatogr.* **125**, 129–156 (1976).
17. *Chimia* **27**, 391–395 (1973).
18. Personal recollections (LSE).
19. V. R. Meyer, *Chimia* **45**, 101–102 (1991).
20. J. Schreiber, *Chimia* **25**, 405–407 (1971).

INDEX*

*Names mentioned only in references at the end of each chapter but not in the main text have been omitted from the index.